CONTENTS

Sudden Death in Opera

To David,
There is more to
opera than meets
the eye,

Michael
June 28th 2023

The co-authors of this book draw on a wealth of experience, knowledge, understanding and, not least, love of opera, aptly defining this our most plural and complete form of artistic creation and expression, to explain the romantic themes of love and death, recurring throughout the operatic canon, and to provide brilliant insights and fascinating forays into the overlapping fields of culture, neuroscience, medicine, philosophy, history and religion. The authors' multidisciplinary research and scholarship have resulted in a work which is both accessible and enlightening for any opera lover, and a particularly valuable source of background information and interpretative inspiration for those making opera, be they working on the stage, in the pit or behind the scenes.

—Ian Ritchie (Artistic Director, The Musical Brain, and freelance curator)

Sudden Death in Opera:

Love, Mortality and Transcendence on the Lyric Stage

By

Michael Trimble,
Robert Ignatius Letellier
and Dale Hesdorffer

Cambridge Scholars Publishing

Sudden Death in Opera:
Love, Mortality and Transcendence on the Lyric Stage

By Michael Trimble, Robert Ignatius Letellier and Dale Hesdorffer

This book first published 2021. The present binding first published 2023.

Cambridge Scholars Publishing

Lady Stephenson Library, Newcastle upon Tyne, NE6 2PA, UK

British Library Cataloguing in Publication Data
A catalogue record for this book is available from the British Library

ISBN (10): 1-5275-9795-4
ISBN (13): 978-1-5275-9795-2

DUCA
Ma dee luminoso
in corte tal astro qual sole brillare.
Per voi qui ciascuno dovrà palpitare.
Per voi già possente la fiamma d'amore
inebria, conquide, distrugge il mio core.

DUKE
So bright a star
should be shedding its brilliance on my court.
You would make every heart beat faster here.
The fires of passion already flare
headily, conquering, consuming my heart.

(Francesco Maria Piave, *Rigoletto*, Act 1)

STIMMEN
Selig sind die Liebenden
Die der Liebe sind nicht des Todes,
Und auferstehn werden,
Die dahingesunkenen sind
um Liebe, um Liebe, um Liebe,

VOICES
Blessed are they that love.
Those who have loved shall not die.
And those who have died for love
shall rise again.

(Erich Korngold, *Das Wunder der Heliane*, Act 3)

LIST OF FIGURES

Chapter 19

Chapter 20

Chapter 21

Chapter 23

Chapter 25/Coda

ACKNOWLEDGEMENTS

The authors are so grateful to those friends and colleagues who have contributed their thoughts and ideas to our book, and have helped with its production: Jackie Ashmenall, Mark Berry, Paul Dawson-Bowling, Peter Fuller, Michael Graubart, Eddie Gregson, Kousuke Kanemoto, Marco Mula, Mary Robertson, Roger Scruton, Jeffrey Swann, Semir Zeki, translators of the many librettos some of whom we are unable to acknowledge by name, the production team at Cambridge Scholars, and the staff at Nationalarchiv der Richard-Wagner-Stiftung Bayreuth.

PROLEGOMENON

NIGEL OSBORN

The colloquialism "It ain't over till the fat lady sings" sums up a popular view of opera: that operas are full of buxom singers and extreme emotions, and tend to end when women die. In fact, the saying has been attributed historically to baseball games, church services as well as to the final Acts of Wagner's operas *Götterdämmerung* and *Tristan and Isolde*.

As it happens, the nature of and reasons for Isolde's collapse and subsequent demise at the end of *Tristan* go to the heart of this extraordinary book on love and unexplained death in opera, elegantly penned by musician and cultural historian, Robert Letellier, epidemiologist Dale Hesdorffer and distinguished neurologist Michael Trimble. It is a wonderful journey through the emotional landscapes of our cultural history, as expressed in the most empathetically communicative of our art forms - opera. The authors trace the operatic and medical history of *Liebestod* (love-death) from Fouqué's novella *Undine*, much influenced by Paracelsus in a combination of the new scientific medical spirit and Neo-Platonism, through E.T.A. Hoffmann's opera/melodrama, to Lortzing's *Undine* and Wagner's own *Tannhäuser*. Both Tristan and Isolde die of a "love potion" But whereas Tristan kills himself by ecstatically removing the dressings from his wounds, Isolde collapses in a way reminiscent of disorders of the autonomic nervous system. In fact, as Michael Trimble has pointed out in his excellent writings on music, the experience of music is in itself closely related to autonomic activity.

The authors also explore the philosophical and literary highways and byways of this colourful journey. The theme of love and unexplained death is traced from the Bible, through the stories of Ananias and Sapphira, Enoch and Moses to *Pyramus and Thisbe, Romeo and Juliet*, Goethe, Scottish ballads, Emily Brontë and Tennyson. As the authors point out, Wagner's own death was intimately intertwined with the loves of his life and his planned article "The Eternal in the Feminine" - the diagnosis of heart disease was kept from him. And as they conclude:

"Humans are distinguished from other animals by several fundamental behaviours, especially the six L's of life: Language, Lying, Laughter, Lacrymation, Lyric and Love. For Wagner, and the rest of us, death is always the present-future—without death, there is no life; without life there is no love; without love there is no art; without art there is no music, and a world without music would be inhuman".

PROLOGUE

> Opera is a song of love and death, of conditions which bypass rational understanding. This linked it to those ancient mysteries which the Renaissance revived; its practitioners remain devotees of the mysterious transformations probed by religion.[1]

The association of love with death is a universal phenomenon in art and in the world's religions. And since art has emerged as a way of seeking to articulate or express the universal truths enshrined in faith and mythology, it is hardly surprising that it is central to that most complex fusion of drama and song, the lyric genre of opera. Peter Conrad's book *A Song of Love and Death: A Meaning of Opera* is one of the only comprehensive texts to explore the unbelievably believable world of one of the greatest art forms of human creation. This seems somewhat surprising, in the sense that so many famous operas involve death at some point, even in apparently comic circumstances. In this book, we have attempted to closely examine what we have termed *Unexpected Death,* those that seem to occur without perhaps an obvious cause, so-called *Unexplained* Death.

All who write about the history of opera start with the Orpheus myth. One of the very first operas, Claudio Monteverdi's (1567-1643) *Orfeo* (1607), tells of the origins and mysteries of music and the importance of Apollo (the father of Orfeo), bringer of light and order. The story is well-known. Orfeo is to marry Euridice, and when she dies from a snake bite, he resolves to rescue her from Hades. Pluto hears Orfeo's beautiful music and grants him permission to take Euridice back to the world on the condition that he does not look at her as they ascend. He looks, she disappears, Apollo appears, and Orfeo ascends with him to heaven, where he will see Euridice in the stars.

As we follow opera's development, Orphic laments are renewed on the stage until today, yet from early times the energies of Dionysus, with the charge of Eros, complemented by Thanatos, galvanized and emboldened the myths and music. For about the last 250 years a relatively stable 'canon' of 'classical' music which 'harnesses' human nature has developed, notably in the Western world: a tradition, upheld by our responses to music, so often intense and emotional.[2] Opera charts the latter. The 19th century saw a rise in the number of deaths with transcendental dimensions and themes of redemption. It is widely accepted that such developments characterised the operas of Richard Wagner (1813-1883), who wanted to 'redeem' not only his protagonists but opera itself. Music needed drama and vice-versa, delivered with mighty sounds and glorious visions exulting final moments of epiphany.[3]

There are many well-explained causes of death in opera (for example, by murder, poison, illness etc., but it is the *Unexpected Deaths* which intrigued the authors. It seemed to be one of the *topoi* linked with the rise of Romanticism, and we wanted to explore how they may intertwine with such concepts as Redemption, transcendentalism, states of consciousness and religion. We set about this task by reading many librettos, examining the associated musical styles, and looking at the socio-historico-cultural circumstances of the plots. Between us we have seen and written about many operas, explored such themes in discussions with friends and colleagues, and asked ourselves, as others we later note have implied, are those *Unexpected Deaths* unexplained*?* Indeed, sudden death is a commonly reported experience, but even today, with our advanced medical knowledge, sometimes causes cannot be found. Watching characters suddenly expiring on stage, so often females and at the end of the drama, asked for a resolution.

In this book, to we have tried to examine that challenge by combining our understanding and intimate knowledge of music (Robert Letellier) with our studies of medicine (Michael Trimble), and epidemiological expertise on causes of death in various circumstances (Dale Hesdorffer). To complement our disquisition, we explore some philosophical perspectives of those who have ventured to step into the murky waters of music, look at the meanings of transcendentalism, tried to navigate from the erotic mysteries of *Tristan und Isolde* to the spiritual transcendence of *Parsifal*, and constructively deconstruct the metaphysical dialectic between love and death. The latter brings us to the famous *Liebestod* (Love-Death), the long history of which we uncover: it was not simply a Liszt-Wagner neologism. We consider in some detail the compositions and writings of Wagner. The *raison d'être* of our work should not be perceived as yet another study of Wagner, but he is germane to many of our ideas, and the principal articulator of the issues involved. His ideas were kindled early. At the age of 19 he sketched out the libretto of his first opera, *Die Hochzeit* (*The Wedding*), inspired by an E. T. A. Hoffmann (1776-1822) tale. A bride (Ada) is attracted to a stranger (a magnetism referred to by Johann Wolfgang von Goethe, 1749–1832, as elective affinities) whom she sees at her wedding procession. She shudders at his sight. Later, he climbs to her bedroom, where she awaits her bridegroom (Arindal). She fights with him and hurls him into the courtyard below, where he dies. The bride's father wants the murdered man to be buried in front of the whole clan because a divine sign will reveal who is guilty of the murder. The bride shows signs of madness, locks herself in her room, but appears at the funeral and, as the body is carried past, sinks lifeless, falling onto his dead body.

Even before *Die Hochzeit*, as a teenager Wagner had written sketches for stories with gruesome endings; such metaphysical paradigms that came to dominate his later works clearly had early beginnings.[4]

The opera was never performed in Wagner's lifetime, and he immediately went on to compose *Die Feen* (1833).[5] Unlike *Die Hochhzeit*, the latter (which we discuss in Chapter 19), has many of the themes and tropes that can be found in one form or another in all Wagner's later operas, especially the idea that the power of love can overcome death. *Die Feen* moreover has an Orpheus and Euridice-inspired ending in which unreconcilable worlds are reconciled with redemptive poignancy. In later chapters, we discuss Wagner's obsession with death, and his views on the feminine—*Die Ewig-Weibliche*, the eternal feminine.

The first Part of this book deals with the origins of opera and the emergence of Romantic ideology (Chapters 1 and 2), transcendentalism and phenomenology (Chapters 3 and 4), and out-of-body experiences (Chapter 5). We open several lines of enquiry which lay out the foundations of our journey to explore the *Unexpected*, introducing some philosophical ideas that are emergent and necessary to penetrate into cultural and musical developments of the 19th century. It is difficult to explore themes of death in opera from the specific perspective we have chosen without noting well-reported experiences in contemporary literature and medical circles of alteration of mind and body, such as twilight states, trances, dreams and bodily transmigrations. At the same time we need to emphasise matters of the heart, metaphorical and literal which are embedded in literature from very early days.

After that lead-in, Part 2 reviews some Romantic *topoi.* We briefly meet Orpheus again, but then note the dramas of several operas with death as a prominent theme (Chapter 6). This is followed by biblical, poetic and dramatic accounts of sudden death, introducing Love-Death, the *Liebestod* itself (Chapter 7). These themes echo from ancient sources, but we follow their representation in opera as a prelude to various forms of Love-Death

(Chapter 8). In these, the apparent cause of death derives from the drama (such as by illness, poisoning or execution) and may be referred to as explained. The chapter ends with a comment on the following two chapters which expand on important concepts which reverberate in later sections of our book. These concern the Elemental Being (from early origins to her resurrection, especially with Hoffmann's Undine) and the transcendental symbolism of water and flames (Chapter 9). In chapter 10 we draw attention to a variant of love-death which is rarely discussed, the masculine *Liebestod.*

Before embarking on our analysis of the operas that we identify with *Unexpected Death,* we first return (in Part 3) to the development of music and opera from the late 18th to the late 19th century, picking up from our introduction in Chapter 2. We have split the dramatic progression of operatic exposition into two sections to better enable the co-occurrence of musical themes, especially following the French revolution(s), with singing (as the castrati lost favour), scenery and lighting (Chapter 11). Chapter 12 looks at the fate of the unbroken musical line, the development of the aria, the growing influence of the librettist and the rise of French Grand Opera.

In Part 4 we get to the heart of the matter, the kernel of our exposition to try to understand the so-called *Unexpected Deaths.* As an introduction to our methods we return to important medical themes which were progressing in the 19th century, and include a brief excursion into what is referred to as the *Autonomic Nervous System* (ANS). We introduce physiology, and aspects that we consider have been seriously neglected in discussions of these deaths, pivoting on the heart as the 'seat' of the emotions, and the widespread somatic effects of autonomic disturbance. We thought that these might play a role in explaining those *Unexpected Deaths,* and one way to find out was to read the relevant librettos. To identify the latter, we inductively used our knowledge of many operas, and singled out as many as we could find with such deaths and noted references to signs and symptoms of ANS dysfunction (Chapter 13).

There is a brief intermission, and the operas that are our primary database are shown in List 1. Part 5 is our summary of 50 operas, in which one or more of the protagonists suddenly die. We briefly describe the story, quote sections of the libretti which hint at autonomic instability, set the deaths in contexts of the social/political circumstances of composition, and review the music.

To help our venture, we have divided the libretti into six chapters. Chapters 14, 15 and 16 deal with females and males separately in whom the ANS could be implicated in the sudden death.

We singled out some operas for special consideration, analysed in Chapters 17, 18 and 19. They do not all reference to autonomic events, but we selected them because their special themes are relevant to ideas fostering Romantic opera (such as *La Muette di Portici*), because they adopt specific settings (such as the Undine and Faust operas, or *Pélleas et Mélisande*), or because they were composed by Wagner. The reason to look in detail at the latter is their special relevance to our concerns: List 1 revels that he was the composer whose operas most involved *Unexpected Death,* especially in females.

In Chapter 20 we analyse specific aspects of the operas from List 1, with a further brief medical interpolation before we document important themes and variations that infuse Romantic operas with meaning, so often elided in productions or discussions of the genre.

In Chapters 21, 22 and 23 we delve into associations intertwining transcendence, Redemption, Love and the *Liebestod*, and in Chapter 24 we offer reflections on our quest into the unexplained: we finish with a coda.

We cite the titles of all works, especially operas, in the original language with English translation. We use English-language versions of the libretti for immediacy of access for readers. Some terms are traditionally known in their original medium: *Liebestod*, *Ewig-Weibliche*, but are used interchangeably with their English translations: 'Love-Death', 'the Eternal Feminine'. *Liebestod* is often, but not always, associated with the musical form of this *topos* because of its occurrence in *Tristan und Isolde*. Our book is devoted to exploring the meaning of these terms.

A key understanding of the love-death, a recurrent *Liebestod* motif, is the death of a protagonist (usually female) through illness in a situation which highlights the fragility of the human condition and underscores the tragedy of a doomed love that can find fulfilment only in death.

In all these matters the questions of song (as Apollonian) and inebriation/ madness (as Dionysian) are operatic phenomena intimately associated with the core issues of transcendence, Redemption and their relation to the key concepts/metonym of *Unexpected Death*. In this respect the Mad Scene from Donizetti's *Lucia di Lammermoor* (1835) and Isolde's *Liebestod* in

Tristan und Isolde (1865) attain iconic status where music, love and transcendent death coalesce perfectly.

We have surveyed a small collection of operas from the vast corpus that has been composed over time, and it must be the case that we have missed many which may have ANS references in the libretti to explain an *Unexpected Death*. We have tried to be inclusive but have been somewhat limited by the 'canon' of works that have held critics and audiences' attention over time. We assume that this is so because these operas, their themes and music, and the effects they have on us, contain and retain special features and emotional resonances which resound within the human spirit.

Whether we succeed in explaining the *Unexpected Deaths,* only further examination of this theme by others will tell. In any case, we hope that our embrace of the theme of *Unexpected Death* within cultural, scientific and political events related to the composers and their compositions will help those who want to explore further some of the greatest works of music ever composed. Opera, bringing life even to death, renders comprehension about what it means to be a human being, offering an unfolding of self-knowledge and feelings perhaps greater than any other art form.

Notes

[1] Conrad, P., *A Song of Love and Death: A Meaning of Opera* (New York, Poseidon Press, 1987), 356.
[2] This history is nicely outlined by Mauceri, J. *For the Love of Music: A Conductor's Guide to the Art of Listening*. (London: Weidenfeld & Nicolson, 2019).
[3] An excellent text, with special emphasis on German music and opera is: Eichner, B. *History in Mighty Sounds: Musical Constructions of German National Identity, 1848-1914* (Woodbridge: The Boydell Press, 2012).
[4] *Die Sarazenin* (Saracen Girl) and *Die Bergwerke zu Falun* (The Mines of Falun, from a Hoffmann story.)
[5] See for example: Glasenapp, C. F., *Das Leben Richard Wagners*, trans. Ashton Ellis, W.M. (London, Kegan Paul, Trench, Trübner & Co, Ltd 1900), 142-3. The links from *Die Hochzeit* to *Die Feen* include old Germanic and Norse names, end-rhyme and alliteration, and an early example of a *Leitmotiv.* Not much music was written, but his sister Rosalie thought it was terrible, and told Wagner to destroy it.

PART 1:

ORIGINS

CHAPTER 1

INTRODUCTION TO DEATH AND OPERA

THROUGH the soft evening air enwinding all,
Rocks, woods, fort, cannon, pacing sentries, endless wilds,
In dulcet streams, in flutes' and cornets' notes,
Electric, pensive, turbulent artificial,
(Yet strangely fitting even here, meanings unknown before,
Subtler than ever, more harmony, as if born here, related here,
Not to the city's fresco'd rooms, not to the audience of the opera house,
Sounds, echoes, wandering strains, as really here at home,
Sonnambula's innocent love, trios with Norma's anguish,
And thy ecstatic chorus Poliuto;)
Ray'd in the limpid yellow slanting sundown,
Music, Italian music in Dakota.

While Nature, sovereign of this gnarl'd realm,
Lurking in hidden barbaric grim recesses,
Acknowledging rapport however far remov'd,
(As some old root or soil of earth its last-born flower or fruit,)
Listens well pleas'd.

(Walt Whitman, *Italian Music in Dakota*)[1]

The Theme

"And the secret of love is greater than the mystery of death" [*Und das Geheimnis der Liebe ist grösser als das Geheimnis des Todes*]. In 1901, the composer Richard Strauss (1864-1949) came across the German translation of Oscar Wilde's play *Salomé*, and the opera, with the libretto by Hedwig Lachmann (1865-1918), was premiered in Dresden in 1905. The play was a cynosure of the prevailing *fin-de-siècle* art, closely entwined with late Gothic symbolism, decadence, and aestheticism. Wilde probably was inspired by the novel *À Rebours* (1884) by Joris-Karl Huysmans (1848-1907) in which the decadent Des Esseintes slavers over two paintings by Gustave Moreau (1826-1898) depicting Salome. One, *The Apparition* (1876), shows the head of John the Baptist hovering in front of her, eyes staring down on the near naked dancer, dripping blood surrounded by a halo.

The story of the opera is well known even if it seems to have no biblical equivalent. The imprisoned Jokanaan (John the Baptist) rails against Salome's mother and her incestuous marriage to Herod. Salome persuades the captain of the guard, Narraboth, to let her see Jokanaan, and he is brought from prison. She is besotted by him, he refuses to kiss her, she develops a passion which rapidly becomes an obsession. Herod's own lust for Salome leads him to ask her to dance, for which he will offer her anything she desires. After the seventh veil has fallen, it is the head of Jokanaan on a silver dish. She kisses the lips of the severed head, his lips and blood tasting bitter, the taste of love. [**Fig. 1.1** Moreau, *Salome*]

Not to be confused with Mary Salome, present at Christ's crucifixion (Mark 15:40-41), here Salome is portrayed as "the symbolic incarnation of undying Lust…the accursed beauty…like the Helen of ancient myth".[2] The music is intoxicating, and in this short story there are three deaths (Narraboth, Jokanaan and Salome). The story also includes a suicide, incest, a curse, religion, a fascination with the East, obsession, madness and altered mental states, eroticism, necrophilia and a *Liebestod.* The encounter between Salome and Jokanaan functions at a deep Biblical level. There are sacred and profane implications, along with Salome's erotic language and imagery of the Song of Songs.[3] Salome's obsession with Jokanaan is paralleled by Narraboth's and Herod's for her. Her descent into madness activates her already overexcited body. Salome is swiftly executed; there is only the bitter sweet of tragedy to savour, no redemption or salvation—but there is the music.

The closing scene of the opera, with the Dance of the Seven Veils, the fatal request, the execution of Jokanaan, the presentation of his head, and the extended soliloquy as Salome indulges her lust, conjure up a claustrophobic atmosphere, a sultry intensity of torrid heat and moral torpidity. The atmosphere is sustained by the tonal language, with its subdued intensity, its constant background of shimmering violin tremolos and a reiterated three-note motif, quiet but obsessively insistent. It generates unease and fearfulness, representing as it does the nauseous fixation of Salome's perverted desire.

Salome is one epigone of a *femme fatale*, which fascinated the artists of the 19th and 20th centuries, with a historical resonance going back to Eve, Circe, Cleopatra, and many others, whose existence hovers between reality and mythology, often with supernatural overtones. In the 12th and 13th centuries, courtly love within the Troubadour tradition vies with eroticism and the fearful embrace. Passion undermining social conventions is a theme well represented in poetry and literature, especially reinvigorated in the Romantic era. John Keats's (1795-1821) *La Belle Dame sans Merci* and *Lamia,* Gustave Flaubert's (1821-1880) *Salammbô,* Anatole France's (1844-1924) *Thaïs,* compete with Algernon Swinburne's (1837-1909) *Dolores* and *Faustine*, and images portrayed in the paintings of Dante Gabriel Rossetti (1828-1882) and Moreau. Such depictions provide dramatic potential and find expression in opera. [**Fig. 1.2** *Salome*]

Wilde's Salomé is a reflection of the moon, pale and ever-loitering, a daughter of Babylon. Along with others such as Frank Wedekind's (1864-1918) *Lulu* "…created for every abuse,/To allure and poison and seduce,/ To murder without leaving trace",[4] they portray myths which segue with the more Romantic concerns of the female and powers of love and loss which emerged with opera of the 19th century. This book is concerned with the different physical and psychological congeries which abound in Strauss's *Salomé*.

As a younger composer, Strauss had described himself as a complete Wagnerian. But even if later he made "a detour round him", the ideas and works of Richard Wagner and the *Liebestod,* quite different from some of the other songs of love and death that we discuss later, will be central to the themes of our work.[5]

A Song of Love and Death

Peter Conrad, in *A Song of Love and Death* (1987), refers to opera as "an art devoted to love and death".[6] From the earliest operas, portraying the plight of Orpheus and his loss of Eurydice, death by one means or another, so often forms part of the drama. Orpheus—the son of Oeagrus the Greek river god (or Apollo) and the muse Calliope—was a singer of holy songs, performed with a lyre or cithara, mastered harmony, charmed the flora and fauna and fell in love with Eurydice. She was bitten on the ankle by a serpent, when alone and possibly running away barefoot from the shepherd Aristaeus, who was chasing her (according to Vergil), perhaps before the marriage to Orpheus was consummated. The virgin could not be revived by the songs of Orpheus, and died.[7]

The idea of drama being performed with music emerged in Florence in the 1590s and the first opera to survive intact was Jacopo Peri's (1561-1633)

Euridice (1600, libretto by Ottavio Rinuccini, 1552-1621). The score was probably known to Claudio Monteverdi (1567-1643) whose opera *La favola d'Orfeo* (1607) is considered as the first great opera. Here was a statement of the power of music to arouse emotion, reflecting on the glorious past of Classical Antiquity [**Fig. 1.3-4**, Orpheus, mosaics]

At the end of Peri's *Euridice*, Orpheus and Eurydice embrace. Monteverdi has Apollo descending, assuring Orpheus that he will be immortal and see Eurydice again among the stars. Orpheus wants his lamentations to be heard by all.

The music of Christof Willibald Gluck's (1714-1787) *Orfeo ed Euridice* (1762) has him mourning by the side of Lake Avernus, a portal to the underworld. As in other versions of the story, he descends to Hades to recover Eurydice (in some to redeem her) and to recover his lost self. After all, he was not the first to go to Hades (the *katabasis*): Hercules had been there. But Pluto laid down the law and stipulated that Orpheus could take Eurydice with him as long as he did not look back at her until he had returned across the Avernus. Gluck's Eurydice implored him to turn to her; Monteverdi pitted love against death, but he looked, and she disappeared [**Fig. 1.5-10**, Orpheus in the Underworld, *Orpheus and Eurydice*].

Luigi Rossi's (1597-1653) *Orpheus* (1647) and Gluck's *Orfeo ed Euridice* gave us happy endings. Orfeo's lament in Gluck's music *"Che farò senza Euridice"* (What will I do without Eurydice) is of such beauty that Amor appears and restores her to life. Orfeo and Eurydice return from the underworld and there is general rejoicing. In Joseph Haydn's (1732-1809) *Orfeo ed Euridice* (1791), on the ascent, Eurydice places herself directly in front of Orfeo and removes her veil. He sees her 'gracious face', but the Furies seize Eurydice. A bacchante (follower of Dionysus) offers Orfeo a drink—the nectar of love, which is a poison—and he dies. In Ovid's tale, Orpheus was subject to *sparagmos,* being torn asunder by the Maenads on account of rejecting women.[8] The happy endings rather give way to a developing Romanticism and the destructive influences of Dionysus. However, whichever version of the myth is adopted, the ability of music to elevate souls and overcome death is poignant. Conrad insists that "Orpheus is opera's founder, and he presides over it throughout its subsequent history".[9] [**Figs. 1.7-10,** The Death of Orpheus]

The Birth of Tragedy

Friedrich Nietzsche's (1844-1900) *Die Geburt der Tragödie aus dem Geiste der Musik* (*The Birth of Tragedy out of the Spirit of Music*, 1872) has been read in many ways. It can be seen as a diatribe against the decline of German culture from the high standards of Classical Greek ideals, within which is an appeal for a regeneration of the former, focussed on works of poet and composer Richard Wagner. It is surely an encomium to Wagner, who at the time of its publication was a confidant as well as a reflective mirror for Nietzsche's ideas, musical and classical. However, Nietzsche's book is also a serious attempt to examine the birth and then decline of Tragedy as an art form. It is one opening to examining the question of why Tragedy is somehow pleasing to us as an aesthetic art-form. What is the pleasure we get from seeing Tragedy on the stage, in the opera house, at the cinema or on the television? Just why do we have the response we have to Tragedy, which often involves crying, and a consolation of the falling tears? [**Fig. 1.11** Nietzsche]

Three aspects of Nietzsche's theories are pertinent to the ideas opening up in him and this book. Firstly, the full title—*The Birth of Tragedy, out of the Spirit of Music*—affirms music as central both to his thesis but also to the questions posed above. Secondly, he introduces the interplay between the two iconic Greek gods, Dionysus and Apollo. These cannot be taken literally, and are used by him as emblems for his theories. We will meet them again later. Thirdly, and much less commented on, is Nietzsche's use of physiological and psychological ideas that were current at the time of his composition.[10]

Nietzsche was a scholar of Greek culture, and as a frequent visitor to Wagner's house at Triebschen he was well acquainted with Wagner's ideas. The latter was incorporating Greek theatrical concepts and models into the plots of his operas. Music was played and philosophy discussed. The intellectual *Zeitgeist* had heralded a return to the ideas of Ancient Greece, inspired by despair at the lack of a defined German theatrical and cultural tradition.

The Desires of Eros

Ever since our species began to tell stories, the awareness of, and later the conflict between the world of nature and that of human consciousness became one avenue of exploration. So did the clash between consciousness and social conventions. The enclosed intellectual world of the Middle

Ages, dominated by Scholasticism, setting Aristotelian and Neo-Platonic thinkers at odds, struggled with Eros.

Most people do not immediately associate Plato (428/7?–348/347? BC) with the erotic, but Jill Gordon, in her book *Plato's Erotic World,* quotes *Symposium, Phaedrus, Charmides, Lysis* and *Alcibides* as the 'erotic dialogues' supporting her view that "In actuality, Plato's entire world is permeated with eros" [11]

The split between the soul and the body, the transition between life and death, and concepts of beauty, are colligated with the divine cosmological *Ursprung* of eros as elements of the human soul. Vision and beauty, the latter an object of eros, are linked to our insight and inner wisdom, which can be guided towards understanding (*nous*). *Nous* can see above the heavens, to the gods and their realm, the sun and the moon, to the divine, which guides human souls.

> *Nous* in Plato is not mere intellect…it is reason or thought moved by desire, by the desire of the soul for that which is akin to it, the desire to know and enjoy its object in that complete union which the great mystics have sought to describe, and which Plato himself so often describes in terms of sexual imagery…where the mystical aspect of his philosophy is prominent…[12]

But this desire is not free, because the alienated, earthbound human soul is beset by appetites (*epithumia* from *thymos*), which have a physical connotation. Gordon notes Plato's references to the swelling of the passions, the pounding heart, the activity of the lungs to cool the heart, and appetites for eating and drinking. His tripartite image of the soul (rational, appetitive, and passionate) allows for the *logos* to be disturbed by the senses. Famously in the *Phaedrus*, he portrays *logos* as a charioteer, Logos driving two horses, Eros and Thymos. The soul is a compound of three components: a charioteer (*Reason*), and two winged steeds: one white and one black, the later, concupiscent desire, posing the problems.

The word *epithumia* is taken up in the New Testament, implying desire but with both positive and negative connotations, yet warning against evil and worldly perturbations. The question of appetite took on a new and prominent meaning and would occupy an intensified role in the concept of moral theology emerging especially from the writing of Paul of Tarsus (?-64 AD) and his enthusiastic emulator Augustine of Hippo (354-430) some three centuries later. In Paul's Letter to the Galatians (5:16-24) he presents two ways determined by the concepts of 'the flesh' and the Holy Spirit.

16 But I say, walk by the Spirit, and do not gratify the desires of the flesh.

17 For the desires of the flesh are against the Spirit, and the desires of the Spirit are against the flesh; for these are opposed to each other, to prevent you from doing what you would.

18 But if you are led by the Spirit you are not under the law.

19 Now the works of the flesh are plain: fornication, impurity, licentiousness,

20 idolatry, sorcery, enmity, strife, jealousy, anger, selfishness, dissension, party spirit,

21 envy, drunkenness, carousing, and the like. I warn you, as I warned you before, that those who do such things shall not inherit the kingdom of God.

22 But the fruit of the Spirit is love, joy, peace, patience, kindness, goodness, faithfulness,

23 gentleness, self-control; against such there is no law.

24 And those who belong to Christ Jesus have crucified the flesh with its passions and desires.[13]

The way of the Spirit is manifested in the Fruits, qualities that distinguish the enlightened and balanced path of life.

Additional dangers of body/soul imbalance and a clash between appetites are the onset of madness, eros pinioned between the divine and the human. Within that space the hero resides, sired by erotic encounters between humans and immortals, and urged by eros to seek objects and experiences beyond our mundane human desires. This ineffable metaphysical realm of noetic value, seen as a revelation, allowing experiences that are non-rational, intuitive and insightful, but not purely subjective, link traditional Greek Tragedy to the operatic dramas which are the major subject of this book.

Nietzsche, Wagner and Schopenhauer

Nietzsche and Wagner were both at one point influenced by the philosophy of Arthur Schopenhauer (1788-1860). He viewed music as very special among the arts, an expression of what he referred to as the Will (*die Wille*), in its rawest manifestation. Music, not dealing with things in the real world and being independent of the world, was able to reveal essential truths about the intrinsic nature of things, and thence life [**Fig. 1.12** Schopenhauer].

René Descartes (1596-1650) in his philosophy had argued that it was possible to doubt the existence of everything, and with such doubt, through logic, to arrive at the certainty of his own existence (*Cogito ergo sum*—I think therefore I am), and also the certainty of God. But what kind of 'I', begs the very question of what the individual ego actually is, and an understanding of us as *being* in our world, as members of a society and the human race.

The distinction between phenomena and noumena goes back to the philosophy of Emmanuel Kant (1724-1804). Phenomena are things in the world as they appear to us, and noumena are things 'in themselves' to which we have access only through our perceptions and which we cannot know directly. The world we see (in our perceptual field), is only a representation of what is actually out there. Both Descartes and Kant put limitations on human knowledge, what we could ever know, yet both ignored how our emotions interacted with cognitions: What the mind seeks is not intellectual understanding (*Verstand*) but meaning *(Vernunft)*. Mythical and logical thought are not the same, since many aspects of myth are inaccessible to logic, and the truths of logic are without precedent in myth.

Art, aside from music, cannot, in Schopenhauer's scheme of things, get underneath the illusion and behind the appearances—to the thing-in-itself, to touch what he rather unfortunately called 'the Will'. Music, seen by Nietzsche as Dionysian, in contrast, escaped the phenomenon of individuation.

Tristan und Isolde (1865) is the only Wagnerian opera discussed in *The Birth of Tragedy*. In that opera the influence of Schopenhauer was central, as were the contrasts between light and darkness, day and night, and between passion and societal expectations. The drama of the two lovers, sinking into unconscious bliss, combined with the music full of unresolved discords, suggested to Nietzsche a balance between the Apollonian (words and story line) and the Dionysian reflecting "the mystery of death in life, the unity in duality".[14]

Out of these three influences, namely philosophy, music and the ideas of Wagner, Nietzsche developed a new theory of art, especially Tragedy. For him, beauty, the high-point of Greek art, as also for Johann Joachim Winkelmann (1717-1768), was insufficient to explain the differences between music and the visual arts: "For Greek art has taught us there is no truly beautiful surface apart from some terrifying depth".[15] Nietzsche's philosophy was one of *becoming*, in contrast to *being*, since change, transience and destruction were elemental to the Dionysian. Heraclitus of

Ephesus (535-475 B.C.), the weeping philosopher, he who could not step into the same river twice as things were never in a fixed state, only in flux, was a model. What is important in the context of our book however, is not so much the apparent opposition between Apollo and Dionysus, but their *synthesis*, and their *psychological* as opposed to any Classical or theological meaning.

Some of the words that Nietzsche used relate to the *physiological*, including 'drives', 'wills', 'energies' and 'impulses'. The German word *Trieb* is frequently written, referring to desire and instinct. Apollo and Dionysus are personified as driving creative forces (*Kunsttrieb*), physiological phenomena. Form under the tension of force (the artistic articulations of Apollo shaping Dionysian energy), is embodied in the tragic hero (as individual) being overwhelmed by forces and circumstances, natural and social. Drives are organic, visceral, they shape characters and destiny, and Tragedy is revealed as a narrative of time, of life and of becoming, and as "artistic powers which spring from nature itself". Aesthetics for Nietzsche was "nothing but a kind of applied physiology".[16]

Affairs of the Heart

The association between what are referred to as autonomic disturbances (discussed later in more detail) and passion find a place in the earliest of Western poetry. Sappho (630-580 BC), born on Lesbos, adorer of grace and beauty, leapt to her death into the Ionian Sea after a failed love affair. The sapphic stanza is named after her.

> Like the very gods in my sight is he who
> sits where he can look in *your eyes*, who listens
> close to you, to *hear* the soft voice, its sweetness
> murmur in love and
>
> laughter, all for him. But it breaks my spirit;
> underneath my breast all *the heart is shaken.*
> Let me only glance where you are, *the voice dies*,
> I can say nothing,
> but my *lips are stricken to silence*, underneath
> my *skin the tenuous flame suffuses*;
> nothing shows in front of my *eyes*, my *ears* are
> muted in thunder.
>
> And the *sweat breaks running upon me, fever*
> *shakes my body, paler* I turn than grass is;

> I can feel that I have been changed, *I feel that death has come near me.*[17]

The italics emphasise the bodily effects portrayed in her love experience. Affecting all her senses, envisioned especially through the eyes, she perspires, her body shakes, she is rendered speechless, and has a feeling akin to dying. [**Fig. 1.13** Sappho]

These sentiments are vividly revisited in the life and poetry of Dante Alighieri (1265-1321). At the age of 9, he saw the 8-year-old Beatrice with angelic beauty, dressed in crimson. He fell in love with her, and so started a new life—*Vita Nuova.* Nine years passed before he glimpsed her again, perhaps in a church, and converted this unconventional passion into one of the greatest artworks of all time, the journey and ultimate goals of the human soul in *La Divina Commedia* (*The Divine Comedy*) (c. 1308-c.1320). He feels faint when he anticipates not seeing her. The first sonnet in the *Vita Nuova* describes a dream in which personified Love has Dante's heart in his hand and holds a nearly naked sleeping woman covered in crimson, later identified as Beatrice. She is awoken by Love and eats the heart and ascends towards heaven. He describes his feelings on seeing Beatrice as follows:

> The moment I saw her I say in all truth that the vital spirit, which dwells in the innermost depths of the heart, began to tremble so violently that I felt the vibration alarmingly in all my pulses…[18]

From that point on, Love ruled over his soul. Immediately after this experience he became wan, weak and debilitated for a time. Thereafter, the *Vita Nuova* contains many references to his bodily feelings, of fear, heart tremors, throbbing pulsations in his left chest, trembling, loss of voice power and much, much weeping. There are many thoughts dwelling on Beatrice, "Love is encompassed in my Lady's eyes".[19] One day, after considering that he and she would both die, he described dream-like imaginings which included seeing the dead Beatrice, dead birds flying in the air, stars, angels and personified Death, whom he desired to come to him.

As a story about the love of nature and the nature of love, Dante in the *Divine Comedy* explores not only how the Universe is put together, but what lies behind the divine purpose: light contrasted with darkness, the dissolution of time and space, harmony and unity, beauty, Beatrice, the face and eyes, all 'will and desire turned, /as wheels that move in equilibrium, / by love that moves the sun and other stars'.[20] Eyes:

> From my dear Lady's eyes a light doth gleam,
> so clear and noble that, where it doth shine,
> Things are revealed that no artist can define....[21]

The metaphysical nature of the erotic seeking something beyond human experience, linked with pre-embodied memories, hints at predestination, with a seeking for the two parts of the whole to come together.

A late-Medieval writer, exiled from his native Florence, in a culture well-versed in courtly love and chivalry, Dante took the secular love-poetry and surrounded it with a religious aura. He was inspired by the philosophy of St Thomas Aquinas (1225-1274), straddling those of Plato and Aristotle (384-322 BC). Man is matter as well as spirit, sensory experience providing images from which the human mind can abstract valid universals: human reason gaining power from divine truth and revelation. Desire, present in all of us, is immanent, and should lead towards God. The pursuit of happiness is a part of us, seeking ecstasy [**Fig. 1.14** Dante]:

> The mind, which is created quick to love,
> when roused by pleasure into conscious act
> will tend towards such things as give us delight. [22]

The deepest parts of Hell are not hot, but cold and even tears are frozen. But no one is beyond repentance, which is not the same as forgiveness for those who can put themselves on the *right* path (in the *Inferno* the various turns during his descent are to the left, but in the ascent in the *Purgatorio* they are to the right). In the *Inferno* the souls are lost, but not necessarily sinners (there are courtesans and whores in *Paradiso* [e.g. cantos 8 and 9], and several popes in *Inferno* [e.g. Canto 19]). Francesca da Rimini is in the second circle of the *Inferno*, reserved for the lustful. But it is not the love between her and Paulo il Bello which is the transgression, but the kiss, aroused when reading together the love story between Lancelot and Guinevere, which was a breach of courtly order (Canto 5). They are swept continually by violent winds, emblematic of their wayward passions. Pyotr Ilyich Tchaikovsky's (1840-1893) tone poem about the couple, a fantasy after Dante (1876), captures the fraught futility of their adulterous passion. In the opera by Riccardo Zandonai (1883-1944, 1914) their secret meetings, marked by wine and kisses, are discovered, and both are killed in a type of Love-Death. Other operatic characters found in *Inferno* are Semiramis, Helen of Troy, Dido, Cleopatra, and Tristan. [**Fig. 11.15-18** Paolo and Francesca]

At the end of Canto 5, having heard of the sad sufferings of Francesca, Dante records

> …I, in such great pity,
> fainted away as though I were to die.
> And now I fell as bodies fall, for dead.[23]

There are many episodes in the *Divine Comedy* echoing those in the *Nuova Vita* of his attacks of fainting, suggestions of loss of consciousness, and an equivalence of Love and Death.[24]

Unlike John Milton (1608-1674), who wrote *Paradise Lost* (1667) to justify the ways of God to men, Dante wanted to bring people into the Christian fold. His *Divine Comedy* can be viewed an overarching tale, an epic journey, from being lost in a dark wood at the midpoint of his life, to a state of bliss and eternal light, guided at first by Vergil, and thereafter by Beatrice. Johann Wolfgang von Goethe's *Faust,* Shakespeare's sonnets, or Marcel Proust's (1871-1922) *À la recherche du temps perdu* (*In search of Lost Time)* represent similar artistic endeavours of lives enduring journeys towards some kind of understanding and redemption. Later we will sense the same trajectory in the epic adventures of Wotan, Brünnhilde and Siegfried's in Wagner's operatic cycle *Der Ring des Nibelungen* (1876).

In the Renaissance, the philosophical swings between various forms of Platonism and Aristotelianism hardly resolve their tensions. The mathematician/philosopher Alfred North Whitehead (1861-1947) said that all Western philosophy was a footnote to Plato, yet it was the cataloguer and systemiser Aristotle who had provided the rival earth-bound scheme.[25] According to the historian Arthur Herman (1956-), the idea that the rise of modern science involved a struggle between reason and religion is not merely wrong but also misleading. The real struggle has been between Aristotle and Plato. Both were concerned with things eternal and immutable, but they also needed to deal with what was transitory and changing. For Plato the idealist, wisdom came not through the senses, but through contemplation and reason; while for Aristotle knowledge was derived *a posteriori,* from experience.[26] Other dualisms, notably active versus passive, reason versus passion, Apollo versus Dionysus, right versus left, and religion versus science are parts of this continuing debate. [**Fig. 1.19** Raphael, *The School of Athens*]

Michelangelo Buonarotti (1475-1564) was well acquitted with the work of Dante, and a sonnet to him ends "Ne'er walked the earth a greater man than he".[27] This was concerned with how spiritual love (Platonic) emerges from physical love (sensual): and eyes have a most important place in this transformation.

I let my eyes become my poison's gate
When they let sharp arrows pass them freely,
And for soft glances made my memory,
which never will decrease, a nest and crypt,
Violent passion for tremendous beauty
Is not perforce a bitter mortal error,
If it can leave the heart melted thereafter,
So that a holy dart can pierce it quickly.[28]

Michelangelo gets beyond Platonic idealism, as night and day, dawn and dusk, the feminine in the masculine and Eros and Thanatos all found greater expression in his sculpture. No better image of the holy dart can be seen than in the statue by Gian Lorenzo Bernini (1598-1680) of the *Ecstasy of Saint Teresa* (sometimes referred to as the *Transverberation of Saint Teresa*) in the Cornaro Chapel, Santa Maria della Vittoria, Rome. [**Fig. 1.20** Bernini, *St Teresa*]

William Shakespeare (1564-1616) expresses similar ideas: Sonnets 24, 46 and 47 are referred to as the 'eye-heart' sonnets.

Sonnet 24
Mine eye hath played the painter and hath steeled,
Thy beauty's form in the table of my heart;
My body is the frame wherein 'tis held,
And perspective that is best painter's art.
For through the painter must you see his skill,
To find where your true image pictured lies,
Which in my bosom's shop is hanging still,
That hath his windows glazed with thine eyes.
Now see what good turns eyes for eyes have done:
Mine eyes have drawn thy shape, and thine for me
Are windows to my breast, where-through the sun
Delights to peep, to gaze therein on thee;
Yet eyes this cunning want to grace their art,
They draw but what they see, know not the heart.

Sonnet 47
Betwixt mine eye and heart a league is took,
And each doth good turns now unto the other.
When that mine eye is famished for a look,
Or heart in love with sighs himself doth smother,
With my love's picture then my eye doth feast

And to the painted banquet bids my heart.
Another time mine eye is my heart's guest,
And in his thoughts of love doth share a part.
So either by thy picture or my love,
Thyself away are present still with me;
For thou no farther than my thoughts canst move,
And I am still with them, and they with thee;
Or if they sleep, thy picture in my sight
Awakes my heart to heart's and eye's delight.

The Metaphysical Poets were especially concerned with such conceits, as in John Donne (1572-1631) to a lover:

Our eye-beames twisted, and did thread
Our eyes, upon a double string…
And pictures on our eyes to get
Was all our propagation.

Their souls

…(which to advance their state,
Were gone out), hung 'twixt her and me.
Soe soule into the soule may flow,
Though it to body first repaire.[29]

Religion was much embedded in his poetry, as in his cry "Batter my heart, three-personed God".[30]

These metaphysical aporia met the challenge of the Renaissance, in which new forms of artistic expression sensed a changing world view: "How beauteous mankind is! O brave new world / That has such people in't!". New continents were explored, as were the stars of the universe and the shadows of the mind: "Tis all in pieces, all coherence gone", as Donne put it.[31]

Psychological penetration of protagonists and antagonists, and metaphysical entanglements "as imagination bodies forth/ The form of things unknown" are expressed in Shakespeare's plays, especially the last ones. E.M.W. Tillyard (1889-1962) referred to "a sense of different planes of reality [which] renders probable a certain amount of symbolism".[32] Prospero acknowledges Caliban as his own progeny, born of a witch, "a thing of darkness", living on Prospero's enchanted island, where Ariel 'sucks' in the same place as the bees. These developments in the arts conflicted with the developing Enlightenment perspectives, and prospered in the literature and music of the late 18th and 19th centuries with the Romantic era.

Romantic Times

Historical epochs achieve their identity only in retrospect and are never perfectly demarcated. Quite when to point to the beginning of a Romantic era fails adequate discussion, yet there is something about the ideas and spirit of the *Weltanschauung* fully emerging in the 19th century which is unmistakable and of central importance to the themes we are developing in this book.

Some might suggest origins going back to the period of Greek heroes with the struggles of Oedipus and Orpheus against fate. Others perhaps can point to Martin Luther (1483-1546), who fostered a revolution that undermined the strictures of Catholic authority and engendered a religious and secular spirit of liberation. Milton's epic *Paradise Lost* (1667-74), spanning eons of time and the whole universe, has Satan as the first character to appear, and William Blake (1757-1827) famously concluded that Milton himself was of the Devil's party. The poem raises awareness of the striving of Satan's consciousness and issues of free will.[33] Like Oedipus, the blind Milton wanted to tell of things invisible to mortal sight: insight since the sighted only see surfaces.

While the Enlightenment period forged progress in the arts and sciences, there was an emerging interest in minds, motivations and memories of individuals. There was a shift away from empirical philosophies towards idealism and it was the era of political as well as industrial revolutions, notably in America, England and France. Further, it engendered the cult of celebrity.

Notes

[1] Whitman, Walt. *Leaves of Grass*. 'Italian Music in Dakota' was published first in the 1881–82 edition of *Leaves of Grass*. This manuscript was likely composed between 1879 and 1881, after Whitman took a trip to the West (though not the Dakota Territory) and before the poem was published. Whitman observed: "But for the opera, I could never have written *Leaves of Grass*". See Brasher, Thomas L. (2008). Judith Tick, Paul E. Beaudoin (ed.). "Walt Whitman's Conversion To Opera". *Music in the USA: A Documentary Companion* (Oxford: Oxford University Press), 207.

[2] Huysmans, Joris-Karl. *Against Nature* (*À Rebours*). Trans. Robert Baldick (London: Penguin Books, 2003), 53. There is reference to the story in Matthew 14: 1-12, and Mark 6: 14-29. The opera was written in French and German, with differences between the two, Strauss preferred the German version.

[3] Letellier, Robert Ignatius. *The Bible in Music* (Newcastle: Cambridge Scholars Publishing, 2017), 359.
[4] Wedekind, Frank. *The Lulu Plays.* Trans. Stephen Spender (London: Alma Classics, 2015), 10.
[5] Kennedy, Michael. *Richard Strauss: Man, Music, Enigma* (Cambridge: Cambridge University Press, 1999), 26, 303.
[6] Conrad, Peter. *A Song of Love and Death* (London: Chatto & Windus, 1987), 11.
[7] There are many references to the life and times of Orpheus. A. Wroe, *Orpheus: The Song of Life* (London: Jonathan Cape 2011) is very comprehensive.
[8] Ovid. *Metamorphosis The Death of Orpheus,* Book IX, 1-66.
[9] Ibid., 19.
[10] Apollo was for Nietzsche the god of light, the god of sculpture, the god of restraint, the god of the line and of the *principium individuationis.* Dionysus aligns with song and dance, increases bonds between people, releasing and freeing the enslaved soul of man. Friedrich Nietzsche, *The Birth of Tragedy: Out of the Spirit of Music.* Trans. S. Whiteside (London: Penguin, 1993).
[11] Gordon, J. *Plato's Erotic World From Cosmic Origins to Human Death* (Cambridge: Cambridge University Press, 2012), 1-3.
[12] Hackworth, R. *Plato's 'Phaedrus'* (Cambridge: Cambridge University Press, 1952), 10.
[13] *The Holy Bible.* Revised Standard Version (1894). Catholic Edition (London: Thomas Nelson & Sons, 1966).
[14] Liébert, G. *Nietzsche and Music* (Chicago: University of Chicago Press, 2004), 150.
[15] Friedrich Nietzsche quoted in Liébert, G. Ibid., 43. Nietzsche altered many of his ideas as his philosophy progressed. The place of music diminished with his dissatisfaction with Wagnerian Romanticism, and he rejected many of Schopenhauer's ideas. However, Dionysus was still there to the end.
[16] Moore, G. *Nietzsche, Biology and Metaphor* (Cambridge: Cambridge University Press, 2002), 85.
[17] Lattimore. Richmond. Trans. from *Greek Lyrics*, ed. Richmond Lattimore (Chicago: University of Chicago Press, 1960).
[18] Dante Alighieri. '*La Vita Nuova*' (London: Penguin Classics). Trans. B. Reynolds (Cambridge: Cambridge University Press 1969), 29.
[19] Ibid. 21:60.
[20] Dante. *Paradiso. The Divine Comedy*. Trans. R. Kirkpatrick (London: Penguin Classics, 2012), Canto 33: 143-145.
[21] Dante. *Canzoniere.* Trans. E. H. Plumptre (London: Sir Isaac Pitman & Sons Ltd), Sonnet 31, 1-3, 6.
[22] Dante. *Purgatorio. The Divine Comedy*. Trans. R. Kirkpatrick (London: Penguin Classics, 2012), Canto 18: 19-21.
[23] Dante. *Purgatorio*. Ibid. Canto 5: 140-142
[24] A discussion of these episodes is given by the Neuropsychiatrist Marco Mula. He considers the possibility that Dante had epilepsy. In terms of the morbidity at his time from fevers, head injury and the like, and the fact that he was a member of the Ars Medicorum et Apothecariorum, he must have been familiar with the

condition. See Mula, M. "Epilepsy in Dante's Poetry", *Epilepsy and Behavior*, 57 (2016): 251-254.

[25] Whitehead, Alfred North. *Process and Reality* (New York: Free Press, 1979), 39.

[26] Herman, A. *The Cave and the Light: Plato versus Aristotle, and the Struggle for the Soul of Western Civilisation* (New York: Random House, 2013), 327. In the famous fresco by Raphael (1483-1520), *The School of Athens,* painted between 1509 and 1511, the philosophies of Plato and Aristotle are contrasted. Plato, on the left, is pointing upwards and surrounded by Pythagoras, Socrates and other thinkers whose ideas linked together idealism. Aristotle is on the right with thinkers allied to empiricism, including Ptolemy and Euclid. Herman summarises thus: "Plato and Aristotle serve as the twin fountainheads of Western reason, intellectual coequals who sum up the entire scope of human knowledge...The painting...captures the dual character of Western culture almost from its start", p. xx. These dualisms have reverberated through neuroscience. See Michael Robert Trimble, *The Intentional Brain: Motion, Emotion and the Development of Modern Neuropsychiatry* (Baltimore: John Hopkins Press, 2016).

[27] Longfellow, Henry Wadsworth. *Poems and Other Writings* (New York: Literary Classics, 2000), 700. The poem was translated into English by Longfellow.

[28] Michelangelo. *The Complete Poems and Selected Letters of Michelangelo.* Trans. C. Gilbert (Princeton: Princeton, University Press, 1980). Sonnet fragment 22: 1-4, p. 16: "*Sonnet to Tommaso Cavalieri*", 258: 1-4, p. 145.

[29] Gardner, H. *The Metaphysical Poets* (London: Penguin Books, 1972): John Donne, *The extasie,* pp. 74-77.

[30] Donne, J. *The New Oxford Book of Seventeenth Century Verse* (Oxford: Oxford University Press, 1991), Sonnet XIV, p. 118.

[31] William Shakespeare, *The Tempest*, 5:1; Sonnets 24 and 47. See Paterson, D. *Reading Shakespeare's Sonnets* (London: Faber and Faber, 2010), 74, 138; *John* Donne, *An Anatomy of the World, First Anniversary* (London: W. Stansby, 1625), lines 206-9,213.

[32] Tillyard, E. M. W. *Shakespeare's Last Plays* (London: The Athlone Press, 1938), 68. He was especially referring to *Cymbeline*, *The Winter's Tale* and the *Tempest*. Quotes: *The Tempest,* 5.1.186-187; 5.1 88: *Midsummer Night's Dream,* 5:1. 15-16.

[33] There was much to this, as the poem, although allegorical, has a vast political backdrop, covering the troubled times of Milton's lifespan. Milton himself was a republican who was only saved from punishment at the time of the Restoration of the monarchy in 1660 on account of his reputation as a poet and his infirmity and blindness.

Satan:
"Which way shall I fly
Infinite wrath, and infinite despair?
Which way I fly is hell; myself am hell" (*Paradise Lost*, 4:73-75).

CHAPTER 2

THE EMERGENCE OF ROMANTIC THOUGHT

Music the fiercest grief can charm,
And fate's severest rage dis arm:
Music can soften pain to ease,
And make despair and madness please;
    Our joys below it can improve,
    And antedate the bliss above.

(Alexander Pope, 'Ode for Music on St Cecilia's Day')[1]

Italy and France

While Italian opera had already split in the 18th century into two genres, *opera seria*, and *opera buffa*, in France there was much academic philosophical cavilling over what was referred to as the *Querelle des Bouffons* (The War of the Buffoons*)*. The importance of reason as opposed to emotion in the composition and experience of music, within the formal structure of the classical plays of Jean Racine (1639-1699) and Pierre Corneille (1606-1684), conflicted with the free spirit and liberty of Jean-Jacques Rousseau's (1712-1778) romantic novel *La Nouvelle Héloïse* (1761), a best-seller which made him famous.

Between 1752 and 1754, in Paris, a troupe known as Les Bouffons performed Giovanni Pergolesi's (1710-1736) archetypal *opera buffa*, *La serva padrona* (1733, *The Servant Turned Mistress*), with a libretto by Gennaro Federico (1726-1744), and other Italian comic works. Their great success, and the growing enthusiasm for the fair theatres, led to this polemical war, a clash, essentially literary in nature, centred on the confrontation between the solemn past (enshrined in Italian *opera seria* and the grand Baroque French *tragédie lyrique*) and the lively topical present (both the farce and sentiment of the newly emerged Italian comic opera and the unsophisticated popular vaudevilles). The Bouffons helped to introduce French composers to the idea that the libretto should serve to enhance the music, whereas in the fair theatres the composer had played a subservient role.[2]

The Bouffons also proved influential on musical content, since many of the *ariettes* they used were initially borrowed popular pieces transferred to new works (in the process of parody), where in turn they served as models for original compositions.

The disputations had a happy outcome in the emergence of a uniquely French form of opera—the *opéra-comique*—an amalgam of the fair theatre performances, the English ballad opera, the German *Singspiel* and the Italian *opera buffa*. While the debate had often been couched in nationalistic terms, subsequent compositions, both by French composers and by a series of resident foreigners, offered excellent examples of this new genre.

The first battle of the Guerre des Bouffons began in 1752, the year when Rousseau staged his one-act comic opera *Le Devin du village* (*The Village Soothsayer*) in the Court Theatre at Fontainebleau. Both the libretto and music were his own. The music had been compiled in the manner of a *pasticcio*, reflecting the popular romances and vaudevilles being heard in the Parisian fairs. The score pleased both sides in the aesthetic debate, appealing to the French in its manner and sentiment, and to the Italians in its employment of recitative and continuously composed form.

Rousseau had hoped to provide a model for the new French comic opera but was not immediately successful. However, his achievement was developed, notably by the Neapolitan composer Egidio Duni (1709-1775), who settled in Paris in 1755 and began first assembling and then composing *opéras-comiques* after the style of Rousseau.

Germany

In Germany the conflict was not so much the new philosophies against Classicism, but against rationalism, the cornerstone of the Enlightenment. Goethe's *Götz von Berlichingen* (1773), a historical play based on the life of an imperial knight-poet, and his novel *Die Leiden des jungen Werthers* (*The Sorrows of Young Werther,* 1774), ushered in what was referred to as *Sturm und Drang*: many of the succeeding texts later found their way into operas. Werther, chiding rationalists, has his senses stretched to breaking on account of his passionate love for the married Lotte. Following the fate of Francesca and Paolo, they experience such joy and tears and intense emotion, leading to kisses, while reading a translation of the works of Ossian, a supposed Gaelic epic written in the third century BC and 'translated' into rhythmical prose by a Scotsman, James MacPherson (1736-1796) in 1761. The relationship is too much for Lotte, who realises that she must separate from Werther, who then shoots himself.

In tales of this genre there was much dying, grief and joy, wars and unhappy love, leading towards the Gothic. Goethe wrote Classical-style dramas such as *Iphigenie auf Tauris* (1779) and straddled what might be called the Classical/Romantic *Wanderweg*. He was quite negative about French poetry and philosophy, and paved the way to a fully-fledged Romanticism with his *Faust* legend (1808, 1832). The movement was further developed by the philosopher-poet Johann Gottfried Herder (1744-1803). The latter, *pace* Kant, insisted that language was central to understanding the spirit of the *Volk,* with ancestral origins waiting to be revived. Herder considered that poetry was the first human language and, along with music, was one of the languages of emotion.[3] **[Fig. 2.1** Goethe]

One consequence was interest in other national epics, such as *Nibelungenlied* in Germany, and stories by the Scottish poet and novelist Sir Walter Scott (1771-1832). Scott was one of the significant originators of Romanticism, and many of his stories found their way into opera.[4] Initially Goethe had considered the Classical 'healthy' and the Romantic 'sick', but after meeting the poet-philosopher Friedrich von Schiller (1759-1805), he confessed that, contrary to his inclinations, he was a Romantic.[5] Goethe was fascinated by archetypes (*Urtypen*), and some philosopher scientists at the time, the *Naturphilosophen,* sought causes within forces that could explain archetypes and their progressive variations. This led to organic, quasi-biological metaphysical theories of nature, in contrast to those based on empirical Newtonian principles. For them, time, evolution, and *becoming* were embedded in the concept of the development of *Homo sapiens* as well as of the self; the less-organised appearing before the complex successor meant the whole must be considered as more than the sum of the parts.[6]

Schiller's *Über naïve und sentimentalische Dichtung* (1795) was a watershed in the development of the Romantic era. He opined that art, as well as civilisation, was at a crisis point, stemming from a century before. This was another spat like the *Querelle des Anciens et des Modernes.* In Schiller's essay naïve poetry, associated with the Ancients, was concerned with the immediate: it is object possessed (he cites, for example, Homer and Shakespeare) and deals with the surface of life, nature represented directly.

This was unable to express the modern sentiment (hence *sentimentalische*), in which the poet himself becomes involved in seeking the essence of nature, an ideal to be sought for, interposing his personality on the works. This meant tackling the gulf between himself and his *Umwelt*, between the ideal and the real, a psychological ingredient missing with the Ancients.

There were poets transitioning between the two (such as Goethe). But while Schiller himself wanted to preserve much of the ancient traditions, his poetry, plays and theories have become memorialised in music. This includes Ludwig van Beethoven (1770-1827) in his 9th Symphony (*An die Freude*—Ode to Joy, 1824), and in several operas: *Don Carlos* (1867, Giuseppe Verdi, 1813-1901), *Maria Stuarda* (1835, Gaetano Donizetti, 1797-1848), *Die Braut von Messina* (1882/3, Zdeněk Fibich, 1850-1900), *Luisa Miller* (1849, Verdi) and *Guillaume Tell* (1829, Gioacchino Rossini, 1792-1868), to name a few.

Novalis (Georg Philipp Friedrich Freiherr von Hardenberg, 1772-1801) knew Schiller and other philosophers who were advancing Romantic ideas, stimulating an interest in the discipline of aesthetics. They included Karl Friedrich Schlegel (1772-1829), Kant, Johann Gottlieb Fichte (1762-1814) and Friedrich Wilhelm Joseph Schelling (1775-1854).

In 1794 Novalis met and fell in love with 13-year-old Sophie von Kühn. They were engaged in 1795, but she died of tuberculosis two years later. He eulogized night and death as times of ecstasy, and visualized Sophie in Paradise, where they would be united in the presence of Christ, a mystical and loving union with her and the universe as a whole after his own death. He gave to Romantic literature the image of the *die blaue Blume* (the blue flower) and wrote of the indwelling of infinity in the finite.[7]

Novalis expressed his grief in *Hymnen an die Nacht* (1800; Hymns to the Night), six prose poems interspersed with verse celebrating night, or death. They are a mixture of prose and poetry, avowing transcendent feelings which encode day and night—the latter embracing sleep, death and love, a union between two souls. The realm of the night was timeless and infinite, creative; but also in his writings the erotic was tangible, even when he was at the side of Sophie's grave [**Figs. 2.2 a & b** Novalis].

> Once when I was shedding bitter tears, when, dissolved in pain, my hope was melting away, and I stood alone by the barren mound which in its narrow dark bosom hid the vanished form of my life—lonely as never yet was lonely man, driven by anxiety unspeakable—powerless, and no longer anything but a conscious misery. —Out of the blue distances—from the

> hills of my ancient bliss, came a shiver of twilight—and at once snapped the bond of birth—the chains of the Light. Away fled the glory of the world, and with it my mourning—the sadness flowed together into a new, unfathomable world.—Thou, Night-inspiration, heavenly Slumber, didst come upon me.—The mound became a cloud of dust—and through the cloud I saw the glorified face of my beloved. In her eyes eternity reposed—I laid hold of her hands, and the tears became a sparkling bond that could not be broken. Into the distance swept by, like a tempest, thousands of years. On her neck I welcomed the new life with ecstatic tears. It was the first, the only dream—and just since then I have held fast an eternal, unchangeable faith in the heaven of the Night, and its Light, the Beloved.[8]

These ideas resound in some of the paintings of Caspar David Friedrich (1774-1840): the natural within a supernatural, physical presence with dream-like visions, images of Christ the redeemer, as in Blake's 'To see the World in a Grain of Sand/ And a Heaven in Wild Flower'.[9] The ordinary needed a higher meaning, as the philosopher Wilhelm Dilthey (1833-1911) expressed it: [**Fig. 2.3** Caspar David Friedrich]

> On the basis of this culture there arose a poetic world created by Goethe, Schiller, and Jean Paul, which has been developed further through Novalis and Hölderlin. The entire spiritual development of Europe was influenced by this new world-historical force.[10]

This is reflected in themes developed in this book, especially in the Romantic musical tropes of the 19th century.

Dreams and Death

The Romantics had a particular interest in dreams and states of mind that hover between the conscious and the unconscious, either sleep or drug-induced reveries. Interest in dreams has a long philosophical tradition. For Plato and the Neo-Platonists, the dream was a *vacatio animae*, moments of loss of the self, and hence reason, but such moments liberated the soul and could lead to poetic or divine inspiration. Prophecy and dream interpretation are part of a continuing history, and dream interpreters existed long before Sigmund Freud (1856-1939). As we shall see, in the context of different layers of experience, and contact with another world, dreaming, along with sleeping, is relevant not just as a theme in many operatic tales, but perhaps also in inspiring composers towards their compositions. The equation of sleep with death, and between both these states and an alternative transcendental world, will be explored in later chapters. It is a central feature of the Gothic imagination.

Love and Death in Opera

We now wish to unwrap some recurrent themes encountered in the operatic corpus as we move from the 17th century towards Romantic times, especially the growing emphasis on the psychological conflicts of protagonists. The Renaissance saw a development of music itself.

The Sonata Principle

One key development in the history of music during the middle of the 18th century was the definitive formulation of the sonata principle, where a form of musical construction encapsulated a concept of dramatic locution that effectively changed, by innate structure, the nature of musical discourse. The triple movement of an exposition of two contrasting themes, their variation, juxtapositioning, interplay and development, with their eventual reconciling resolution, provided a wordless mode of dramatic interaction that stirred the pulses of musical creativity from that time on. The Italian composer Leonardo Vinci (1690-1730), prompted by the extraordinary gifts of the castrato singer Farinelli (Carlo Broschi, 1705-1782), began to differentiate ever more sharply the central section of the operatic aria, elaborating the contrast with the opening and closing sections, and so strongly influencing the development of the sonata principle. The emergence of the early symphony by Johann Stamitz (1717-1757) and the Mannheim School, by Carl Philipp Emanuel Bach (1714-88) and by Joseph Haydn (1732-1809), established the form decisively, and fed straight into the mood and self-expression of the intensely emotional *Sturm und Drang*, the harbinger of Romanticism.

There was also the ascendency of the value of music over words as interactions between humans, rather than between gods and humans, drove the drama along. The textures of human love, the eternal return and Goethe's concept of *das Ewig-Weibliche* (the Eternal Feminine) coincided and collided with salient conceits, as metaphysical and transcendental ideas and experiences became of increasing concern in the arts and philosophy, finding expression through music, especially opera.

While it may seem that Orpheus lost his voice towards the end of the 19th century as the 'realism' of the operas of Jules Massenet (1842-1912), Giacomo Puccini (1858-1924), Ruggero Leoncavallo (1857-1919) and Pietro Mascagni (1863-1945) dampened Romantic aspirations, and as the 20th century saw music itself lose its begotten foundations, his echo is still with us. Orpheus, after all, had human qualities, as did his Eurydice. They possessed that most human of emotional needs: Hope, and its companion Faith, and experienced Love and Loss. Neither trusted themselves and neither trusted the other: he looks, she undergoes a second death. In some versions she is in the real human underworld of crooks and hoodlums (Jacques Offenbach, 1819-1880); in others they are reunited on earth (Gluck). In variants of the theme, they will be together in heaven or some other ethereal realm of togetherness.

Contrasts and conflicts between dark and light, night and day, reason and inspiration, are thematic as redemption through love becomes epiphanic. As Roger Scruton (1944-2020) put it: “the redemption offered by the *Liebestod* is no illusion. It offers the very thing that redemption is, namely a transcendence of the world of appetite into the world of values”. The mysteries of our human condition are revealed through music.[11]

Mozart and the late Classical operas

Pre-Romantic but important hints of this development are features of Wolfgang Amadeus Mozart’s (1756-1791) operas **[Fig. 2.4]**. While in many ways epitomizing the style and rational equilibrium of the Classical Age (as perfectly embodied in his *opere serie Idomeneo,* 1781, and *La Clemenza di Tito,* 1791), with the inspirational help of the librettist Lorenzo da Ponte (1749-1838), Mozart developed insights into the human predicament almost unequalled in the Classical mode. In the genre of *opera buffa* (*Le nozze di Figaro,* 1786, and *Cosi fan tutte,* 1790) his works acknowledge a growing revolutionary social sensibility stimulated by the subversive drama of Pierre-Augustin Caron de Beaumarchais (1732-1799). But in his operas *Don Giovanni* (1787) and *Die Zauberflöte* (*The Magic Flute*) (1791) there is a new mood and strong surging power that begins to move away from the rational realism of the Classical. With myths of modern times (Don Juan) and the magic of folk theatre (Emanuel Schikaneder, 1751-1812), Mozart begins to explore the irrational powers and symbolism associated with the advent of Romanticism. This was paralleled in the rediscovery of folk poetry by Bishop Thomas Percy (1729-1811) and Herder, and the emergence of the Gothic Novel (1764-1821), with its worlds of bold adventure, wild nature, dark passions, night and dreams. The power of an avenging fate or providence (in the Statue of the Commendatore and the damning of Don Giovanni) and the detailed binary imagery and mystic ordeals of *Der Zauberflöte,* with its shaping use of day and night and the incipient ramifications, demonstrate a deep and irresistible compulsion breaking out of Classical certainties, finding new world of bold new and challenging spiritual exploration.

Der Zauberflöte is a type of rescue opera, featuring a fundamental struggle between light and darkness which are taken through various levels and elaborations. These are basically determined by Sarastro and his realm of the sun, and the Queen of the Night and her empire of the moon and stars. The struggle and rescue focus on the destinies and development of the young couple Tamino and Pamina. Both are taken on developmental journeys from being isolated and ignorant individuals to become mature members of their social and intellectual community, symbolized in their marriage union. The *Enwicklungs- geschichte* (development history) is highlighted in the vow of silence that Tamino must embrace and Pamina suffer, and the ritualized trials by fire and water they undergo together when, protected by the Flute, they walk unharmed through the elements, rites of initiation and cleansing, birth and death. They undergo a type of Love-Death in their ordeal, but in typical Enlightenment mode, Mozart's interest is in life here and now.

The process represents an archetype of the nature of mankind. The reconciliation of the male and female principles is embodied in the emergence of Tamino and Pamina as man and wife, symbols of a new age of hope illumined by Sarastro's light of reason, and eschewing the Queen of the Night and her realm of the unconscious. This was both an appeal to Masonic principles of clarity and order, and a topical political message, advocating the new order of sensible, rational politics under the Emperor Joseph II (1741-1790), as opposed to the perceived Catholic obscurantism of the Empress Maria Theresa (1770-1780). However, Schikaneder's parable rises above its own intentions, and in celebrating the triumph of rational Enlightenment, also looks enticingly at the mysterious fluidity and seductive attraction of the nocturnal realms.[12]

Sudden Unexpected Death in Opera

Above we have given a brief compendium of causes of death that form the apotheoses, the high points of the dramas in many operas, notably tragedies, and noted some significant musical developments. The subject of operatic deaths is a topic often discussed, but our intention is to highlight causes of death which we refer to as *Unexpected Death,* apparently unexplained, and of concern to authors such as Catherine Clément, who laments some women die "of nothing, just like that". In a discussion of female roles of 18th-century opera, the director Barrie Kosky pondered:

> In Greek theatre the women characters are the most interesting. ...It's Electra, Antigone and Clytemnestra. ...It's the same with Shakespeare's women, or the women in Baroque plays and operas. There is Mozart too—his women characters are incredible. But then something went terribly wrong. It's difficult to say exactly when. But throughout the 19th century, women were pigeon-holed—it's as if they were shackled. ... In 19th-century operas the women are suddenly either sick, crazy, dying or all three. It is impossible to imagine finding a woman like Agrippina in any German, Russian or French opera of the 19th century.[13]

The selection of operatic deaths cited above follow expected causes, integral to the story. But we are interested and will explore those deaths which occur apparently without obvious explanation, yet form a vital part of the operatic corpus, and remain popular with audiences.

We have identified 50 operas where such events occur. We have examined the librettos, profiled the sex and age of the *dramatis personae*, and the context within which the tragedy happens. The historical times when the operas were composed, and their composers, form a part of our analysis.

Specific themes emerge. These include a preponderance of females, especially in the 19th century, who die mainly at the end of the operas, and all in the context of tragedy. We explore the librettos for evidence that the composer has perhaps identified a weakness or instability of the protagonists' emotional states and the moments of demise, including the musical themes surrounding the final moments of the expiration.

It is not possible to begin to understand our aims without an exposition of the flux of philosophical, medical and social ideas that abounded in the 19th century. In view of the transcendental implications emerging from the tragic implications of the plots, in the next chapter we examine, in philosophy, religion and music especially, transcendental experiences and transfiguration (*Verklärung*). These occupy the borderlands between mind and body, with the growing awareness in the medical sciences of the unconscious forces driving human behaviour, and the liminal mental states and trances that appear in the operas, sometimes brought about by magic or potions. Composers were influenced by the underlying philosophical ideas, in particular those stemming from Kant and Georg Wilhelm Friedrich Hegel (1770-1831), through to the later works of Schopenhauer and Nietzsche. This was notably true of Wagner, with whom rapture/love- death is a main theme.

Notes

[1] Pope, Alexander. 'Ode for Music on St Cecilia's Day'. Emily Fragos, *In Music's Spell* (New York: Alfred Knopf, 2009), 43.
[2] The *Théâtre de la foire* refers to the theatres that put on performances at annual summer and winter fairs at St Germain and St Laurent in Paris.
[3] Herder emphasized the distinction between *Zivilisation* (Civilization) and *Kultur* (Culture), the latter being the temperament of the *Volk* (a historic-psychological entity). This had much to do with artistic, social and religious affairs, as opposed to Civilization, which emphasised the rational requirements of the civil: in other words belonging to the development of the city and civilization (*civitas*—citizens united by law). Herder collected poetry and stories from the Germanic past, tasks taken up by others such as the Brothers Grimm (Jakob Ludwig Karl, 1785-1863, and Wilhelm Karl, 1786-1859), Ludwig Achim von Arnim (1781-1831) and Clemens Brentano (1778-1842), whose collections influenced Wagner. For much on the importance of such ideas for German history see: P. Watson, *The German Genius: Europe's Third Renaissance, The Second Scientific Revolution and the Twentieth Century* (London: Simon Schuster, 2010).
[4] Scott's popularity was huge and his works were translated into French and most other European languages. A six-volume edition of his works published in the 1820s sold some one-and-a-half million copies. See Pittock, Murray (ed.). *The Reception of Sir Walter Scott in Europe* (London: Bloomsbury, 2006).
[5] Eckermann, J .P. *Gespräche mit Goethe in den letzen Jahren seines Lebens*, 3rd ed. (Berlin: Aufbau-Verlag, 1987), 286, 350.
[6] A superb development of these ideas is given in R. J. Richards, *The Romantic Conception of Life: Science and Philosophy in the Age of Goethe* (Chicago: University of Chicago Press, 2002).
[7] Novalis's exact meaning of the Blue Flower is unclear. It appears in his unfinished novel *Heinrich von Ofterdingen* (1802). From his notes, Novalis intended to finish with Ofterdingen picking the blue flower, but if this related to the infinite, or the eternal beyond death, is speculation. Nonetheless it became an important symbol of German Romanticism.
[8] Novalis. *Hymnen an die Nacht.* Ed. P. Kluckhohn and R. Samuel. Ibid, Third Hymn, 1:135. Trans. R. J. Richards, Ibid., 33.
[9] Blake, William. *Auguries of Innocence* (1789). See *Selected Poems*. Ed. P. Butter (London: Everyman, 1993), 121.
[10] Dilthey, Wilhelm. *Poetry and Experience: Selected Works, Volume 5.* Ed. R.A. Makkreel and F. Rodi (New Jersey: Princeton University Press, 1985), 253.
[11] Scruton, Roger. *Death-Devoted Heart: Sex and the Sacred in Wagner's Tristan and Isolde* (Oxford: Oxford University Press, 2004), 192.
[12] Cf. Spaethling, R. "Folklore and Enlightenment in the Libretto of Mozart's Magic Flute". *Eighteenth-Century Studies*, 9:1 (1976): 45-68.
[13] Clément, Catherine. *Opera: The Undoing of Women* (London: Virago Press, 1989); Kosky, Barrie. *Agrippina* (London: Royal Opera House, 2019), 13-14.

CHAPTER 3

TRANSCENDENTAL EXPERIENCES

> What I felt is indescribable, and if you will deign me not to laugh, I will try and translate it for you … To be immediately felt carried away, under a spell ... I often experienced quite a strange feeling, the pride and enjoyment of understanding, of being engulfed, overcome, a really voluptuous sensual pleasure, like riding into the air or being rocked on the sea. In general those deep harmonies remind me of those stimulants which accelerate the pulse of the imagination … There is everywhere something elevated and everlasting, something reached out beyond, something excessive, something superlative … this would be if you like, the final paroxysm of the soul …
>
> (Charles Baudelaire, *Les Fleurs de mal*)[1]

Introduction

The words of Charles Baudelaire (1821-1867), poet and art critic, addressed to Richard Wagner after he had attended a performance of *Tannhäuser* in Paris in February 1860.[2] He was describing the immense feelings that music could arouse in him, and its connections to some sort of epiphanic experience.

In his book *Wagner Moments*, James Holman collected together favourite Wagner experiences of 107 people, not all musicians or artists. Responses included that of the tenor Plácido Domingo, someone who rarely let emotion get in the way of his performing technique, who said that when performing as Siegmund in *Die Walküre,* as he lay dying,

> … tears well up in me that I cannot suppress, because I realise that my death was caused at the instigation of my own father, the God Wotan. The reason for this overpowering emotion is the combined genius of Wagner as a librettists and composer.[3]

The composer Max Reger (1873-1916), after hearing *Parsifal*, wrote that he cried for two weeks; and Hugo Wolf (1860-1903), after hearing the same opera observed:

> This is without doubt the most beautiful and sublime work in the whole field of art. My whole being reels in the perfect world of this wonderful work, as if in some blissful ecstasy. I could die even now.[4]

Willa Cather (1873-1947) in her short story *A Wagner Matinee,* describes an evening when an elderly farm-living aunt is taken to a concert in New York by her nephew. Several Wagner pieces are played.

> Soon after the tenor began the Prize Song, I heard a quick-drawn breath, and turned to my aunt. Her eyes were closed, but the tears were glistening on her cheeks, and I think in a moment more they were in my eyes as well. It never really dies, then, the soul? It withers to the outward eye only, like that strange moss which can lie on a dusty shelf half a century and yet, if placed in water, grows green again. My aunt wept gently throughout the development and elaboration of the melody.

The concert closes with Siegfried's Funeral March.

> My aunt wept quietly, but almost continuously. I was perplexed as to what measure of musical comprehension was left to her, to her who had heard nothing but the singing of gospel hymns in Methodist services at the square frame school-house on Section Thirteen. I was unable to gauge how much of it had been dissolved in soapsuds, or worked into bread, or milked into the bottom of a pail.
>
> The deluge of sound poured on and on; I never knew what she found in the shining current of it; I never knew how far it bore her, or past what happy islands, or under what skies. From the trembling of her face I could well believe that the Siegfried march, at least, carried her out where the myriad graves are, out into the gray, burying-grounds of the sea; or into some world of death vaster yet, where, from the beginning of the world, hope has lain down with hope, and dream with dream and, renouncing, slept.[5]

An account as an audience participant at Bayreuth after a performance of *Parsifal* given by Virginia Woolf

> Here at Bayreuth, where the music fades into the open air, and we wander with *Parsifal* in our heads through empty streets at night, where the gardens of the Hermitage glow with flowers like those other magic blossoms, and sound melts into colour, and colour calls out for words, where, in short, we are lifted out of the ordinary world and allowed merely to breath and see—it is here that we realise how thin are the walls between

one emotion and another; and how fused our impressions are with elements which we may not attempt to separate.[6]

Paris

The above extracts are but a small selection of special experiences that people have had singing, listening to or writing about Wagner's operas.

In the 19th century, Paris was the cultural centre of Europe. Honoré de Balzac (1799-1850) in his cycle of novels *La Comédie humaine* refers to scenes from the opera: the *soirée à l'Opéra,* a social event par excellence was 'a metaphor for Parisian society', it could change everything. Gustave Flaubert's (1821-1880) heroine Emma Bovary had her passions re-awakened after attending a performance of *Lucie de Lammermoor,* linked with passion, tragedy, madness and suicide. In Gaston Leroux's (1868-1927) *Le Fantôme de l'Opéra* (1909-10), the first strange accident happens during a performance of Gounod's *Faust,* one of the most popular operas on the Paris stage for nearly half a century.[7]Marcel Proust (1871-1922) was able to use a *théâtrephone* at his home to listen to live opera from Paris opera houses. In his great novel sequence *À la recherché de la temps perdu* [**Fig. 3.1**]. Wagner is mentioned more than any other composer. The *petit phrase* (little phrase) of Vinteuil, his fictional composer, appears in several episodes. It takes on particular significance for Swann and his changing feelings for his lover Odette, yet for Marcel the psychological intensity changes with time as Vinteuil's sonata for piano and violin began to evoke more profound sensual experiences. The sonata and the music of Wagner, especially *Tristan und Isolde,* are related in special ways in Marcel's meditations.

> In this way Vinteuil's phrase had, like some theme from *Tristan,* for example, which may also represent to us a certain emotional accretion, espoused our mortal condition, taken something human that was rather touching. Its destiny was linked to the future, to the reality of our soul, of which it was one of the most distinctive, the best differentiated ornaments. Maybe it is the nothingness that is real and our entire dream is non-existent, but in that case we feel that these phrases of music, and these notions that exist in relation to our dream, must be nothing also. We will perish, but we have for hostages these divine captives who will follow us and share our fate. And death in their company is less bitter, less inglorious, perhaps less probable.[8]

In *La Prisonnière* (*The Prisoner*) Marcel is playing music from *Tristan und Isolde* on his piano, and he likens the special phrase to Wagnerian

Leitmotifs, referred to by Wagner as *Gedächtnismotives*. Marcel's mediations on music and its inseparability from the sensuous, the erotic, love, jealousy and other emotions were a turning point in his life, leading him to become a writer, with Time as the great artist: memory freed from the burden of time and habit. As he played Wagner, he could hear the latter "exalting, inviting me to share his joy". Only in a work of art was finding lost time again possible, as a metaphor, "Our slightest desire, while it is a unique chord, contains the fundamental notes out of which our whole life is built".[9]

Philosophical Interlude

Wagner was one of the few composers who was much influenced by philosophy. The writings of the anarchists, socialists and materialists such as Pierre-Joseph Proudhon (1809-1865) and Ludwig Feuerbach (1804-1872) were a part of the intellectual *Zeitgeist* of his earlier years, as were the influential ideas of Hegel: Wagner had much exposure to the latter's philosophy at the University of Leipzig.

Parsifal, in Act 1 of the opera, faints, having learnt from Kundry that his mother is dead. When Parsifal asks Gurnemanz "Who is the Grail?", the enigmatic answer given is "Time here becomes space".

Time and space, the two co-ordinates of our lives, are forever a philosophical conundrum. Plato had argued for two kinds of things in the universe. Forms were mind-independent (non-visible, non-changing) and distinct from ordinary things in the world (visible and changeable). The soul shared the property of Forms, but forever in motion; immortal, free of desire and home of the *logos* (that which gives order and intelligibility to the world). This implied ideas of an *a priori,* pre-scient (from the Latin *prae scire*), the pre-existence of ideas, a re-collection. Similar ideas are found in other philosophies, such as Hinduism and Buddhism.

Eastern Influences

Many classical philosophies, such as those from India, Greece and China, emphasize the beginnings of existence with a unity, with ideas of the infinite nature of an enduring *All*. In most there is a very close binding with religion (from the Latin *ligare*), particularly so in India. One very old metaphysical concept which remains elemental in Hindu, Sikh and Buddhist cultures revolves around *karma* (literally action, which determines the cycle of re-birth). The belief in reincarnation, linked either to one's behaviour in this world, or on cosmic designs and intentions, has implications for the future. Almost *à la* Newton, every action has a reaction, *karma* as a law of cause and effect.[10]

The sacred Tantra texts (*tan*, 'to weave') described rituals transgressing religious and social boundaries, sensual yet divine, and the passionate union between a god and goddess bringing enlightenment and compassion.[11]

In some works, such as those in the *Upanishads* of Indian philosophy, knowledge is gained not through learning or sight, but by meditating to attain the ultimate, that which is without parts. In some variants the role of 'reason' is paramount. Trance-like states are one way that supernatural and mystical knowledge can be attained. [**Fig. 3.2** Tantra]

In contrast, the Japanese orientation is more related to exteriority and less to interiority. Experience, the grasping of nature, especially with the eye, and the awareness of time and the fragility of all things, are overriding concerns. "Religion in Japan is not about belief, instead [it] is about consciousness transformation." It is based much more on aesthetics, exemplified by the tea ceremony and the annual cherry blossom display.[12]

Julian Baggini, in his excellent review of many Eastern philosophies, calls attention to the importance of taking a comprehensive and overarching perspective of them when considering the development in the West of Kant's Copernican revolution.[13] He suggests *phenomenological metaphysics* for such an enquiry, and quotes Kant: "There will always be metaphysics in the world, and what is more, in every human being, and especially the reflective ones".[14]

Continental Philosophy

The concept of a self, an immortal agent with free will, is fundamental to much Eastern and Western philosophy. It lies at the heart of the developed Christianity stemming from earlier Platonic resonances. In the 17th century Descartes, by doubting everything, sought mathematical-like certainty for the non-mathematical. He considered that humans are composed of two substances: *res cogitans,* the thinking mind, and *res extensa,* the body; the mind and the body simply did not directly connect.

The late 18th and early 19th century saw intellectual changes which had considerable importance for the development of Romanticism and the philosophy of aesthetics. Moving away from religious attachments, and having some regard to scientific developments, the origins of modern Continental philosophy can perhaps be identified by *Phänomenologie des Geistes* (*The Phenomenology of Spirit*) (1807), Hegel's complicated journey from the rise of consciousness, via Reason, to Absolute Spirit. This was for him a historical process, linked with what is referred to as 'the Hegelian dialectic'. In simple terms, any positive that comes to exist (*thesis*), stimulates its opposite (*antithesis*), and in combining the two they are then raised into a higher unity (synthesis) in which both are preserved. The latter then becomes a new *thesis*, so the historical time-based algorithm is repeated, but with different levels of attainment towards ultimately a transcendent *synthesis*. This *Aufheben* was a radical shift for Western philosophy in which the principles of Plato and Aristotle had held sway for so long. Newtonian physics had established fixed laws of movement and space, but the latter were now revealed as subject to time and change. The dialectic makes things move forward—very troubling for Platonic ideas or for those of an Aristotelian cast that wanted to categorize things that existed based on essential properties. *Aufheben* has at least three meanings: 'to raise up', 'to cancel out', 'to preserve'.[15]

A considerable problem for those of non-German-speaking empirical leanings was with the word *Geist*—translated variously as spirit, mind, wit, vitality—which is abstract rather than concrete in meaning. In the penumbra of *Geist* was *Weltgeist* (world-spirit), and *Volksgeist* (national spirit), terms used by earlier German-speaking philosophers and not to be confused with Pantheism (that God is everywhere in nature, which is itself an expression of God's nature). These ideas forwarded an agenda for idealistic philosophy, which has always been at the background of intellectual thinking, especially when mind and its relationship to the body and to the soul have been involved.[16]

Hegel shifted the Cartesian 'I' to an ontological rather than epistemological concern: in other words, to asking 'what is the nature of that which exists?', in contrast to 'how do we have knowledge of that which exists?' Kant was more concerned with the latter. What the senses provided was quite incomplete (*a posteriori*) and 'understanding' ordered perceptions by means of *a priori* (before experience) actions, the Categories. These basic forms of understanding were necessary for making our experience possible, giving synthesis and unity of mental images. For Kant, the self is the source of agency; the 'I' (*the transcendental ego*) requires the synthesis of the manifold, the necessary unity of apperception in one consciousness.

Kant referred to *transcendental idealism,* much adrift from the naïve realism of the empiricists, the latter implying that all that we know derives only from experience (the infant mind being a *tabula rasa*). But experience requires sensory stimuli, and the transcendental possibility of ordering things in time and space, *a priori* and not only dependent on sensation, a *transcendental synthesis*.

In the mind there are not objects but representations, namely phenomena, which are the basis of knowledge in Kant's philosophy. The noumena are inaccessible to us, since they are only perceived by us as part of the apperception. We give form to the knowable world.[17] Appearances are not 'things in themselves' but representations within the bounds of time and space, the latter being *a priori* conditions of our cognition.

There are many criticisms of Kant's ideas. Free from contingent empirical limitations, there may be a tangential glide towards a Cartesian disembodied ego, and/or an alternative anti-metaphysical argument adding layers of 'reality' to our experiences. Naturalism, which posits that the laws which operate in the universe govern the nature of our world, is a broad philosophical/scientific church which tends to empiricism, rejecting non-physical causal explanations. Yet the representations are within us, and they relate in some causal way to things outside us, although the inner realm must also be accountable. Such 'idealist' philosophies "imply a coherent, expansive, and multilevel realist metaphysics".[18]

Cosima Wagner noted in her diary on the 21 December 1877 that her husband commented: "Today I have set a philosophical precept to music: 'Hence space becomes time'".[19]

Time Moves On

Time moved on for philosophy, and philosophy moved on with time.

> Time present and time past
> Are both perhaps present in time future
> And time future contained in time past.
> If all time is eternally present
> All time is unredeemable.
> What might have been is an abstraction
> Remaining a perpetual possibility
> Only in a world of speculation.
> What might have been and what has been
> Point to one end, which is always present.

Thomas Stearns Eliot (1888-1965) whose early life ('Prufrock' days) was living in an unreal city, a 'waste land', in which "death had undone so many", in 1927 underwent a conversion to Anglo-Catholicism and took vows of chastity. His poem *Ash Wednesday* was the voice of a man in search of God. Not hoping to turn or know again, he pleads:

> With a new verse the ancient rhyme. Redeem
> The time. Redeem
> The unread vision in the higher dream …
> And let my cry come unto thee.

Deeply influenced by Dante, having travelled through *Inferno*, he seeks *Paradise* unfolding in *The Four Quartets. Burnt Norton* challenges the puzzle of unredeemable time:

> Desire itself is movement
> Not in itself desirable;
> Love is itself unmoving,
> Only the cause and end of movement,
> Timeless, and undesiring
> Except in the aspect of time
> Caught in the form of limitation
> Between un-being and being.

At *Little Gidding*:

> What we call the beginning is often the end
> And to make an end is to make a beginning.
> The end is where we start from.

This a Ring, from which

> … all shall be well and
> All manner of things shall be well
> When the tongues of flame are in-folded
> Into the crowned knot of fire
> And the fire and the rose are one.[20]

We introduced Schopenhauer in Chapter 1, and it is well known that his philosophy was of considerable influence on Wagner. *Pace* Kant, Schopenhauer was interested in the noumena, of which music was a manifestation. Brian Magee puts it as follows: "For it means that music is not arbitrary in the sense language is. Language is entirely a human creation; but music is rooted in the nature of things".[21] Schopenhauer's *Wille* is arguably the pulsating energy that invests all creation, "the essence of everything in nature … eternal becoming, endless flux …"[22] Music escaped the phenomenon of individuation.

I pant for the music which is divine,
My heart in its thirst is a dying flower;
Pour forth the sound like enchanted wine,
Loosen the notes in a silver shower;
Like a herbless plain, for the gentle rain,
I gasp, I faint, till I wake again.[23]

The Schopenhauerian influence was central in *Tristan und Isolde*, as were the contrasts between, between light and darkness and between day and night. Schopenhauer's philosophy led him to the conclusion that the only escape from life's suffering and striving could come from annihilation of the *Will*, hence with death, bound up with Eastern Buddhist ideals. The drama of the two lovers, sinking into unconscious bliss, combined with the music full of unresolved discords, suggested to Nietzsche a balance between the Apollonian and the Dionysian reflecting "the mystery of death in life, the unity in duality".[24]

Wagner had written about Apollo, the slayer of the Python (serpent-chaos) in his book *Art and Revolution*. Dionysian images filter through the philosophy of Schopenhauer and other writers, including Hegel and the philosopher-poet Friedrich Hölderlin (1770-1843).

In *The Birth of Tragedy* Nietzsche recalled that "… art derives its continuous development from the duality of the *Apollonian* and *Dionysian*":[25] Greek or Attic Tragedy emerged out of the spirit of music. Apollo, the god of boundaries, reflected by the epic in poetry and Doric sculpture, is contrasted with Dionysus, the god of melody, lyric poetry and the dance. Dionysus, the god of wine, of ecstasy and of rapture, also presides over the rupture of boundaries. Tragedy came from fusion of Apollonian beauty with Dionysian sadness. In Western art they come together, as do the two sexes, bringing unification.

Nietzsche portrayed psychological interpretations of the two gods, and in later work wrote with some dismay about the Hegelian redolence of his youthful text. But his philosophy proposed new ways of viewing the relationship between perception and emotion in which the emotions dominate, and a psychology based not on the dualism of the Cartesian cataract but on Apollo and Dionysus. These basic drives and tensions are organic, and in John Richardson's view, are closely linked with *Rausch* (intoxication)—they enhance sensitivity to sensations, "quicken the organism ... aesthetic experience as characterised by this 'visceral' excitement or heightening".[26]

In fact no clear view of Tragedy emerges from the book, but Nietzsche's polarity of the two gods was an approach taken up by many others, notably Thomas Mann in *Death in Venice* (1912) and *Dr Faustus* (1947); Hermann Hesse (1877-1962) in *Goldmund and Narcissus* (1930); and Henrik Ibsen (1828-1926) in plays such as *Emperor and Galilean* (1873) and *Hedda Gabler* (1891).

Camille Paglia (1947-), giving but little credit to Nietzsche, contrasts Apollo, the law-giver and representative of sculptural integrity, with Dionysus, god of fluids, of *sparagmus* (the fate of Orpheus) and of pleasure-pain; objectification and identification respectively. Her thesis is that the Apollonian and the Dionysian were two great principles governing the sexual personae in art, but also in life. She continues:

> Apollo is the hard cold separatism of Western personality and categorical thought. Dionysus is energy, ecstasy, hysteria, promiscuity, emotionalism —heedless indiscriminateness of idea or practice. Apollo is obsessiveness, voyeurism, idolatry, fascism—frigidity and aggression of the eye, petrifaction of objects. Human imagination rolls through the world seeking cathexis. Here, there, everywhere, it invests itself in perishable things of flesh, silk, marble, and metal, materialisations of desire. Words themselves the West makes into objects. Complete harmony is impossible. Our brains are split, and brain is split from body.[27]

These themes are core integrants of the main stories and operas we highlight in this book, summed up by the philosopher Jacques Derrida: "Darkness and light, the founding metaphor of Western Philosophy".[28]

Nietzsche's philosophy calls for a 're-evaluation of all values' (*Umwertung aller Werte*), yet he was very anxious to avoid nihilism. One of his controversial ideas was that of the Eternal Return, linked to Dionysus:

> What was it that the Hellene guaranteed himself by means of these mysteries? *Eternal* life, the eternal return of life … the triumphant Yes to life beyond all death and change; *true* life as the overall continuation of life through procreation, through the mysteries of sexuality … [29]

The thought of the eternal recurrence is the highest ideal, for one who would wish to continue to re-live his life, over and over, human all too human. This is a main theme in Nietzsche's philosophical tale *Also sprach Zarathustra: Ein Buch für Alle und Keinen* (*Thus Spoke Zarathustra: a book for all and for no-one)* (1896). He appears to have come to this idea from Heinrich Heine (1797-1856), and while it is a thought experiment, it was meant as something positive, that would emphasise the value of life, a recurrence of all that is wished for. Hence Nietzsche's *Ja Sagen,* saying yes to life, encompassing redemption: "To redeem what is past and to transform all 'it was', that I willed it—that is first redemption for me![30]

Jouissance is a surplus of enjoyment and pure pleasure, an emotion unrelated to logic and "through music, even our passions can enjoy themselves".[31]

O man, take care!
What does the deep midnight declare?
'I was asleep—
From a deep dream I woke and swear:
The world is deep,
Deeper than day had been aware.
Deep is its woe;
Joy—deeper yet than agony:
Woe implores: Go!
But all joy wants eternity—
Wants deep, wants deep eternity.'[32]

Notes

[1] Baudelaire, Charles. *The Flowers of Evil. Poison.* Trans. James McGowan. (Oxford: Oxford University Press, 2008).
[2] Barth, M., Voss, H. and Dietrich, E. *Wagner, a Documentary* (Oxford: Oxford University Press, 1975), 193.
[3] Holman, J. K. *Wagner Moments: A Celebration of Favourite Wagner Experiences* (New York: Amadeus Press, 2007), 54.
[4] Quote from J. K. Holman (Note 2), xviii.
[5] Cather, W. S. "A Wagner Matinée'. *Everybody's Magazine*, 10 (1904): 325-328. Alex Ross, in his book on Wagnerism writes about the involvement of Wagner operas and music in Cather's life and novels. Ross, A. *Wagnerism: Art and Politics in the Shadow of Music* (London: 4th Estate, 2020), 322-351.
[6] Woolf, Virginia. "Impressions at Bayreuth", *The Times* (August 1909).
[7] See for more of these associations Newark, C. *Opera in the Novel from Balzac to Proust* (Cambridge: Cambridge University Press, 2011), 43.
[8] Proust, Marcel. *The Way by Swann's*. Trans. L. Davis (London: Penguin Books, 2002), 352-353.
[9] Proust, Marcel. *The Fugitive*. Trans. Bersani. In: *Marcel Proust: The Fictions of Life and Art*. 2nd ed. (Oxford: Oxford University Press, 2013), 91. Ross, A. ibid. has a superb section on Proust's adoration of music, 391-399.

[10] The various themes and variations of these Eastern philosophies and their associated religious attachments are subtle and complex for the Western mind. Much help on this comes from J. Baggini, *How the World Thinks: A Global History of Philosophy* (London: Granta, 2018). He notes that *karma* is one of the earliest philosophical concepts in human history, so important because it determines one's rebirth.
[11] Ranos, I. "Sacred Transformations". *British Museum Publications* (Spring/Summer 2020): 24-27.
[12] Okakura Kakuzō's *Book of Tea* (1906) remarks: 'Teaism ... is founded on the adoration of the beautiful among … the sordid facts of everyday existence ... a tender attempt to accomplish something possible in this impossible thing we know as life ...' Baggini, J. Ibid., 50, 299.
[13] A development of Kant's philosophy in which the mind provides a systematic structure of representations.
[14] Baggini, J. Ibid., 169.
[15] Note this is a different use of the term dialectic from the Socratic dialectic. The latter was a form of critical oppositional argument, an inductive method to reveal the truth of a proposition, and a way to the logos. For an excellent account of Hegel's logic etc; see https://plato.stanford.edu/entries/hegal-dialectics.
[16] Excellent reviews of these philosophers are many: J. McCumber, *Time and Philosophy A history of Continental Thought* (Durham: Acumen, 2011). The relationship between Wagner and philosophy cannot be bettered than by Brian Magee in *Wagner and Philosophy* (London: Allen Press, 2000).
[17] Yovel, Y. *Kant's Philosophical Revolution: A Short Guide to the Critique of Pure Reason* (Princeton: Princeton University Press, 2018). For an excellent guide to some of these complexities, see M. Kuehn, M. *Kant a Biography*. (Cambridge: Cambridge University Press, 2001).
[18] Ameriks, K. 'On reconciling the Transcendental Turn with Kant's Idealism'. In Gerdner, S. and Grist M. *The Transcendental Turn* (Oxford: Oxford University Press, 2015), 45.
[19] Wagner, C. *Cosima Wagner's Diaries,* 2 vols. *Volume One, 1869-1877*. Trans. G. Skelton (London: St James's Place, 1978), 1007 [21 Dec.1877]. The second draft of the libretto of *Parsifal* was completed in 1877, and the composition of the music started the same year.
[20] Eliot, T. S. *Burnt Norton. Little Gidding. Four Quartets* (London: Faber & Faber, 1940), 7, 42, 44; *Ash-Wednesday* (London: Faber & Faber, 1930), 13, 20, 25.
[21] Magee, B. Ibid. 1997, 187.
[22] Schopenhauer, A. *The World as Will and Idea* (London: J. M. Dent, 1995), 43, 85.
[23] Shelley, P. B. *Music. The Complete Poetic Works of Percy Bysshe Shelley* (Oxford: Oxford University Press, 1914), 651.
[24] Liébert, G. *Nietzsche and Music* (Chicago: Chicago University Press, 2004), 150.
[25] Nietzsche, F. *The Birth of Tragedy*. Trans. S. Whiteside (London: Penguin) 1993), 14.

[26] Richardson, J. *Nietzsche's New Darwinism* (Oxford: Oxford University Press, 2004), 229.
[27] Paglia, C. *Sexual Personae: Art and Decadence from Nerfertiti to Emily Dickinson* (New Haven: Yale University Press, 1990), 96.
[28] Derrida, J. *Writings and Difference* (Oxford: Routeledge, 2001), 31
[29] Nietzsche, F. *Twilight of the Idols*. Trans. W. Kaufmann. In *The Portable Nietzsche*. Ed. W. Kaufmann (Cambridge Companion 54) (New York: Viking, 1954), 561.
[30] Nietzsche, F. *Also Sprache Zarathustra. Das Lexicon der Nietzsche Zitate*. Ed. Prossliner, J. V. and Ross, W. (Munich: Deutsche Taschenbuch Verlag, 1892), 196. *Die Vergangenen zu erlösen und alles 'Es war' umzuschaffen in ein 'So wolte ich es!'— das heiße mir erst Erlösung!*
[31] Nietzsche, F. *Beyond Good and Evil*. Trans. M. Fabe (Oxford: Oxford University Press, 1998), 63.
[32] Nietzsche, F. *Zarathustra's Roundelay* [*Zarathustras Rundgesang*]. *Also Sprach Zarathustra*, I-IV. Trans. W. Kaufmann (Munich: Dünndruck-Ausgabe/ De Gruyter, 1988), 404.

O Mensch! Gib acht!
Was spricht die tiefe Mitternacht?
„Ich schlief, ich schlief—,
Aus tiefem Traum bin ich erwacht:—
Die Welt ist tief,
Und tiefer als der Tag gedacht.
Tief ist ihr Weh—,
Lust—tiefer noch als Herzeleid:
Weh spricht: Vergeh!
Doch alle Lust will Ewigkeit—
—will tiefe, tiefe Ewigkeit!

CHAPTER 4

GOETHE, THE GOTHIC AND MEDICAL MINDS

On the death of God: "Who gave us the sponge to wipe away the whole Horizon?"

(Friedrich Nietzsche, *The Gay Science*)[1]

Urtypus

One of the important tenets of Romanticism was, and still is, the discerning of inner directing mechanisms of nature. Life is a continuum of the impetus of evolution, and the abiding principle is feminine, Mother nature. Goethe was in search of archetypes, the *Urtypus* from which various forms in nature emerged. The image of the Homunculus in *Faust II* is one metaphorical image, a prototype from which living forms will evolve. Goethe was a friend of the great explorer Alexander von Humboldt (1769-1859), who, with a broader sweep, developed his image of *Naturgemälde* in which all of nature was connected. [**Fig. 4.1** Humboldt, *Naturgemälde*].

Goethe's considerably interest in rocks, fossils and plants led to the book *Versuch die Metamorphose der Pflanzen zu erklären* (*Experimental Observations to Elucidate the Metamorphosis of Plants*) (1790), in which he argued for a primordial form for all plants, existing ones and fossils representing metamorphic variants. He extended his theories to the animal kingdom, emphasising the principle with at least some influence from Kant, that surface phenomena could not reveal the archetypes.[2] He found his ideal *Urpflanz* celebrated in his poem *Gingo biloba,* a two-leaved symbol of love from the Japanese Ginkgo tree [**Figs. 4.3-4** *Urpflanze*].

By the East this tree's entrusted
To my garden, and its leaf
Has to an edifying secret
They can savour who can read it.

It is a single living essence
That divides within itself?
It is two that chose each other,
Whom we recognise as One?

I can answer such a question.
I've discovered its true sense.
Can't you feel it in my poems
I am double, I am One?[3]

Goethe was in Rome between 1786 and 1788, retrospectively described in his *Italienische Reise* (*Italian Journey*, 1816-19). He became invigorated by Italian and Classical art, and even more enchanted with the beautiful young women. He composed a cycle of twenty poems with a provisional title *Erotica Roma,* in addition to the twenty-poem collection *Römische Elegien* (*Roman Elegies*, 1786-88).

Finding *Amor* (Cupid or Eros) in Rome, he brought together poetry, ideal biological forms, the erotic and his adoration of female beauty, instantiating *die Ewig-Weibliche* [the Eternal Feminine] of his later creativity. In the first draft of *Faust*, Faust gazes at Gretchen and sees her beauty. Mephistopheles gives him a magic potion and then declares: "with this potion, in love, you will soon see Helen in every woman".[4]

Goethe's novel *Die Wahlverwandtschaften* (*Elective Affinities*, 1809), explored the idea of chemical attraction of the elements. It is about a foursome who change partners: Eduard and Ottilie

> If they found themselves in the same room, it didn't take long before they were sitting or standing together ... it required no glance, no word, no gesture, no touch only the purest proximity. Then they were not two persons, it was only a single person in unconscious, utter contentment ... for them life was a riddle whose solution they found only in each other.

They both develop headaches (he on the right side, she on the left) and they both have considerable guilt. They do not marry. Both die: "... peace hovers above their resting place, cheerful, familiar images of angels look down on them from the ceiling, and what a happy moment it will be when, one day, they awake together".[5]

The "transcendental leaf" was the elemental kernel, the *Ewig-Weibliche* which *Zeit uns hinan*—eternally draws us onward, towards heaven.

Transcendental Ascent

For some two millennia, the view that the world was structured from an unalterable reality held prominence. For the likes of Schlegel and Novalis,

> … above all for ETA Hoffman, the tidy regularities of daily life are but a curtain to conceal the terrifying spectacle of true reality, which has no structure, but is a wild whirlpool, a perpetual *tourbillion* of the creative spirit which no system can capture: life and motion cannot be represented by immobile, lifeless concepts, nor the infinite and unbounded by the finite and the fixed. A finished work of art, a systematic treatise, are attempts to freeze the flowing stream of life; only fragments, intimations, broken glimpses can begin to convey the perpetual movement of reality.[6]

Thus Isaiah Berlin's summing up of the progress of the crooked timber of humanity from which, quoting Kant, man is made, and nothing entirely straight can be carved. The sway between subject and object, soul and body, idealism and materialism became unsteady towards the end of the 18th century, reflected in the work of William Blake who was appalled with Newton's breaking up of light, on a similar path, if from a different approach to that of Goethe. Blake's painting of Newton measuring up an image on a scroll with a compass, circumscribing nature into numbers, emphasised the break-up of nature. The very idea that the purity of light could be broken by a prism was an anathema—light could not contain something darker than itself. Undoing the unity of Apollo could only lead to unknown consequences: it was an *Urphenomenon*. [**Fig. 4.2** Blake, *Newton*]

Science Enfolding

Emanuel Swedenborg (1688-1672) was a mystic philosopher with considerable interest in the spiritual world. He led an eccentric solitary existence, writing extensively, his eight-volume *Arcana Coelestia* (1749-56) being more than two million words long. He became paranoid, with religious hallucinations and delusions, ultimately declaring that Jesus Christ had his Second Coming through him. He believed that the spiritual world was a special universe with a linked spiritual community; he founded the Church of the New Jerusalem. He studied medicine and was especially interested to gain a better understanding of the soul by investigating the nervous system.

Kant had read his magnum opus, and wrote *Träume eines Geistersehers, erdurch Träume der Metaphysik* (*Dreams of a Spirit Seer, as Elucidated*

Through Dreams of a Metaphysician). Although sceptical of much of Swedenbourg's thesis, within it he found a metaphysical introduction to a transcendental world, with moral implications.[7] Swedenbourg paved the way not only to exploring transcendental possibilities, but also to appealing to neuroscientists to look to the brain's associations with the soul. For Goethe, who discussed such matters with Swedenborg, "transcendence is never directly apprehended by means of some rare revelation, but rather immanent only in empirical reality, a deepening of empirical experience".[8]

Concepts related to transcendental idealism, harking back to Kant's ideas, continued to occupy philosophy, and were part of the developing study of aesthetics, poetic sensibilities, literature and education. These topics were taken up by Fichte, Friedrich Schlegel, Friedrich Wilhelm Joseph Schelling (1775-1854), Friedrich Daniel Ernst Schleiermacher (1768-1834), Schiller and Goethe, who reflected on the infinite potential within human nature to the transcendental, central to the Romantic agenda. Dichotomies between self and nature, subject and object, idealism and realism, were brought together along with forces of attraction and repulsion, the nature of love, and love of nature.

These transcendental philosophies were not simply a move to solipsism but were closely allied to Phenomenology, reasserting the a-temporal nature of the human mind. Edmund Husserl (1859-1938) posited a science of pure phenomenology and referred to 'transcendental consciousness' and himself as a transcendental idealist. He wanted to clarify the essence of the stream of consciousness, transcendental subjectivity as an absolute independent realm of direct experience (*Ein-sicht: in*-sight).[9]

Transfiguring the dualism of heaven and earth into a spiritual unity however ran into dangers, since not all spirits were benevolent: some could be devilish.

Gothic Turns

The Goths were a Germanic tribe that settled in Scythia on the shores of the Black Sea, and were a part of the migrations that eventually led to the downfall of the Roman Empire.[10] The very terms 'Goths' and 'Huns' evoke stories of carnage and political collapse, but embedded in the myths was the supernatural. Applied to architecture, the Gothic style developed in France in the 13th century and became dominant in France, England and the European diaspora. The Gothic could be contrasted with Greco-Roman classicism (with the classic rounded arch), and the development of pointed arches and flying buttresses was viewed with distaste by some, especially in the Italian Renaissance. In the 17th and 18th centuries the term 'Gothic' became a byword for 'barbaric', devoid of 'classical elegance' and hinting at the 'dark' ages, which from an artistic point of view were by no means dark.[11]

In 1764, Horace Walpole published *The Castle of Otranto*, generally considered to be the first Gothic Novel.[12] Suspense and sensation, supernatural with shrouded ghosts, castles and collapse all became the backdrop to the genre referred to as Gothic. Several Gothic Novels followed, notably Ann Radcliffe's *The Mysteries of Udolpho* (1794), Matthew Lewis's *The Monk: A Romance* (1795), through to Charles Robert Maturin's *Melmoth the Wanderer* (1820), with Jane Austen's *Northanger Abbey* (1817) providing a wry look at the vogue and readership of this fiction.

Radcliffe wrote a text on the supernatural in poetry (in praise of Shakespeare's plays such as *Macbeth*, *Hamlet* and *Cymbeline*) and distinguished terror from horror. Edmund Burke (1729-1797), in his *Philosophical Enquiries into the Origin of Our Ideas of the Sublime and the Beautiful* (1757), implied such fiction could bring aesthetic pleasure. Space, especially architectural settings such as ruined castles, (un-)natural boundaries (e.g. forests) or crossing territories, along with sleep and nightmares, lost women, and romantic love/death/separation, became the core themes. They were allied to chivalrous tales, with giants, champions, knights and romance: Ghosts ('who's there?') and monsters (*Frankenstein* 1818, *The Vampyre* 1819, *Dracula* 1897) were important personae.

Many poets took up these themes, like the Graveyard Poets—Thomas Gray (1716-1771), Edward Young (1683-1765), Robert Blaire (1699-1746) and, importantly for opera, Sir Walter Scott (1771-1832).

American Gothic

There were many resonances of European literature in American poetry and stories, but there were distinct differences: after all, Americans could not boast ruined churches and castles. But they had a strong Protestant ethic, and with it a fear of the devil, witchcraft and 'savages'—at war with the civilized white settlers. Indians became demonised as devil-worshippers as Romanticism segued into materialism and naturalism.

American novelists, especially Nathaniel Hawthorne (1804-1864), Herman Melville (1819-1891) and Edgar Allan Poe (1809-1849), played on irrational and unconscious impulses, undercutting the phantom of rationality and the myth that 'all men are created equal', the emphasis shifting to 'perhaps all men are created evil'.

The trail of the Gothic emerged with the works of Charles Brocken Brown (1771-1810), the first 'professional' novelist (*Edgar Huntly; or Memoirs of a Sleepwalker*, 1799) and continued with Poe. The somnambulistic Huntly goes into a forest, eats a panther and kills Indians (written after a yellow-fever epidemic in Philadelphia). Poe's story of Roderick Usher and the collapse of his house (*The Fall of the House of Usher*, 1839) and Hawthorne's *The Scarlet Letter* (1850) involve secrets, sin and guilt: the letter in the latter is A for adultery. Melville's *Moby Dick* (1851) tells the story of the primordial gargantuan whale and the deeply flawed Captain Ahab, who gets around with the jawbone of a sperm whale to replace his lost leg: secrets from the past, not revealed. Themes of light and dark permeate the novels and poems of Melville whose books have much to do with the sea. The mystical whiteness of Moby-Dick is terrifying yet spiritual. The polarities within his unfinished tale *Billy Budd,* (published posthumously in 1924) can be interpreted by seeing Budd as a Christ-like figure brought down by the forces of the evil Claggart. We will meet with Billy Budd later in the opera by Benjamin Britten (1913-1976).

The alcoholic Poe suffered episodes of depression, and was well acquainted with loss and death, especially the deaths of women (his mother; his foster mother; Helen Stannard, mother of a friend; and his young wife, Virginia Clemm). He idolized and idealised them: the notion of beauty, contrasted with mess and decay. Interestingly, consumption (tuberculosis) provided for him a type, the wan and pale women with reddened cheeks, and yellow or hyacinth hair. His females' eyes are bright and glowing (Annabel Lee, Eulalie, Eleonar), Helen disappearing except for her eyes which "remain? ... they lead me through the years".[13] There is the embodiment of temporal loveliness approaching an eternal beauty, but lacking any individual development. Most of his women are almost silent, yet they have possibilities of life beyond the grave. His aestheticism insisted on the role of the 'effect' of an artwork, seeking beauty in the creation, itself elevating the 'soul' and not primarily the intellect.[14]

Gothic metaphors abound in his tales, the *Pit and Pendulum* (1842) and *The Descent into the Maelstrom* (1841) being allegorical descents into the depths of the soul. The house of Usher has "vacant eye-like windows", doors as mouths, in a "mansion of gloom"—Roderick Usher being an artist gone mad. In portraying Roderick, Poe said: "I mean to imply a mind haunted by phantoms—a disordered brain". The Gothic themes of madness, grief, forebodings of hopelessness, and personification of death held his creative attention.

Annabel Lee, Poe's last poem (1849), was composed for his child bride, Virginia Clemm, who had died some two years earlier. In poetry permeated with transcendent moments, accentuated with his ever-skilful rhythms and rhymes, the melancholy beauty of the potential reunion with the lover, soul with soul, is affirmed.[15]

It was many and many a year ago,
In a kingdom by the sea,
That a maiden there lived whom you may know
By the name of Annabel Lee;
And this maiden she lived with no other thought
Than to love and be loved by me.

I was a child and *she* was a child,
In this kingdom by the sea,
But we loved with a love that was more than love—
I and my Annabel Lee—
With a love that the wingèd seraphs of Heaven
Coveted her and me.

And this was the reason that, long ago,
In this kingdom by the sea,
A wind blew out of a cloud, chilling
My beautiful Annabel Lee;
So that her highborn kinsmen came
And bore her away from me,
To shut her up in a sepulchre
In this kingdom by the sea.

The angels, not half so happy in Heaven,
Went envying her and me—
Yes!—that was the reason (as all men know,
In this kingdom by the sea)
That the wind came out of the cloud by night,
Chilling and killing my Annabel Lee.

But our love it was stronger by far than the love
Of those who were older than we—
Of many far wiser than we—

And neither the angels in Heaven above
Nor the demons down under the sea
Can ever dissever my soul from the soul
Of the beautiful Annabel Lee;
For the moon never beams,
without bringing me dreams
Of the beautiful Annabel Lee;
And the stars never rise,
but I feel the bright eyes
Of the beautiful Annabel Lee;
And so, all the night-tide,
I lie down by the side
Of my darling—my darling—my life and my bride,
In her sepulchre there by the sea—
In her tomb by the sounding sea.

The sea, death, unity with the beloved, and pursuit of beauty expressed through poetic brilliance that Poe himself knew gave rise to tears and ineffable feelings:

> … through a certain, petulant, impatient sorrow at our inability to grasp now, wholly, here on earth, at once and forever, those divine and rapturous joys, of which through the poem, or through the music, we attain to but brief and indeterminate glimpses.[16]

Ulalume is considered one of his finest poems, set in forest and fields, on a gloomy autumn evening. The composer Daniel-François-Esprit Auber (1782-1871) is named in the verse. Auber's opera *Le Lac des fées (The Fairy Lake*, 1839) takes place in the Harz Mountains in the 15th century. There is a lake surrounded by high rocks, and in the cloudy realm of the fairies lies Rodolphe's castle with a magnificent Gothic hall inside. Albert, a young student, has fallen in love with a fairy, Zélia: she has been forced to live on earth because Albert has stolen her veil. At the last moment, however, she regains her veil, and disappears to her fairy sisters. To welcome her back, the Fairy Queen allows Zélia a wish: but she chooses to renounce immortality, and returns to Albert on the earth. In the closing transformation scene (Zélia's return to earth)

> ... we see her passing through the clouds, Albert, alone in his room and weighed down with despair, is about to kill himself—He raises his eyes and sees Zélia who is coming down on a cloud towards him, holding out her arms. He runs to her, and the curtain falls.[17]

Gothic tales and sentiments are found in many operas around this time, including Carl Maria von Weber's (1786-1826) *Der Freischütz* (*The Free*

Shooter, 1821); Heinrich Marschner's (1795-1861) *Der Vampyr* (1828) and *Hans Heiling* (1833); Meyerbeer's *Robert le Diable* (*Robert the Devil*, 1831); Gaetano Donizetti's *Lucia di Lammermoor* (1835); and Wagner's *Der fliegende Holländer* (*The Flying Dutchman* (1843).

Tennyson

Tennyson's mystic trances and transcendental experiences are reflected in his poetry.

Tennyson's poetry implies an escape from the real through dreams or madness, or an escape into some higher truth through vision and the quest. This entails an inner rather than an outer awareness, from the life of the imagination rather than from a sense of responsibility to society. This conflict, between the appearance of the external world and the reality of the individual consciousness, is found to be the underlying theme of the *Idylls of the King* and runs through nearly all the more serious poems of Tennyson's later period. The vision of the Holy Grail is referred thus:

> "Let visions of the night or of the day
> Come, as they will; and many a time they come,
> Until this earth he walks on seems not earth,
> This light that strikes his eyeball is not light,
> This air that smites his forehead is not air
> But vision—yea, his very hand and foot—
> In moments when he feels he cannot die,
> And knows himself no vision to himself,
> Nor the high God a vision, nor that One
> Who rose again: ye have seen what ye have seen."
> So spake the king: I knew not all he meant.

By taking mythological themes, Tennyson was able to explore the darker depths of consciousness, the Dionysian impulse closely linked with creativity, as in *Demeter and Persephone.* This is described as a domestic drama of the hearth, but it is a story of bereavement and descent to the underworld, a mysterious realm that Persephone must periodically visit, reflecting the fragility of human stability, since Persephone must always return.

For Tennyson inner awareness is allied with an idealistic philosophy and artistic creativity. The source of all true inspiration is within the intuitional consciousness of the individual. This sense of the immateriality of exteriority, of the strangeness and mystery that always lie just under the appearance of things (phenomenology), of the 'unknowableness' of the human mind in its instinctual perceptions (Kant), is the kernel of nearly all of Tennyson's best poetry. The imagination not only conceives, it also shapes: 'it is the Human existence itself'.

Coleridge opined: "But oh! Each visitation/Suspends what nature gave me at my birth,/My shaping spirit of Imagination". He called this creative power *esemplastic,* that which shapes the whole.[18] He was much influenced by German philosophy, and could not accept that the mind was a blank slate. It was an active principle that shaped the world of individual experience, and he reframed Descartes's principle to *sum quia sum* (I am because I affirm myself to be). He introduced us to the words unconscious and psychosomatic, and the interesting idea that a stomach sensation was attached to his thoughts, thought being motion.

American Transcendentalism

At this point it is worth noting the rise of the transcendental movement in America, especially through the works of Ralph Waldo Emerson (1803-1882) and Henry David Thoreau (1817-1862). Originating in New England, the movement was an attempt to move away from British influences, and looked towards the philosophies of the East. The term transcendental was taken from the writings of Kant.

Emerson's *Nature* (1836) and Thoreau's *Walden* (1854) are founding texts, but there were many other others in the movement. Originating from Unitarianism, but with an emphasis on the philosophy of Kant, and against the empiricism of John Locke (1632-1704), they extolled the higher qualities of the human mind and the infinitude of the individual.

The American Transcendental movement emphasized inner spiritual values, and laid the grounds for the naturalist works of Walt Whitman (1819-1892) and his collection of poems, *Leaves of Grass* (1855). Whitman's poetry reveals his concern with 'cosmic consciousnesses' and a sense of oneness with others, especially those who suffer. He considered he would never have written *Leaves* if it were not for opera. He adored Italian *bel canto*, and is quoted as saying 'he was told that Wagner is *Leaves of Grass* done into music'.[19] The American Transcendental movement influenced composers such as Edward Alexander MacDowell (1860-1908) and Charles Ives (1874-1954).

Mesmerism, Magnetism and Medicine

The rise of transcendental ideas and philosophies was associated with studies of transient alterations of the mental state, especially dreams, but also as encountered with mental illness, the taking of drugs, and with therapeutic benefits from psychological interventions. In France, Franz Anton Mesmer's (1734-1815) use of "animal magnetism" for treatment was popular, and Mesmerism spread to England, linked with an ever-growing interest in somnambulism, cataplexy, *déjà vu* experiences and automatic writing.

The French neurologist Jean-Martin Charcot (1825-1893) and his associate the psychologist Pierre Janet (1859-1947) wrote extensively on hysteria, its symptoms and causes. The latter's theories involved a failure of psychological synthesis, amnesia and constriction of the field of consciousness. Subconscious parasitic ideas, concealed in the deeper layers of the mind and related to traumatic events, acted as *idées fixes*, and were a target for treatment. His concept of psychological automatisms was very influential, leading to later ideas of dissociation, and to concepts of hysterical amnesia, multiple personalities and the subconscious. The latter was able to influence motor somatic expression. The increased emphasis on psychological as opposed to neurological causes as a basis for hysteria heralded a watershed in the rising field of psychosomatic medicine with the theories of Sigmund Freud. The division between madness and sanity was becoming blurred [**Fig. 4.5** *Les fascines de la Charité*].

This was in part a reaction to the perceived excesses of positivism: a Neo-Romanticism, but Romanticism with more than a tinge of decay and decadence in the air. Although the underlying philosophy was from Germany, the movement was evident not only in medical circles, but also reflected in the arts of the time, notably in Paris and Vienna.

Interest in the mystical and the unconscious led onto surrealism, a seeking of beauty for beauty's sake, and the aesthete, a servant of beauty, exemplified by Oscar Wilde. Yet Dorian Gray could not escape his own natural decadence nor could art save him by beauty alone. Entwined were dandyism, exoticism, eroticism and a lot of Satanism.

Goethe, whose colour theory required both dark and light together, suspicious of Kantian philosophy, and whose last words are reported to have been *mehr Licht* (more light) could not let Faust escape the two souls within him:

There dwell two souls, alas, within my breast!
The one desires a parting from the other.
The one clings to the world with earthly lust,
Hangs on for all its worth, with all its senses.
The other struggles mightily from the dust,
And yearns to reach the realms of lofty forebears.[20]

Notes

[1] Nietzsche, Friedrich. *The Gay Science* (German: *Die fröhliche Wissenschaft*), occasionally translated as 'The Joyful Wisdom' or' The Joyous Science' is a book by Friedrich Nietzsche, first published in 1882 and followed by a second edition, which was published after the completion of *Thus Spoke Zarathustra* and *Beyond Good and Evil*, in 1887. This substantial expansion includes a fifth book and an appendix of songs. See *The Gay Science* (London: Random House, 1972), 209.
[2] It is often argued that Goethe was a rather poor scientist, and that he got many theories quite wrong, most notably his views *pace* Newton about colour. However, for the time he was writing, at the cusp of the scientific revolution, he had many fruitful ideas, and made some significant discoveries, including the intermaxillary bone in humans. The controversy was that if humans did not have this bone it was evidence of their separation from other animals.
[3] Goethe, J. W. *Westöstlicher Divan, Suleika Nameh—Buch Suleika Gingko Biloba. Reich-Ranicki M. 1000 Deutsche Gedichte* (Frankfurt am Main: Insel Verlag, 1994), 367; Safranszki, R. *Goethe: Life as a Work of Art.* Trans Dollenmayer (London: Liveright Publishing Corporation, 2017), 477.
[4] Mason, E. C. *Goethe's Faust: Its Genesis and Purport* (Berkeley: University of California Press, 1967), 209.
[5] Safranszki, R. Ibid. 2017, 442-443.
[6] Berlin, I. *The Crooked Timber of Humanity: Chapters in the History of Ideas.* (London: John Murray, 1990), 232.
[7] Johnson, G. R. "Swedenborg's positive influence on the development of Kant's mature moral philosophy". In *Philosophy, Literature, Mysticism: an anthology of essays on the thought and influence of Emanuel Swedenborg.* Ed. Stephen McNeilly (London: The Swedenborg Society 2013), 25-44.
[8] R. Safranszki, Ibid, 2017, 493.
[9] A useful text to follow some of Husserl's (difficult) ideas: Husserl, E. *Ideas: General Introduction to Pure Phenomenology*. Trans. William Ralph Boyce Gibson (1869 –1935) (London: Routledge Classics, 2012).
[10] The original Goths appear in what is now northern Poland; they migrated south in the 2nd and 3rd centuries, and fragmented into separate Gothic groups under pressure from the Huns. See Heather P., *Fall of the Roman Empire* (Oxford, Oxford University Press, 2006).
[11] For excellent illumination see: Arthur Herman, *The Cave of Light: Plato Versus Aristotle, and the Struggle for the Soul of Western Civilisations* (New York: Random House, 2013); Tom Holland, *Dominion: The Making of the Western Mind*

(London: Little Brown, 2019). The Dark Ages are better referred to as the period of Late Antiquity. See Brown, P. *The World of Late Antiquity* (London: Thames and Hudson, 1971).

[12] Walpole also built his own Gothic pile—Strawberry Hill. Later the 'High Gothic Victorian' became a national style with the rebuilding of the Houses of Parliament associated with the architecture of Augustus Pugin (1812-1852) and the works of John Ruskin (1819-1900).

[13] Poe, E. A. *Sonnet to Science. The Complete Poetry of Edgar Allan Poe* (New York: Signet Classics, 1996), 109.

[14] Poe was deeply affected by the death due to tuberculosis of his wife Virginia Clemm, aged 24, whom he had married when she was 13.

[15] Poe, E. A. *Annabel Lee.* Ibid., 116.

[16] Taylor, M. *Last Works: Lessons in Leaving* (New Haven: Yale University Press, 2018), 150-151.

[17] See Letellier, Robert Ignatius. *Daniel-François-Esprit Auber: The Man and His Music* (Newcastle: Cambridge Scholars Publishing, 2010), 330-342.

[18] Tennyson, A. 'The Holy Grail', *The Works of Alfred Tennyson* (London: MacMillan & Co, 1907), 433. Coleridge, S. T. 'Dejection: An Ode', Wu, W. *Romanticism: An Anthology*, third ed. (Oxford: Blackwell Publishing, 2006), 676. *Esemplastic,* a word that Coleridge created: Coleridge, S. T. *Biographia Literaria* (London: Dent, 1965), 91.

[19] See Ross, A. Ibid. 2020, 156-157. Whitman never saw a completed Wagner opera, but did write an essay *The Poetry of the Future.*

[20] Safranski, R. Ibid., 527.

CHAPTER 5

OUT OF BODY INTO MIND

I dissolve into a grove of women, into tree and lake,
Into spring and lotus, into the open sky.
My soul I had considered sound and stable
takes flight and rises with a thousand wings.
I grow boundless and dissolve
In this lustrous cosmos united with the world.

(Hermann Hesse, *The Seasons of the Soul*)[1]

Out of Body Experiences (OBE) are by no means uncommon. They are often described in such terms as 'leaving my body', sometimes accompanied by a sense of floating and a feeling of looking at someone else. It is an experience of literally being disembodied.

I saw Eternity the other night,
Like a great ring of pure and endless light,
All calm, as it was bright;
And round beneath it,

Time in hours, days, years,
Driv'n by the spheres
Like a vast shadow mov'd;
in which the world
And all her train were hurl'd.

Henry Vaughan (1621-1695) sets down a poetic evocation of a transcendent state in *The World*, Platonic and Dantesque, in which 'the ring' is a central metaphor of eternity and the soul's marriage with Christ in which light triumphs over darkness.[2]

Evidence of such experiences has been sought by those, including mediums, looking for confirmation of a life after death, giving proof that the spirit can leave the body. Many are simply curious to explore or enhance the boundaries of their own mental states. The latter often seek some kind of revelation brought about by artificial physiological experimentation or by taking cerebrally active agents. The possibility of such journeys can seem threatening to certain religious groups, who fear that such untoward adventures into the spirit world might lead to encounters with jinns, or that the soul might get lost and not be able to find its way home.

Susan Blackmore settles for the following definition: "An experience in which a person seems to perceive the world from a location outside the physical body". This is not to be confused with dream-like states, and is distinguished from *Doppelgängers*, autoscopy, *déjà vu*, *déjà vecu* and other variants.[3]

Déjà vu and *déjà vécu* have been described many times in poetry and literature. Famously, Charles Dickens (1812-1870) wrote about his own *déjà vu* experiences and put one such episode in *David Copperfield* (1849-50). He even suggested that they may have a link to reincarnation, and precognition.

> Has this been thus before?
> And shall not thus time's eddying flight
> Still with our lives our love restore
> In death's despite,
> And day and night yield one delight once more?[4]

Doppelgängers are two appearances of one person in different places, while autoscopy refers to the experience of seeing the self as image, dislocated and projected into external visual space, outside the body, (extracorporeal). A variant of the latter is heutoscopy, in which it is unclear which of the two selves is the real one. Franz Peter Schubert (1797-1828), using the poem by Heinrich Heine, perceived a *Doppelgänger*:

> A man stands there too, staring up,
> and wringing his hands in anguish;
> I shudder when I see his face—
> the moon shows me my own form!
> You wraith, pallid companion,
> why do you ape the pain of my love
> which tormented me on this very spot,
> so many a night, in days long past?[5]

In some surveys up to 25 per cent of people report an OBE, but the figure given by Blackmore is 15 per cent. Most OBEs are experienced lying down, while autoscopic images are more often reported standing. Descriptions of leaving the body have some common elements, but also much diversity. They often occur in sleep-related states or in association with intense stress: they can be triggered by accidents, or occur after operations. They are not dreams, however, and are reported as being real, even if someone is on the borderlands of sleep. Some people describe what are called lucid dreams. The sleeper seems awake, knows they are dreaming, yet claim to be able to control the ongoing dream. [**Fig. 5.1** Joan of Arc]

Many people relate a journey, such as travelling down dark tunnels, sometimes with a white light at the end. They report seeing illuminated objects, hearing noises, including singing, and feeling shakes or vibrations. Some claim to visit strange places, including heaven, or meeting angels, spirits or daemons. Some OBEs assume an astral plane to which the astral body projects, where spirits exist. Objects can seem to be surrounded by an aura, and may be linked by a silver chord to the physical body. Generally these kinds of experiences are reported as pleasurable, although not always so. They are frequently peaceful and calm, in contrast to descending into hell.

The Artificial

Alterations of the mental state are frequently reported after a person ingests drugs. Indians in South America and Mexico have used peyotl (from cactus root) for religious inspiration since recorded time. In the 21st century, substances such as Lysergic acid (LSD), cannabis, ketamine (antagonist of N-methyl-D-aspartate, a brain neurotransmitter), dimethyl-tryptamine (DMT), and psilocybin (from magic mushrooms) are popular: these can provoke brain state changes with a variety of startling effects. Investigations confirm consequential neuroanatomical and neuro-physiological changes, beyond the nature and scope of this book.[6]

Interest in such altered states of mind was popular in the 19th century, and operas with potion-induced changes of behaviour are common, as we discuss later. We have noted Baudelaire in this context describing the immense feelings that music could arouse in him and its connections to an epiphanic experience. He described the brain as a “cramped and mysterious laboratory ... there is a nervous shock that makes itself felt in the cerebellum”.[7]

Thomas de Quincy (1785-1859) wrote his *Confessions of an English Opium-Eater* in 1821, and in France Jacques-Joseph Moreau (1804-1884) (referred to as Moreau de Tours), was extolling the virtues of hashish, which he had encountered in the Orient. He was impressed by its power to alter the mind and speculated on its potential to reveal the mysteries of alienation (insanity), and the source of delusions.[8] Moreau was led to the conclusion, not accepted at that time, that delusions were not always pathological, but there was continuity between normality and abnormality which could be explored medically.

The physician Valentin Magnan (1835-1916) investigated the deleterious effects of cocaine, morphine and alcohol.[9] Absinthe, laudanum, ether, nitrous oxide, chloroform, and other substances, including alcohol, were widely considered inspirational. As Baudelaire put it, opium magnified and expanded experience:

> Opium will expand beyond all measures,
> Stretch out the limitless,
> Will deepen time, make rapture bottomless,
> With dismal pleasures
> Surfeit the soul to point of helplessness.[10]

The last half of the 19th century was a time when the mystical and unconscious aspects of the mind were widely discussed, not only in medicine but also in the arts. Accounts proliferated of the psychedelic impact of chemical substances, natural or artificial, and Romantic, especially Gothic, tales spread the news. Robert Louis Stevenson's (1850-1894) *Strange Case of Dr Jekyll and Mr Hyde* (1886) was one of many dealing with drug-inspired transposition. The Salon de la Rose+Croix founded by Joséphin Péladan (1859-1918) linked ideals of beauty with Eastern influences, Christianity and the occult.[11] The idea was to transcend the mundane for a spiritual alterative.

Many have described their travel in astral planes, some drug-inspired. The Rev. Richard Sigmund, in his account *My Time in Heaven,* saw his own name in golden letters engraved on heaven's gates. The neurosurgeon Eben Alexander cites his experiences as proof of heaven. He entered a world of consciousness "completely free of the limitations of [my] physical brain … the place I went to was real".

One of the most famous examples is the mystical experience of the Apostle Paul who claimed to have a vision of heaven:

> 1 I must boast; there is nothing to be gained by it, but I will go on to visions and revelations of the Lord.

> 2 I know a man in Christ who fourteen years ago was caught up to the third heaven—whether in the body or out of the body I do not know, God knows.
>
> 3 And I know that this man was caught up into Paradise—whether in the body or out of the body I do not know, God knows—
>
> 4 and he heard things that cannot be told, which man may not utter.
>
> (2 Corinthians 12:1-4)

Such experiences can be life-changing, and are discussed by William James (1842-1910) in his *Varieties of Religious Experience* (1902) in which he defines the central features of mystical states of consciousness. This is *Ineffability*, an experience which defies expression; a *Noetic* quality, that which gives knowledge and insight into certain truths "unplumbed by the discursive intellect"; *Transiency*, as these states cannot be sustained for long, mostly under an hour; and *Passivity,* related to a superior power. Such states can herald a new plane of existence, what James refers to as "cosmic consciousness".

> We pass into mystical states from out of ordinary consciousness as from a less into a more, as from a smallness into a vastness, and at the same time as from an unrest to a rest. We feel them as reconciling, unifying states.

These can only be described as transcendental, and give challenges to those rejecting a dualism, with a split between mind and body. "The existence of mystical states absolutely overthrows the pretention of non-mystical states to be the sole and ultimate dictators of what we may believe".[12]

The Transcendental and the Arts

Baudelaire believed that there was a unity of the arts (*Gesamtkunstwerk*) which we explore further in Chapter 23, and wrote about his synaesthetic experiences invoked, for example, by listening to Wagner's music. The *fin de la siècle* artists were much interested in associations between sounds, sight and symbolism. This is found in the art of Wassily Kandinsky (1866-1944), Odilon Redon (1840-1916), Moreau, and composers such as Alexander Scriabin (1871-1915) and Olivier Messiaen (1908-1982).[13]

The philosopher Jean-Luc Nancy (1940-), in his book *The Muses,* asks why are there several arts and not just one. The muses were nine in number, and historically appear as far back at least as the epics of Homer.

The Odyssey begins with an invocation to a muse, goddess daughter of Zeus, to tell a story. It was the poet Hesiod (c.750-c. 650 BC) who settled on nine, claiming that these nymphs of Mt Helicon gave him breath which was the inspiration for his poetry. He adumbrated that they were daughters of Zeus and Mnemosyne, the goddess of memory and the supposed mother of Apollo. But according to the poet Theognis (570-485 BC), their mother was Harmonia.[14]

Although there may be a special relationship between an individual muse and a particular form of art, the muses were a group. They were feminine, possibly linked to the act of creation (childbirth), but there was no muse of visual art. Thus, for the Greeks, the latter was thought different from the other arts, to do with the hands perhaps rather than the mind).[15]

The melancholic Neo-Platonic philosopher Marsilio Ficino (1433-1499) brought the nine muses together to service music, love and the soul: the *animae musica* made by the eight heavenly spheres and harmony. We retain the concept of such unity in the word Museum, a place to inspire scholarship through integration of knowledge and the arts.

Roger Scruton has noted the coincidence between the rise and development of German Idealism in the 18th century and the apotheosis of German music, music as the art addressed to the soul. He suggested three different meanings of transcendentalism: the theological, the philosophical and the aesthetic. In drawing our attention to the transcendental nature of music, he noted differences, especially in the German culture and language, between a 'hearing' and a 'listening' culture. The latter is a feature of German Romanticism but, as he puts it, "effing the ineffable" is seen as a challenge. *Hören* means 'to listen', yet *Zuhören* is to give oneself up to the experience and *Zugehörigkeit* means belonging: words stemming with *Gehören*, denoting possession.[16]

Scruton uses the expression "acousmatic space … a space full of movement and fields of force in which nothing actually moves…":

> And of which we ourselves could never be a part. In a mysterious way the order of music transforms sequences of sounds into melodies that begin and end, chords that occupy whole areas and gravitational fields that push and pull in ways of their own.

Far away from Newtonian physics and the three dimensional empirical world, Schopenhauer's *Will*, the 'Thing in Itself', behind the veil of representations, where form and content are inseparable, space and time are or seem like one, lies the transcendental power of music.

> It is as though we could reach, through art, to a perfect version of our imperfect states of mind: a version that 'transcends' the ordinary limitations of the human psyche.[17]

Notes

[1] Hesse, H. *The Seasons of the Soul*. Trans Fischer, L. M. (Berkeley: North Atlantic Books, 2011), 95.
[2] Vaughan, H. *The World. Selected Poems* (London: Society for Promoting Christian Knowledge, 2004), 113.
[3] Blackmore, S. J. *Seeing Myself: The New Science of Out-of-Body Experiences* (London: Robinson, 2013), 26. *Déjà vu*—a feeling of familiarity with a present experience: *Déjà vécu*—already having lived through something present.
[4] Rossetti, D. G. *Collected Works* (London: Ellis & Scrutton, 1886), 1:29. An excellent review of *déjà vu* experiences is H.N. Sno, D.H. Linszen, and F. De Jonghe, "Art Imitates Life: Déjà vu experiences in Prose and Poetry". *British Journal of Psychiatry,* 160 (1992): 511-518.
[5] Franz Schubert, *Der Doppelgänger* is one of the six songs from the song cycle *Schwanengesang, with* words by Heinrich Heine. It was written in 1828, the year of Schubert's death.
[6] For an account of work in this area, see Michael Robert Trimble, *The Soul in the Brain: The Cerebral Basis of Language, Art and Belief* (Baltimore: Johns Hopkins Press, 2007).
[7] Callaso, Roberto. *La folie Baudelaire*. Trans. Alastair McEwan (London: Allen Lane, 2012), 23. The cerebellum is the small part at the back of the brain situated under the cerebrum.
[8] For those now aware of the growing literature on the negative effects of marihuana on the human brain, and its link to psychosis, Moreau de Tours was there a long time before. He observed that hashish could lead to a wide range of abnormal mental states, from the mildest morbid impulse to manic excitation and furious delirium. He was a member of the Club des Hachischins in Paris and began to use hashish combined with rose or jasmine essence to treat patients. Members of the Club des Hashischins, founded in 1844 by Théophile Gautier (1811-1872), included Alexandre Dumas (1802-1870), Baudelaire, Hugo, and Gérald de Nerval (1808-1855). His work may be considered as the beginning of the psychotropic era; exploring the effects of drugs on the brain.
[9] This history from a medical point of view is in Trimble, M. R., Ibid. 2016. A recent account of the development and use of psychedelic drugs over the ages is given in: Pollan, M. *How to Change Your Mind* (London: Allen Lane, 2018). The term psychedelic was introduced by Albert Hoffmann (1906-2008), a chemist at Sandoz Laboratories, Basel, who created and experimented with LSD. He believed that its real value was giving a chemical basis for spiritual contemplation. The

psychiatrist Humphrey Osmond (1917-2004) first gave Aldous Huxley (1894-1963) mescaline and coined the epithet: "To fathom in Hell, or soar Angelic/ You'll need a pinch of psychedelic".

[10] Baudelaire, Charles. *The Flowers of Evil. Poison.* Trans. James McGowan. (Oxford: Oxford University Press, 2008), 99-100. Baudelaire had translated de Quincey's book into French.

[11] The Salon de la Rose+Croix (R+C) was founded in 1891. It was bridged with The Croix du Temple et du Graal, a break away from the Rosicrucians. The R+C held exhibitions between 1892 to 1897. See Vivien Green, *Mystical Symbolism: The Salon de la Rose+croix in Paris 1892-1897* (New York: The Solomon Guggenheim Foundation, 2017). Péladan's first novel was titled *Supreme Vice.* He visited Bayreuth and Wagner featured in several of his novels.

[12] Alexander, E. *Proof of Heaven: A Neurosurgeons Journey into the Afterlife.* (New York: Simon & Schuster, 2012), 9; Sigmund, R. *My Time in Heaven* (New Kensington, PA: Whitaker House, 2009); James, W. *The Varieties of Religious Experience* (London: Penguin Books, 1982 [original 1902]), 380-382, 416, 427.

[13] As we discuss later, many philosophers have rather ignored music in their greater scheme of things. Kant, Hegel, Heidegger and Hans-Georg Gadamer (1900-2002), for example, have little to say about music itself, even if their philosophies do enter the areas of aesthetics. The term aesthetics was introduced to philosophy in 1735 by Alexander Gottlieb Baumgarten (1714-1762). Derived from the Greek *aisthetikos,* it emphasized ideals of beauty.

[14] Nancy, Jean-Luc. *The Muses*. Trans. Kamuf Peggy (Stanford: Stanford University Press, 1996). The number of muses varied, from three to nine. They were Clio (history), Euterpe (flute), Thalia (comedy), Melpomene (tragedy), Terpsichore (light verse and dance), Erato (lyrical choral poetry), Polyhymnia (mime), Urania (astronomy) and Calliope (epic poetry). Plato honoured Sappho as the 10th muse.

[15] *Technê* (Greek), a skill or craft, and not the same kind of activity as the arts ascribed to the muses. The Greek *mousiké* embraced all arts the muses had to offer.

[16] Such distinctions are not appreciated by English. Listening requires effort and concentration as opposed to simply having music floating by the ears.

[17] Scruton, R. *Music as an Art* (London: Bloomsbury Continuum, 2018), 71-84, 109-124.

PART 2:

OPERATIC AND LOVE-DEATH

CHAPTER 6

THE OCCURRENCE OF DEATH IN ROMANTIC OPERA

In sweet music is such art,
Killing care and grief of heart
Fall asleep, or hearing die.

(*Orpheus* [William Shakespeare or John Fletcher] in Ivor Gurney, *Five Elizabethan Songs*)[1]

Introduction

We began with the Orpheus myth, central to so many tales and operas, in which the drama of Eros, love, death and resolution or redemption are motifs with enduring meanings and consequences. Works of art beyond opera, including especially literature and poetry, embracing these ideas, reveal that they are universally encountered, especially in great creations such as those of Dante, Shakespeare, Milton, Schiller and Goethe, which resonate with philosophical and religious perspectives. Plato is always there, as is idealism in one form or another. We have discussed the rise of Romanticism, especially following the Kantian revolution, and then philosophical turns through Schopenhauer and Nietzsche, introducing various shades of transcendentalism. The development of musical style over this era is an essential aspect of our enquiry, and has been touched on in Chapter 2. In the next four chapters we will consider the widening perspective of operatic themes, as we move from 17th century towards our current times, especially with growing emphasis on the psychological conflicts of protagonists. We also advance in more detail themes and variations of topics of concern to our main theme. Nearly all can be embraced under the shadow of death, with love and loss, playing back to the ancient Greek tragedies, Orpheus cast in protean disguises. We then move on to that variety of death in opera which has puzzled many and been little discussed, sudden death: *Unexpected* or *Unexplained*?

Love, Rapture and the Apollonian-Dionysian Duality

Wagner famously quipped that every opera needs to have its prayer and its drinking song. This is indeed very often the case, even in his own operas. It remains an interesting phenomenon of this art-form, recurring genres that lend themselves in attractive musical terms to the consolation of invoking higher powers and the extraordinary inebriating pleasure of sharing in the fruit of the vine. Both are perennial features of human activity from the beginning of time, and both play a special role in the huge success of the operatic form that began with the ideals and experiments of the Renaissance Camerata, meeting in Florence in the last decade of the 16th century, with the aim of recreating drama enriched by music as in Ancient Greece.

The prayer and the drinking song are so different and yet of course intimately related: both of them are concerned with seeking a form of transcendence, whether through the hoped-for ecstasy of contact with the divine ordering of the universe, or through the imbibing of alcohol, one of the great agricultural discoveries of mankind in its infancy. Both forms proffer the possibility of an alternative state of mind that can raise the participant out of or beyond the often dreary confines of daily life, with its inescapable and often sorrowful chores and responsibilities.

Both forms are also related to the sense of inspiration: of the literal indwelling or infusion of the Spirit, the divine emanation holding creation and even the universe in its proper order. They are equally related to the use of poetry—that elevated transformation of language from basic communication to a medium of almost mystical communion with some kind of higher order. The Gifts of the Spirit and the Fruits of the Earth are combined through rapture, whether of the soul's aspiring in hope to a higher reality, or through the agency of a liquid capable of enabling entry into a different reality and perception of life.

The images and vectors are brought together in the sacred concept of the Eucharistic chalice, where through a consecrating prayer 'the fruit of the vine and the work of human hands' becomes a spiritual drink that for many is nothing less than a sharing in the divinity of Christ, the manifestation of the omnipotent celestial love. In opera, as so often in religious rites, song becomes the medium of expression for this invocation of the transcendent. As we have already intimated, motivation for this search for a higher or alternate reality is inevitably the pursuit or celebration of love, that universal emotion of attraction for others, mingling the eroticism of physicality with the ennobling aspiration to a higher union that almost seems to promise a reality beyond the limits of human existence and experience in death.

Love is a marvellous situation, chance, fortune, providence, and affirmation of identity, whereby another is able to value us even when we are no longer able to love ourselves. In this way love seeks to confront the dark reality of our finitude with promise of something more, something that might endure even beyond the confines of human experience. Most poetry, and almost all of opera, is devoted to a celebration of this *fortuna meravigliosa* that defies death. Opera, which uniquely combines poetry and song, in fact, almost in itself, becomes therefore an alternative religion in its own right.

Back to Orpheus

It is perhaps no coincidence that the first opera to survive from the aspirations of the Camerata is Monteverdi's *Orfeo* (1607), the first music drama to endure, enshrining a fable that assumes a mythical importance for the cultural nature of opera and its message. Recall that Orpheus ventured into the infernal regions and so charmed Pluto and Persephone/Prosepina that Eurydice was released on the condition that Orpheus did not look back until they reached earth. He turned back to look at her and she instantly vanished. But there were sequences to his life/death. Grieving, he separated himself from the rest of mankind. After the Thracian women (also known as the Bacchantes or Maenads, with whom he had previously celebrated the rites of Bacchus/Dionysus, the god of wine) had torn him in pieces, the fragments of his body were collected by the Muses and buried at the foot of Mount Olympus. His head, still uttering the name of Eurydice, was thrown into the River Hebrus, and carried into the sea. It was borne to the island of Lesbos, where, in some legends, it was buried. Lesbos was, significantly, the home of the great poetess Sappho.

The Bacchae were the female attendants of Bacchus, known for their extravagant gestures and frenzied rites: the title is derived from the Greek *mainas* or 'madwoman'. The Bacchanalia, or Roman festival in honour of Bacchus, were characterized by drunkenness and wild revelry.

The Apollonian and the Dionysian

The name Apollo, who we have met in earlier chapters so far, derives from an Indo-European root *abol*, 'apple', or from the Semitic *ap olen*, 'universal father'. For the Greeks he was the son of Zeus and Leto, and identified with Helios, the sun-god. He was the brother of Artemis (Diana), half-brother of Hermes (Mercury), and father of Asclepius (the Roman Aesculapius). He was the god of music, poetry, archery, prophecy and the healing art. His plant was the laurel, and he is represented as the perfection of youthful manhood.[2]

In contrast, Dionysus of the Greeks, the son of Zeus and Semele, became Bacchus (perhaps of Thracian or Lydian in origin) as the Roman god of wine. The name was originally an epithet of *Dionysiud.* At first represented as a bearded man, he later appeared as a handsome youth with black eyes and flowing locks, crowned with vine and ivy. In peace his robe was purple, in war it was a panther's skin. The Romans seemed to have confused this god with the ancient Italic deity Liber Pater, god of wine. Dionysus, a god of fertility, a suffering god, who dies and comes to life again, gave us gift that loosens care and inspires man to music and poetry. Hence his connection with the dithyramb, tragedy, and comedy: Elements of ecstasy and mysticism are found in his cult. He is frequently represented drawn in a chariot by tigers, and accompanied by a rout of votaries, male and female (Satyrs, Sileni, Maenads, Bassarids).[3] [**Fig. 6.1-4** Dionysus]

The context of the Apollonian–Dionysian duality is fundamental to the very concept of opera and its celebration of song and the transforming power of music. A survey of some significant scenarios through the history of opera will illustrate the permeation and dynamism of the symbolic principle.

Orfeo (Monteverdi, 1607; Gluck, 1762)

This opera establishes the fundamental pattern and symbolism for so many operatic dramas. There are two distinct worlds: one of life and light, another of darkness and death. The hero must leave the light and certainties of life and enter the shadowy, unpredictable and dangerous realms of the Underworld, where he is confronted by challenges of every kind in trying to regain his lost love. Emerging from this shadowy world is fraught, and can so easily end in tragic loss (as when he looks round against the express condition, and loses Eurydice for ever). In both operas, however, the tragic is averted and the lovers restored to each other by divine intervention—either by the consolation of promised immortality by Apollo or the agency of Cupid.

Don Giovanni (Mozart, 1787)

This opera presents two mental worlds. The first is of orthodoxy, order and law (embodied quite literally in the otherworldly statue of the Commendatore, the personification of righteousness and retribution). Opposed is the amoral/immoral, lawless realm of Don Giovanni, a libertine whose attitude is symbolized in relentless sexual activity, a metonym for wholesale disregard for the norms and standards that help society to function smoothly and fairly. The opera plunges into the vortex of this duality in the opening scene where Don Giovanni's attempted rape of Donna Anna is interrupted by her father who is killed in a dual with the profligate. Sexual licentiousness has moved into coercion, violence and murder, and the social fabric threatens to unravel. Don Giovanni represents a roving, free spirit, bold and enterprising in its destructive forcefulness, and revelling in its personal existential choices, regardless of traditional values and expectations, and casting aside the powerful witness of received wisdom. His refusal to repent, even on the terrifying supernatural visitation of the avenging Commendatore, remains a key statement of the Romantic spirit of revolutionary defiance.

Die Zauberflöte (Mozart, 1791)

The symbolism in this work becomes overt in polarities of light and darkness. The Apollonian realm is represented by the high priest Sarastro and his solar alignment. He stands for rationality, control and enlightenment, the right ordering of society and relationships, and assumes a universal paternal persona. Opposed are the darkness, irrationality and potentially destructive world of the Queen of the Night, a chaotic principle realized in maternal imagery, and also associated with the menace of coercion and violence. The lovers Pamina and Tomino must be drawn away from this negative influence, and enabled to love freely as adults. This is resented by Sarastro's masterful but benevolent interventions, and inauguration of the trials by water and fire that signify the process of initiation into illumination and full emotional integration.

Fidelio (Beethoven, 1805, 1814)

Here the semantic field is moved into concepts of liberty and imprisonment, freedom and constraint, justice and tyranny. The external world is untrammelled, one of light and ideals, of righteousness and full conjugal union. This is embodied in Leonore/Fidelio, who acts in selfless and potential danger to uncover fraud and deception, and sets free her high-principled husband Florestan who is incarcerated by the treacherous and murderous Don Pizzaro in the underworld of a hidden dungeon and chains. She is helped by the jailer Rocco who operates on a good-natured but mundane understanding, and by the minister of state Don Fernando who will implement the liberty that Leonore's self-sacrificing love has worked for her husband. Florestan's emancipation is a movement from darkness into light at every level of the symbolic spectrum.

Der Freischütz (Weber, 1821)

The realms are now couched in the semantics of the supernatural, with an ordinary everyday world of human activity, striving and weakness, counterpoised by a secret world of demonism and black magic. In order to gain abnormal powers to succeed beyond natural limitation, the young and unfortunate hunter Max is tempted by the dangerous Kaspar to seek the magic agency of the enchanted bullets provided by the Dark Huntsman Samiel at the midnight ceremonies held in the haunted Wolf's Glen in the depth of the forest. Layer upon layer of symbolism pushes the scenario into a dualistic landscape of light and darkness, day and night, village and forest, realism and magic, orthodoxy and heterodoxy, freedom and servitude, life and death. The challenge of attaining love is beset with the fraught dangers of mortality, both corporeal and spiritual. Agathe must nearly die through the supernatural agent, but eventually the villain is himself outwitted of his life by the fearful ironies of his dark master. The Hermit appears at the end as the voice of illumined spiritual authority, and with the power to enlighten Prince Ottokar for good.

Robert le Diable (Meyerbeer, 1831)

The symbolic world is sustained and complex, with the supernaturalism of *Der Freischütz* and the familial concepts of *Die Zauberflöte* brought into interplay. Like the former, two opposing worlds are presented, one dominated by the father figure Bertram, at the other by the sainted mother figure and her surrogate/agent Alice. This time the alignment is reversed, with the realm of the mother characterized by light and life, and that of the paternal by darkness and death. As in *Der Freischütz*, two opposing worlds are generated: the everyday world of human striving and disappointment, with the weakness of the protagonist in winning his bride, the cause of his dangerous enticement into the perilous realms of the supernatural. As with Max, tempted by Kaspar to journey into the Wolf's Glen, so Robert is induced by his father to seek a magic branch in the ruined Abbey of St Rosalie, where he is inveigled by the supernatural blandishments of ghostly unfaithful nuns to steal the sacred talisman that empowers him unnaturally. The ascendancy of light over darkness is fought out in the last act, in the maternal-paternal dynamic of *Die Zauberflöte*, with the wavering hero, torn between his father Bertram and foster-sister Alice, saved by the eschatological bell of midnight.

Der fliegende Holländer (Wagner, 1843)

Here the clash of realms of light and darkness are proposed in the intrusion of the supernatural into the natural, as the stricken Flying Dutchman comes ashore on one of his seven-yearly reprieves to find a human being prepared to make the ultimate sacrifice of love to free him from his curse. The wild and unpredictable world of the Dutchman's temporary immortality is full of mystery and menace, and is juxtaposed with the bourgeois ordinariness of Daland's Norwegian village with its spinning womenfolk, bemused by the heroine Senta's powerful attraction (through the recitation of the ballad) into terrifying challenges and vagaries of the spirit-world. The two realms are eventually fused in Senta's self-sacrifice, which marks the triumph of love and the reconciliation of the opposites.

Tannhäuser (Wagner, 1845)

The composer conflated two different Medieval legends in this scenario where the typical Romantic juxtaposing of two opposing symbolic realms establishes the perimeters of the moral discourse. Tannhäuser, like Max and Robert, is torn between the two Orphic worlds: light, order and orthodoxy in the Medieval Catholic world of the Wartburg, presided over by the saintly Elisabeth and the paternal figure of benign authority in the Landgrave; darkness, moral confusion and heterodoxy in the supernatural erotic blandishments of the Venusberg, where the goddess of love herself represents a consuming maternal feminine destructiveness similar to that of the Queen of the Night and the damned nuns of St Rosalie. Not even the traditional repentance and sojourn in the Vale of Tears is enough to free

Tannhäuser until the intercession of Elisabeth. Already effective in saving him from death by outraged society, now in her own grief-stricken passing she becomes an effective voice in the Communion of Saints, and facilitates his salvation represented in the Pope's flowering staff.

Faust (Berlioz, 1846; Gounod, 1859; Boito, 1868; Busoni, 1925)

The legend of Faust presents a blueprint for the Don Juan myth with the cherished freedom of the individual, despite the overwhelming weight and portentousness of received wisdom and tradition. In concourse with Mephistopheles, Faust breeches and broaches the realms of light and darkness, and secures supernatural powers symbolized in restored youth and the promise of erotic freedom (cf. Max, Robert, Tannhäuser). The destruction of the loving Gretchen in this process of self-fulfilment parallels the deaths (or near-deaths) of heroines like Agathe, Alice, Senta, and Elisabeth, and empowers her eventually to become the agent of forgiveness, ascent and transformation in the mystical final scene. Both Gounod and Boito are pertinent in this consideration. Only Boito brings in the second more overtly symbolic elements of Goethe's treatment of the myth. (See further discussion of these operas in Chapter 18.)

Carmen (Bizet, 1875)

The heroine in many ways parallels the character of Don Giovanni. She is of an alien Gypsy race, withits temperament and perspective on life. This is characterized by freedom, lawlessness and superstition. She manipulates the infatuation of a Spanish officer to affect her escape from custody, and after promising him the fulfilment of his sexual passion, callously casts him aside for an affair with a glamorous bullfighter. Her licentious personality and milieu represent the polar opposite of the filial, social, military probity and discipline of Don José's background. Her progressive seduction and ruination of this former man of integrity constitutes the triumph of the Dionysian over that of the Apollonian wholeness and self-control of the officer's life. She asserts the sway of eroticism over him during the promises made in the Seguidilla, and completes the destruction of his way of life, already compromised by imprisonment and demotion, in their meeting at Lilia Pastia's tavern. Her siren call merging in with the trumpet reveille of the *appel* represents the moment of transition from control into irreversible lawlessness. The former brigadier becomes a common smuggler in the Gypsy band, wild in the mountains, sexually enthralled to a woman who no longer loves him. Eventually Carmen will fatalistically give herself over to violent death at his hands.

The Queen of Spades (Tchaikovsky, 1890)

The adaptation of Pushkin's short story captures a psychological interpretation of the pattern of the two worlds. The young officer Hermann loves Lisa, but his passion for gambling opens him to a compulsion that intensifies his neurotic sensibilities, and gradually overtakes his reasonable ordered world, turning him into an obsessive, determined to fathom the superstition of the secret of thc Three Cards, reputedly entrusted to the aged Countess during her youth in Paris. Hermann gradually loses his hold on the conduct of normal life, and becoming more reckless as his obsession develops, inducing the Countess's sudden death by visiting her secretly in her chamber at the dead of night and threatening her with a pistol. She dies without speaking. His preoccupation sees him relinquishing control of his life, increasing enthralled to his delusions. This leads to Lisa's suicide, Hermann's growing dissolution as he is visited by the Countess's ghost, and his own eventual suicide in a gambling den.

Death in Venice (Britten, 1973)

The psychological apprehension of the two worlds of emotional and symbolic dualism is given more modern expression in Britten's adaptation of Thomas Mann's Novelle *Der Tod in Venedig* (1911), and the fate of the author Gustav von Aschenbach. In the midst of a sense of his own futility, he seeks refreshment in a sojourn to Venice during a sweltering summer where there is an outbreak of cholera. Here, the *fin de siècle* city, with its decadent luxury, is the setting for the illicit, potentially scandalous, love the older man develops for a beautiful adolescent. His tranquillity is shattered by his observations and growing sexual passion for Tadzio (Tadeusz), of a Polish family holidaying in the same hotel. While nothing overt comes of this all-consuming preoccupation, the inversion of the standards of respectability, and the threat of possible opprobrium, dominate the perception of the protagonist, who eventually dies by himself on a chair on the deserted beach, his ordered life riven by the overwhelming experience of dark torrential passion.

Bassarids (Henze, 1965)

The most modern opera on the two worlds returns to Antiquity in recounting the story of Pentheus, king of Thebes, who resisted the introduction of the worship of Dionysus into his kingdom. Slowly he is driven mad by the god covertly present in his court. His palace is ruined, and he himself destroyed by the Bacchanals, among whom are his mother and two sisters. There is tremendous danger in ignoring, or trying to suppress the darker wilder side of the psyche—of passion, of desire, of frightening inspiration.

Following these general considerations, in the next chapter we look at sudden unexplained deaths, and move to a consideration of the love-death— the *Liebestod.*

Notes

[1] Harrison N. *The Wordsmith's Guide to English Song: Poetry, Music and Imagination.* Volume 2 (Oxford: Compton Publishing, 2016), 12.
[2] Kerényi, Carl. *The Gods of the Greeks* [1951] (London: Penguin: 1958), 147-152.
[3] Kerényi, Ibid., 220-241.

CHAPTER 7

SUDDEN DEATH: CAUSES, SITUATIONS AND *LIEBESTOD*

In the unity of love and death, in the willing acceptance of death for love's sake, and in the renunciation of self for other we glimpse the meaning of human life.

(Roger Scruton, *Wagner and German Idealism*)[1]

The Biblical and Literary Foundations

Unexplained death is an absorbing topic. Medically it can be caused by a sudden cessation of the heart's function, a stroke, an epileptic seizure, or by physiological challenges related to such conditions. Anecdotally it is associated with stress, especially linked, for example, with loss and bereavement. A common factor seems to be an innate physical weakness, strained unendurably by the onset of an excess of physical tension, pain or stress.

The heart and its ancient association with the emotions is key here. This organ, in mammals (those who have mothers), is situated in the chest and in humans just left of centre. We feel it on the left, and literally and metaphorically it is linked to our emotions. It is the focus of feeling in ancient symbolism, yet sometimes the nature of the emotion is ambiguous.

Among the most notorious of sudden deaths is that of the duplicitous couple Ananias and Sapphira, both of whom independently lie to the Apostle Peter in the idealistic days of the early Christian community in Jerusalem. They pretend to have sold their property and to have given the proceeds to the church, but in fact keep back some for themselves. Each individually tells the lie to Peter, and each is overcome by sudden death at Peter's words of reproof: "When Ananias heard these words, he fell down and died … *Immediately* she fell down at his feet and died" (Acts 5: 5, 10). Peter does absolutely nothing, and both deaths come spontaneously.

Interestingly, Peter observes to the man: "Ananias, why has Satan filled your heart to lie to the Holy Spirit...?" (5:3). It is as though the heart passes its own judgment on the situation, and brings about an auto-sentence.

Another sudden death in the Bible is more mystical. In primordial times we hear of Enoch, a man of great righteousness, who does not die but is 'translated' into heaven (Genesis 5:21-24).

21 When Enoch had lived sixty-five years, he became the father of Methu'selah.

22 Enoch walked with God after the birth of Methu'selah three hundred years, and had other sons and daughters.

23 Thus all the days of Enoch were three hundred and sixty-five years.

24 Enoch walked with God; and he was not, for God took him.[2]

The end of the great Moses is also mysterious. When Moses dies before entering the Promised Land, we are told: "So Moses the servant of the Lord died there in the land of Moab ... and he [the Lord] buried him in the land of Moab ... but no man knows the place of his burial to this day. ... his eye was not dim, nor his natural force abated" (Deuteronomy 33:5-7). The level of mysticism increases with the prophet Elijah, who is taken from the sight of his appointed disciple Elisha: "And as they still went on and talked, behold a chariot of fire and horses of fire separated the two of them" (2 Kings 1:11-12).

9 When they had crossed, Eli'jah said to Eli'sha, "Ask what I shall do for you, before I am taken from you." And Eli'sha said, "I pray you, let me inherit a double share of your spirit."

10 And he said, "You have asked a hard thing; yet, if you see me as I am being taken from you, it shall be so for you; but if you do not see me, it shall not be so."

11 And as they still went on and talked, behold, a chariot of fire and horses of fire separated the two of them. And Eli'jah went up by a whirlwind into heaven.

12 And Eli'sha saw it and he cried, "My father, my father! The chariots of Israel and its horsemen!" And he saw him no more. Then he took hold of his own clothes and rent them in two pieces.

The deaths of these three figures are clearly sudden and inexplicable, and have mystical implications: both Moses and Elijah appear speaking with Jesus on the Mount of Transfiguration (in Matthew 17:1-9, in Mark 9:2-10 and in Luke 9:28-36). Clearly something to do with their disposition of heart, with the intensity of their spiritual perceptions and love, drew them by osmotic process into their natural heavenly metier. Their 'deaths' become a transposition of sacrifice propelling them into their rightful place.

So the ancient collocation of love and death takes on an inner dynamic that has overwhelming physical consequences. This is true of the negative manifestation of judgment (in the case of Ananias and Sapphira) and of the positive rhapsody of mystical union (with Moses and Elijah). Sudden death becomes an end, a punishment of separation, or a metaphysical bridge, a transcendental transition to full integration.

Introduction to Love-Death

The poetic world of opera, operating by the laws of love and death, of Orphic rapture and violent rupture, sees the symbolism of sudden death as a key indication of the Romantic path of attainment. The guilt of illicit passion, the intensity of pure love perceived, push the subject both physiologically and socially to an intensity of strain that can find outlet only in sudden death, and an incipient translation to a new world of union and fulfilment. The symbolism comes to the fore in early Romanticism, with the compulsion of erotic love, a metaphor for mystical union that can no longer be opposed by a hostile, conventional world that cannot understand the irresistible passions of the heart. *Liebestod* as a literary term (from German *Liebe*, 'love' and *Tod*, 'death') refers to the theme of erotic death or 'love- death', meaning the couple's consummation of their love in death or after death.

There are influential literary precedents. As with the Scriptures, these go back to Antiquity and include the story of Pyramus and Thisbe in Ovid's (43 BC-17 AD) *Metamorphoses*.

> Pyramus and Thisbe are two lovers in the city of Babylon who occupy connected houses, forbidden by their parents to be wed, because of their parents' rivalry. Through a crack in one of the walls, they whisper their love for each other. They arrange to meet near Ninus's tomb under a mulberry tree and state their feelings for each other. Thisbe arrives first, but upon seeing a lioness with a mouth bloody from a recent kill, she flees, leaving behind her veil. When Pyramus arrives he is horrified at the sight of Thisbe's veil which the lioness had torn and left traces of blood behind, as well as its tracks, and assumes that a wild beast has killed her. Pyramus kills himself, falling on his sword in proper Babylonian fashion, and in turn splashing blood on the white mulberry leaves. His blood stains the white mulberry fruits, turning them dark. Thisbe returns, eager to tell Pyramus what had happened to her, but she finds Pyramus' dead body under the shade of the mulberry tree. Thisbe, after a brief period of mourning, stabs herself with the same sword. In the end, the gods listen to Thisbe's lament, and forever change the colour of the mulberry fruits into the stained colour to honour the forbidden love.[3]

The story of *Pyramus and Thisbe* appears in Giovanni Boccaccio's (1313-75) *On Famous Women* as biography Number Twelve (sometimes Thirteen) and in his *Decameron* (1348). In the 1380s Geoffrey Chaucer (1333-1400) in his *The Legend of Good Women*, and John Gower (1330-1408), in his *Confessio Amantis*, were the first to tell the story in English. Gower altered the story somewhat into a cautionary tale. John Metham's *Amoryus and Cleopes* (1449) is another early English adaptation.

The tragedy of *Romeo and Juliet* (1594-95) ultimately derives from Ovid's story. Here the star-crossed lovers cannot be together because Juliet has been engaged by her parents to another man and the two families are caught up in an ancient feud. As in Pyramus and Thisbe, the mistaken belief in one lover's death leads to consecutive suicides. The earliest version of Romeo and Juliet was published in 1476 by Masuccio Salernitano (1410-1475), while most of its present form was settled when written down in 1524 by Luigi da Porto (1485-1529). Both Salernitano and Da Porto are thought to have been inspired by Ovid's and Boccaccio's writings. Shakespeare's famous adaptation is a dramatization of Arthur Brooke's (-1563) 1562 poem *The Tragical History of Romeus and Juliet* (1562), itself a translation of a French translation of Da Porto's novella.

The 17th century Scottish song *The Ballad of Bonny Barbara Allen* is a fine example of love-death:

It was in and about the Martinmas time,

When the green leaves were a-falling,
That Sir John Graeme, in the West Country,
Fell in love with Barbara Allan.

He sent his man down through the town,
To the place where she was dwelling:
“O haste and come to my master dear,
Gin ye be Barbara Allan.”

O hooly, hooly rose she up,
To the place where he was lying,
And when she drew the curtain by:
“Young man, I think you’re dying.”

“O it’s I’m sick, and very, very sick,
And ’tis a’ for Barbara Allan.”
“O the better for me ye s’ never be,
Though your heart’s blood were a-spilling.”

“O dinna ye mind, young man,” said she,
“When ye was in the tavern a drinking,
That ye made the healths gae round and round,
And slighted Barbara Allan?”

He turned his face unto the wall,
And death was with him dealing:
“Adieu, adieu, my dear friends all,
And be kind to Barbara Allan.”

And slowly, slowly raise she up,
And slowly, slowly left him,
And sighing said, she could not stay,
Since death of life had reft him.

She had not gane a mile but twa,
When she heard the dead-bell ringing,
And every jow that the dead-bell geid,
It cried, “Woe to Barbara Allan!”

“O mother, mother, make my bed!
O make it saft and narrow!
Since my love died for me to-day,
I’ll die for him to-morrow.”

They buried her in the old churchyard,
And Sir John's grave was nigh her.
And from his heart grew a red, red rose,
And from her heart a brier.

They grew to the top o' the old church wall,
Till they could grow no higher,
Until they tied a true love's knot—
The red rose and the brier.[4]

Sir John and Barbara Allan's love continues to grow after death and the rose bush entwines a love knot. The tolling of bell makes Barbara Allan understand what has happened; she dies of shock.

Goethe's ballad *Die Braut von Korinth* (*The Bride of Corinth*, 1797) is a story about a young man who has been promised a bride, but she has been sent to a cloister and died. She appears to the man in the dark of his chamber, and they spend the night together. The girl belongs among the undead and carries the young man off with her in a *Liebestod*: she wants to be cremated with him.

"From my grave to wander I am forc'd
Still to seek The God's long-sever'd link,
Still to love the bridegroom I have lost,
And the life-blood of his heart to drink;
When his race is run, I must hasten on,
And the young must 'neath my vengeance sink.

"Beauteous youth!
no longer mayst thou live;
Here must shrivel up thy form so fair;
Did not I to thee a token give.
Taking in return this lock of hair?
View it to thy sorrow!
Grey thou'lt be to-morrow,
Only to grow brown again when *there*."

"Mother, to this final prayer give ear!
Let a funeral pile be straightway dress'd;
Open then my cell so sad and drear,
That the flames may give the lovers rest!
When ascends the fire
From the glowing pyre,
To the gods of old we'll hasten, blest."[5]

One-sided examples occur in Goethe's *The Sorrows of Young Werther* (1774) and Robert Browning's (1812-1889) poem *Porphyria's Lover* (1836), his first short dramatic monologue and the first of his poems to examine abnormal psychology. Browning uses this mode of exposition to describe a man who responds to the love of a beautiful woman by killing her. The joint suicide of the novelist and dramatist Heinrich von Kleist (1777-1811) and his lover Henriette Vogel (1780-1811) in 1811 is often associated with the *Liebestod* theme.

Elizabeth Barrett Browning (1806-1861) in her epic *Aurora Leigh* (1856) has Aurora speak of the transcendence within her love for the blind Romney:

> "O dark, O moon and stars, O ecstasy
> Of Darkness! Oh great mystery of love …"

which "enlarges rapture … in a hurtle of united souls". For him she is "My morning star" and it is in

> the hour of souls;
> That bodies, leavened by the will and love,
> Be lightened to redemption.[6]

The fate of the knight Hugo at the end of Friedrich de la Motte Fouqué's novella *Undine* (1811) and later the fates of Cathy and especially Heathcliffe in Emily Brontë's (1818-1848) novel *Wuthering Heights* (1849) are of particular significance for detailed consideration of the topos. When Heathcliffe visits Cathy in her illness prior to her death, the violence of their mutual recriminations assumes an almost elemental character, and a startling self-identification of one with the other. The tenor of their exchanges is born out of an obsessional concern with life and death, and the supreme imperative of union that can find its natural fulfilment only in a post-mortal situation, with all the inevitable concerns, obligations and petty emotions of this life obviated.

> Heathcliffe had knelt on one knee to embrace her; he attempted to rise, but she seized his hair, and kept him down.
>
> 'I wish I could hold you,' she continued, bitterly, 'till we were both dead! I shouldn't care what you suffered. I care nothing for your sufferings. Why shouldn't you suffer? I do! Will you forget me? Will you be happy when I am in the earth? ... 'Oh, you see, Nelly, he would not relent a moment to keep me out of the grave. That is how I'm loved! Well, never mind. That is not my Heathcliffe. I shall love mine yet; and take him with me: he's in my

soul. And,' added she musingly, 'the thing that irks me most is this shattered prison, after all. I'm tired of being enclosed here. I'm wearying to escape into that glorious world, and to be always there: not seeing it dimly through tears, and yearning for it through the walls of an aching heart: but really with it, and in it …

Heathcliffe's response is just as elemental.

You loved me—then what right had you to leave me? What right—answer me—for the poor fancy you felt for Linton? Because misery and degradation, and death, and nothing that God or Satan could inflict would have parted us, you, of your own will, did it. I have not broken your heart—you have broken it; and in breaking it, you have broken mine. So much the worse for me that I am strong. Do I want to live? What kind of living will it be when you—oh, God! would you like to live with your soul in the grave?'[7]

The implications of the mysterious condition of the Love-Death are even more fully explored in the closing section of the novel, where Heathcliffe suddenly seems to be able to bring about a confluence of his natural condition and the self-willed psychosomatics of death. The narrator Nelly Dean witnesses the unbidden, sudden inexplicable change in Heathcliffe.

Heathcliff stood at the open door; he was pale, and he trembled: yet, certainly, he had a strange joyful glitter in his eyes, that altered the aspect of his whole face.

'Then you are not afraid of death?' I pursued.

'Afraid? No!' he replied. 'I have neither a fear, nor a presentiment, nor a hope of death. Why should I? With my hard constitution and temperate mode of living, and unperilous occupations, I ought to, and probably shall, remain above ground till there is scarcely a black hair on my head. And yet I cannot continue in this condition! I have to remind myself to breathe—almost to remind my heart to beat! And it is like bending back a stiff spring: it is by compulsion that I do the slightest act not prompted by one thought; and by compulsion that I notice anything alive or dead, which is not associated with one universal idea. I have a single wish, and my whole being and faculties are yearning to attain it. They have yearned towards it so long, and so unwaveringly, that I'm convinced it will be reached—and soon—because it has devoured my existence: I am swallowed up in the anticipation of its fulfilment. My confessions have not relieved me; but they may account for some otherwise unaccountable phases of humor which I show. O God! It is a long fight; I wish it were over!'[8]

> 'It is not my fault that I cannot eat or rest,' he replied. 'I assure you it is through no settled designs. I'll do both, as soon as I possibly can. But you might as well bid a man struggling in the water rest within arms' length of the shore! I must reach it first, and then I'll rest. Well, never mind Mr. Green: as to repenting of my injustices, I've done no injustice, and I repent of nothing. I'm too happy; and yet I'm not happy enough. My soul's bliss kills my body, but does not satisfy itself.'[9]

> Having succeeded in obtaining entrance with another key, I ran to unclose the panels, for the chamber was vacant; quickly pushing them aside, I peeped in. Mr. Heathcliffe was there—laid on his back. His eyes met mine so keen and fierce, I started; and then he seemed to smile. I could not think him dead: but his face and throat were washed with rain; the bed-clothes dripped, and he was perfectly still. The lattice, flapping to and fro, had grazed one hand that rested on the sill; no blood trickled from the broken skin, and when I put my fingers to it, I could doubt no more: he was dead and stark![10]

Heathcliffe sought eternal life with Cathy, she perhaps the same. At the end, the ghosts of both are seen walking on the moors. *Wuthering Heights* has been the subject of two operas: by Bernard Herrmann (1911-1975) (1951),[11] with libretto Lucille Fletcher; and by Carlisle Floyd (1926-) (1958).[12]

Undine we return to shortly.

The same mixture of romance motif and Gothic gloom fixed on the male perspective are found in the French novelist Gaston Leroux's (1868-1927) *Le Fantôme de l'Opéra* (1910). The plot revolves around a beautiful soprano, Christine Daaé, who becomes the obsession of a mysterious, disfigured musical genius who lives in the subterranean labyrinth beneath the Paris Opéra House. Christine sings so exquisitely in Gounod's *Faust* as to bring the chandelier down. The Phantom, Erik, hides her in the depths of his concealed world and teaches her the secrets of his musical craft. He dies in his chambers of illusion; Christine will sing no more.

The Phantom has resemblance to other monsters featured in Gothic novels such as Frankenstein's Creature, Stevenson's Dr. Jekyll, Wilde's Dorian Gray and Stoker's Count Dracula, as well as with Heathcliffe, DU Mautier's Svengali, and Dr Miracle in *Les Contes d'Hoffmann*. Although the Phantom is really just a deformed man, he has ghost-like qualities in that no one can ever find him or his lair. People are frightened by him because of his deformities and the acts of violence he commits. The novel features a love triangle between the Phantom, Christine and Raoul. Raoul is seen as Christine's childhood love: she is familiar with and has affection for him. He is rich and offers her security as well as a Christian marriage. The Phantom, on the other hand, is dark, ugly and dangerous, and therefore represents a forbidden, reckless love. However, Christine, who has analogies with Marguerite in *Faust*, with Antonia in Offenbach's opera, and with Trilby in George Du Maurier's (1834-1896) novel of that name (1895), is drawn to the Phantom (as is Cathy to Heathcliffe). Raoul, the hero of Meyerbeer's *Les Huguenots*, is the safe but more conventional option. But Christine perceives the Phantom as her Angel of Music, and she pities his existence of loneliness and darkness. While she does not die, Erik's passing is a type of Love-Death associated with his adoration of Christine and the art of music. The tale has taken on a new and spectacular life in the musical *The Phantom of the Opera* (1986) by Andrew Lloyd-Webber and Richard Stilgoe. The musical affirms that the mystery of love is still with us, as is the *Liebestod*.

Tennyson's vastly extended elegy *In Memoriam* (1850) for his dead friend Arthur Hallam is a huge meditation on the issues of life and death. The 131 cantos provide perspectives on the sorrows and hopes attendant on both, contrasting the desolation of bereavement with the hopes of finding some meaning in and after death. This is integral to the reality of love and what it means in the mystery of life. Some ideas provide trajectories for the transcendence that is concomitant with the Love-Death.

XLIII
If Sleep and Death be truly one,
And every spirit's folded bloom
Thro' all its intervital gloom
In some long trance should slumber on;

Unconscious of the sliding hour,
Bare of the body, might it last,
And silent traces of the past
Be all the colour of the flower:

So then were nothing lost to man;
So that still garden of the souls
In many a figured leaf enrolls
The total world since life began;

And love will last as pure and whole
As when he loved me here in Time,
And at the spiritual prime
Rewaken with the dawning soul.[13]

The Operatic Realization of Love-Death, the *Liebestod*

In opera, the first major instance of this topos is the Love-Death that occurs at the end of E.T.A. Hoffmann's opera *Undine* (1816), an adaptation of Friedrich de la Motte Fouqué's Novelle of the same name—and the benchmark of this Romantic experience. The mystery of the water-sprite, the hopeless love of an elemental being for a human, and the irresistible hold of this being on the human heart, can only be realized by the kiss from the sprite that initiates the precipitate death of the human protagonist (the kiss as a metaphor for Sudden Death), and ushers him into a world of transfigured integration (as reflected in the postlude to this opera, set in the watery submarine world). The Undine-theme is very interesting in its recurrence and provides a strong masculine variant to the usual feminine subject. Albert Lortzing (1801-1851) would use the same scenario in his *Undine* (1845), and the ubiquity of the concept recurs in the Slavonic versions of the myth, in the *Rusalka* operas of Alexander Dargomyzhsky (1813-1869, 1856) and Dvořák (1901). *Undina* (sometimes *Undine* or *Ondine*) is a fragmentary opera in 3 acts by Tchaikovsky. The work was composed in 1869. The libretto was written by Vladimir Sollogub (1813-1882), and is based on Vasily Zhukovsky's (1783-1852) translation of Fouqué's *Undine.* We enter a fuller discussion of these operas in Chapter 9. [**Figs. 7.1-2**]

The powerful realization of the *Liebestod* in the undine/rusalka kiss that leads to the death of the 'hero' is analogous to the fatality of something venomous, like Cleopatra's asp that brings on rapturous death (a variant on the Judas kiss?). In the same way the poison tree in Meyerbeer's *L'Africaine* (a variant on the Tree of Life in Eden) and the poisoned bread in Humperdinck's *Königskinder* (a magic blasphemy of the Eucharist) are really 'surface' vectors for a truth of love and fulfilment realized in the 'deep' action. But they all lead to the same Love-Death situation.

The terminology and very concept of *Liebestod* are given definitive expression in the final, dramatic aria from Wagner's opera *Tristan und Isolde* (1859/1865). Indeed, Wagner's operas, which we will explore later, make a much fuller use of the concept of sudden death than the work of any other composer.

In his first opera, *Die Feen* (1834), Wagner added an authentic human touch, finishing the story with a catharsis, similar in some ways to a *Liebestod*, a theme which is fundamental to the development of his art. In *Rienzi* (1842) the siblings Irene and Rienzi are not separated, but united in death in loving devotion when they (more or less willingly) give themselves to perish in the conflagration of the Capitol (the vector).

This is also the case with Isolde, and Kundry in *Parsifal* (1882), where no vector is used. With Isolde it is the death of Tristan which induces her mystical state; with Kundry it is the advent and apprehension of her salvation, the flow of grace crystallized in the exposition of the Grail by Parsifal, that initiates her translation. The same could be said for the deaths of Elisabeth and Tannhäuser, the death of Elsa, and to some extent that of Senta (who copies Fenella in Auber's *La Muette de Portici*, 1828). Each is promoted by a crisis in the situation of love that has imploded fatally. These instances are also particularly interesting because they share equally in the centrifugal pull between the spiritual and erotic elements that are endemic to all manifestations of the Love-Death.

The fullest statement is in *Tannhäuser* (1845) where Elisabeth (modelled on St Elisabeth of Thuringia) dies suddenly and inexplicably, where this death is expiatory, and initiates a process of intercession (the Communion of Saints) that has a bearing on the erring hero's salvation. He too dies suddenly on hearing of Elisabeth's passing, and we understand this to be a Love-Death, gaining him access to Paradise when we hear the returning pilgrims bearing the news that the Pope's staff has flowered, like Aaron's Rod (Numbers 17:8). That is, life has come from death.

In *Lohengrin* (1850) Elsa dies suddenly at the end of the opera, when the beloved hero is called away, having restored her enchanted brother to life. The intensity of Elsa's foiled love demands her sudden death, but this time it would seem that there is no suggestion of a heavenly transition. Conversely, at the end of *Parsifal,* the former temptress Kundry, baptized and transformed, is able to pass effortless into a new mode of existence through the salvific agency of the redeeming hero (who himself as administrator of grace must benefit from its bestowal).

The most famous of the sudden deaths is Isolde's at the end of *Tristan und Isolde*. Tristan, who has always linked night and death as the appropriate circumstances for his illicit love of his uncle's bride, hastens his own death by pulling the bandages from his mortal wounds. Isolde, however, experiences the essence of the osmotic process of transition and seems, in ecstasy, effortlessly to wish herself from this life into a new spiritual union with Tristan.

This is at the very heart of the Romantic experience. In the other archetype of tragic love, Donizetti's *Lucia di Lammermoor* (1835), the deceptions and betrayals lead the heroine into madness, now seen not as rupture or illness, but a special state of mind where heightened perceptions see the realities of love and can live the union denied by the factual situations of life. Whereas Edgardo (like Tristan) will end his own life to be with her, Lucia (like Isolde) dies, the sudden death consecrating the foiled union, and serving as the transition to an untrammelled sphere of union. It is always the intensity of emotion, of love, that enables the mystical entrée to this special way of dying. Isolde's enraptured melos, and Lucia's flute, become the vectors of a process of liberation.

The most important late Romantic variant on this theme is Britten's opera (and Visconti's film), *Death in Venice* (1973). Aschenbach's sudden death comes alone in his deckchair, with the epidemic contagion heavy in the air. The plague itself becomes a metaphor of the transition Aschenbach must undergo to be free of human complexity and censure, and so attain the liberation of his loving heart.

There are variations on this pattern, especially emanating from France. The heroines of three French literary works by Prosper Merimée (1803-1870) (*Carmen*, 1845), Henri Murger (1822-1861) (*Scènes de la Vie Bohème*, 1847-49) and Alexandre Dumas (*fils*, 1824-1895) (*La Dame aux camellias*, 1852) have come to represent for many the archetypal experience of opera. These heroines, however, undergo a type of death that is a remove from the kind of sudden death of greatest interest to us. Although all die unexpectedly, death is the result of external factors: murder and illness. These deaths do, however, focus on the relentless portrayal of women as the chief agents of change and death for love, carrying as they do salvific overtones.

The heroine Manon (in the Abbé Antoine-François Prévost's [1697-1763] *nouvelle Manon Lescaut*, 1731) is more subtle. Set by three composers (Auber 1856, Massenet 1884 and Puccini 1893), the heroine is torn between a life of luxury and ease and the call of passionate and dangerous love. She chooses the former, but in her death reveals her true nature. The depiction by Auber captures her sad dilemma better than the more famous settings: Manon dies suddenly, not so much of illness or exhaustion, but out of love for the wavering young Des Grieux.

The French heroine who comes closest to the Biblical and Romantic prototype is Marguerite in Gounod's *Faust* (1859). Here the heroine, betrayed in love and condemned to death for infanticide, dies suddenly amidst a vision of angels. This is the nearest anyone comes to the Fiery Chariot and Horsemen of Elijah's translation to the realms of saving love.

Other variants include the mystery of sleep, the memento and harbinger of death. The sleep of the children Hansel and Gretel in Humperdinck's setting of the fairy tale (1893) comes close to sudden death in their abandonment in the forest. Flights of angels also come to them in their dreams and protect them. It is less happy for the fated couple, the Goose Girl and the Prince, in the same composer's *Königskinder* (1910). Here the young lovers die poisoned in a parody of the Eucharist from the envenomed bread given to them by a witch. Their transition into death comes in the mystery of sleep, with the poison again a vector of the untoward death that brings liberation and integration.

More dramatic is the joint assumption of death by hero and heroine. Bellini's Norma and her illicit but repentant lover, the Roman consul Pollione, go to the funeral pyre together to confront and overcome death in their freely given union (1831). The end of Meyerbeer's *Le Prophète* (1849) follows the same pattern, where mother and son give themselves to each other in a freely embraced death that betokens transformation of the spirit and the promise of a liberated world view. The same is true of the finale of Wagner's *Götterdämmerung* (1876), where the murdered Siegfried is exalted and consecrated by the freely adopted death of the self-sacrificing Brünnhilde. The sudden death is self-assumed, but the trajectory of both, sudden death by love, is held aloft.

The death of the young girl Antonia in Offenbach's *Tales of Hoffmann* (1881) captures another dimension of the sudden death syndrome. In the plot the evil genius, or inescapable Shadow, Dr Miracle, forces the consumptive girl to sing, ostensibly causing her death. It is really much more in the nature of Lucia and Isolde who pass to another mode of existence through the medium of music and its association with Romantic yearning and fulfilment.

The courtesan Thaïs in ancient Alexandria, in Massenet's opera (1894) of that name, is the focus of the reforming religious zeal of the coenobite monk Athanaël. He resists her erotic blandishments and takes her to a convent where she undergoes deep personal conversion. When Athanaël has to confront his own overwhelming erotic passion for her and returns to the convent, he finds her mortally unwell. She dies in his bereft arms, a sudden death that becomes the ladder to a new transfigured existence, as well as a transferred symbol of the redemption she now grasps and mystically embodies. The melody of the famous *Méditation* is exactly in the nature of Lucia, Isolde and Antonia's transforming medium.

The great historical drama of Mussorgsky's *Boris Godounov* (1873) ends with an edict being read against the pretender Dimitri. The anguished Tsar breaks in distraught, and collapses dead for no apparent reason. Like the sudden demise of the dishonest Ananias and Sapphira in the Acts of the Apostles, the Tsar's unbidden death is an external manifestation of an inner self-judgment that must express itself inexorably and unavoidably. His death scene is a dark counterpart to the radiance of Isolde's Love-Death. In both instances, one is not considering illness, poison, or murder, but sudden *Unexpected Death,* our principal theme.

Notes

[1] Scruton R. *Wagner and German Idealism.* BBC Radio Interview (20 May 2013), 151.
[2] *The Holy Bible*. Revised Standard Version (1894). Catholic Edition (London: Thomas Nelson & Sons, 1966).
[3] Ovid, *Metamorphoses*, Book 4: 'The Pathos of Love' containing: 'The daughters of Minyas', 'Pyramus and Thisbe', 'The Sun in love', 'Salmacis and Hermaphroditus', 'The Daughters of Minyas transformed', 'Athamas and Ino', 'The transformation of Cadmus', 'Perseus and Andromeda'. The first English translation by William Caxton was published in 1480.
[4] "Barbara Allen" in *The Oxford Book of English Traditional Verse. Chosen and Edited by Frederick Woods* (Oxford: Oxford University Press, 1983), 122-123.
[5] *Die Braut von Korinth* in *Goethe. Selected Verse, with an Introduction and Prose Translation by David Luke* (The Penguin Poets) (London: Penguin, 1964), 159-168; Safranzki, R. Ibid. 2017, 359.
[6] Barrett Browning, E. *Aurora Leigh and Other Poems.* Ed. John Bolton and Julia Holloway (London: Penguin Classics, 1995), Book 9: 814-816, 908, 939-941.
[7] Brontë, E. *Wuthering Heights* [1847]. Ed. David Daiches (London: Penguin Books, 1965), Ch.15:194-196.
[8] Ibid., 33:354-355.
[9] Ibid., 34:362-363.
[10] Ibid., 34:364-365.
[12] Carlisle Floyd's *Wuthering Heights* was produced at the Santa Fé Opera in 1958, with the soprano Phyllis Curtin as the heroine.
[13] Tennyson, A. Lord. Ibid. 1907, 258-259.

CHAPTER 8

THE VARIOUS FORMS OF LOVE-DEATH

Death makes the lovely soul to wander under the sky.
Death opens unknown doors. It is most grand to die.

('By a Bierside', John Masefield, set to music by Ivor Gurney)[1]

The examples considered in the last chapter suggest that the notion of Love-Death takes on multifarious forms and infuses the nature of the operatic genre. We look now at the variety of these, opening up further the potentials explored in operatic drama.

1. A Single Death for Love

Handel's masque *Semele* (1744) uses the variant on the theme of Zeus/Jupiter's illicit love affairs. He ravishes Semele in a shower of fiery gold in a love that consumes her. This instance encapsulates the potent combination of sexual passion and spiritual aspiration, vital ingredients empowering the subject/victim to enter the mystic realm.

Another trope from Antiquity focuses on the Babylonian queen *Semiramide*, a story with incest, madness and murder which Voltaire (1694-1778) turned into a tragedy (*Sémiramis*, 1748). Semiramide and her lover Assur have murdered her husband, the king Ninus. She later falls in love with the young soldier Arsace, who is in fact her son. She receives a mortal blow from Assur intended for Arsace, who then kills Assur and becomes king. The Queen's demise is the result of a confluence of various strands of illicit love and passionate violence, resulting in her death for love. Metastasio's libretto was extremely popular in the 18th century and received remarkable 19th expression in the adaptations by Count Piosassco Feys (1773-1848) and Gaetano Rossi (1774-1855) for Meyerbeer (1819) and Rossini (1823) respectively.

Gilda, in Verdi's *Rigoletto* (1851), loves the reprehensible, amoral Duke of Mantua with such passion that she is prepared to die for him, even though he has raped and discarded her. She substitutes herself as a victim for the assassin Sparafucile in place of the Duke in order to save his life, even though this means abandoning her disabled and fraught father.

2. Situations Where Both Lovers Die Together (by choice, by circumstance, by coercion)

In Bellini's opera (1831), Norma and Pollione voluntarily decide to die together. Norma publicly confesses her transgression in taking a lover, and her act so overwhelms her erstwhile and perfidious beloved, that he too is moved to heroic self-awareness, admits his transformed heart, and accompanies her to the funeral pyre. The fire will be their passage to transcendence.

In Donizetti's *Lucia di Lammermoor* (1835), both hero and heroine are deceived into false assumptions that lead to death. Lucia murders her unwanted husband and becomes mad before dying, to be followed by the despairing Edgardo, who takes his own life to be with her, away from the world of human anger and passion ("Tu che a Dio spiegasti l'ali").

In Meyerbeer's *Les Huguenots* (1836) the loving protagonists, separated by religious differences and misunderstanding, find themselves together in the darkest night of massacre, free of all social obligation and able to make existential decisions of their own. While Raoul is not able to give up his partisan alliance, Valentine sees their love rising above such abstract, mental constraint and agrees to marry him in the hour of death. They both willingly give themselves over to the executioners, together with their mentor Marcel, having hailed the advent of death as a martyrdom, a gateway to transcendence.

Donizetti's *Poliuto* (1838) presents a similar situation where love, religion and personal loyalty interact and are resolved only in shared death determined by love. Polyeucte has become a secret convert to Christianity and is arrested and condemned to death. His wife Pauline, although in love with Severus, the Roman Proconsul, is overwhelmed and decides to join her husband and die with him in an ecstatic vision of martyrdom that becomes a Love-Death.

A similar concatenation of personal love and public or national loyalty determine Verdi's *Aïda* (1871), where the Ethiopian princess Aïda, held captive in Egypt, falls in love with the hero Radamès but is induced by her father to trick Radamès into betraying vital military secrets. He is condemned to be buried alive, but is overwhelmed to find Aïda waiting in the darkness of the tomb to die with him. Their duet bidding farewell to life on earth expresses perfect love in the face of death ("O terra, addio").

In Verdi's *Otello* (1887) Otello's rabid jealousy and suspicion leads him to murder the one he loves most. He realizes the truth too late and then takes his own life to die upon a kiss, united with Desdemona in death. The role of betrayal and the fatal harmatia of the hero compound the tragedy and lead to a pattern similar to the separate deaths in *Lucia*. The protagonists are united by the suicide of the remaining beloved.

The same pattern characterizes the scenario of Puccini's *Tosca* (1900), where the duplicity of the lustful police chief, Baron Scarpia, means that the promised mock execution of Tosca's beloved Caravadossi is in fact real. Tosca flings herself from the battlements of Castel Sant'Angelo not only to escape arrest for the death of Scarpia, but to be united with her murdered lover. Caravadossi initiates their final duet: "O dolci mani".

3. A Death Caused by Illness, but Nonetheless Dying for Love

A third variant on the Love-Death, the *Liebestod* motif, is the death of a protagonist (usually female) through illness in a situation which highlights the fragility of the human condition and underscores the tragedy of a doomed love that can find fulfilment only in death.

In Wagner's *Tannhäuser* (1845) the stress and agony of the problematic love between the saintly Elisabeth and the troubled Tannhäuser leads to the offstage death of the former, and the onstage demise of the exhausted minstrel on his return from his apparently fruitless pilgrimage to Rome to find forgiveness. The death of Elisabeth is the symbolic moment for the advent of grace for Tannhäuser. The separate deaths of both in succession is a consecration of love.

Verdi's *La Traviata* (1853) provides the classic instance of this kind of *Liebestod*. The love of Alfredo and Violetta is doomed by societal disapproval and misunderstanding. Violetta is identified immediately: her precarious life-style as a courtesan is underscored by her fatal wasting disease. This becomes worse as the story progresses, and it is only in the last stages of her life that reconciliation with Alfredo is possible: but it is at the price of her life. She dies in his arms.

This pattern is repeated in other popular operatic scenarios, especially in the various versions of the adaptations of the Manon story. Her privations see her die in the arms of her lover Des Grieux, whether the action is set in Le Havre or Louisiana.

Puccini's *La Bohème* (1896) more closely echoes *La Traviata* than any other opera. The poor seamstress Mimi is identified as suffering from consumption, and her stormy relations with the student Rodolfo are resolved only by her tragic death in the end leaving her erstwhile estranged lover heartbroken in this classic commingling of love and death.

4. Death Caused by Execution

Associated with sudden death are the instances of execution in opera. These tragic climaxes are also invariably related to the mystery of love. In Donizetti's *Anna Bolena* (1830), Henry VIII's queen, Anne Boleyn, is condemned for adultery and is sent to her death with those accused alongside her. Her last moments are a mad scene in which she recalls her past, the times of happiness, and the betrayal of her love. As the marriage of her rival Jane Seymour to the King is announced by pealing bells, she collapses before she is led to her death. Donizetti dealt with a similar situation in *Roberto Devereux* (1837), with gender roles reversed. The Earl of Essex, although loved by Queen Elizabeth, is himself in love with Sarah, Countess of Nottingham. Robert is accused of treason and sentenced to death by the Queen who still loves him deeply.

Rachel in Halévy's *La Juive* (1835), a Christian foundling brought up as a Jewess, is in love with the young Samuel who turns out to be the disguised Prince Leopold, and who is about to marry Princess Eudoxia. Rachel exposes this deception and is sentenced to be executed in boiling oil. She loves Leopold still and helps to have his opprobrium lessened. She refuses to convert to Christianity and goes bravely to her death as one who has dared to love.

Manrico in Verdi's *Il Trovatore* (1853), is the son of a Spanish nobleman, abducted as a baby and brought up by Gypsies. His love for Donna Leonora brings him into fatal rivalry with the Count di Luna who is really his brother. When Manrico and his supposed mother, Azucena, fall into the Count's power, di Luna has Manrico executed thus fulfilling the Gypsy curse on the Count. Manrico dies soon after Leonora, who has taken poison while pretending to give herself to his rival. Their deaths are a classic *Liebestod* that looks forward to the finale of *Tosca* where suicide and execution characterize the passing of the lovers into transcendence.

Love-Death is even more fully realized in Giordano's *Andrea Chénier* (1896), an opera of the French Revolution, where analogies with *Il Trovatore* are close. Madeleine de Coigny loves the poet Chénier, but is herself loved by Gérard, a servant of her family, who becomes an important figure in revolutionary Paris. He writes an indictment of his rival Chénier who is sentenced to death. Madeleine joins Chénier in his cell and they go to the guillotine together, celebrating their love and its fulfilment in death. The libretto by Luigi Illica (1857-1919) has analogies with the finale of *Aida*.

5. Death Caused by Murder, often Poison (*crime de passion*)

Many violent deaths in opera are brought about either by the stealth of assassination or outright murder. Both are also invariably bound up with the expression of love, sometimes with a familial rather than erotic emphasis.

We have seen how Valentine, Raoul and Marcel in *Les Huguenots* (1836) die united in the face of the bullets of partisan massacre, caught up in an ecstasy of martyrdom perceived as the gateway to transcendence.

The hero Masaniello in Auber's *La Muette de Portici* (1828), based on the historical events in Naples in 1647 when the fishermen rose in revolt against their Spanish oppressors, is poisoned by his disgruntled followers for political reasons. His death impels his mute and despairing sister Fenella to throw herself from the battlements into the lava flowing from the eruption of Mount Vesuvius. It is an act of love, a search for union with her adored brother.

In Donizetti's *Lucrezia Borgia* (1833) the heroine is suspected by her husband Alphonso d'Este of having an affair with the young Gennaro. He is actually Lucrezia's son, although only she knows that. When he is arrested she arranges his escape. She plots the death of some young men who have taunted her, by poisoning the wine at their party, only to discover that she has also killed her son among the revellers. She is heartbroken at this loss.

In Verdi's *Luisa Miller* (1849), based on Friedrich Schiller's tragedy *Kabale und Liebe* (1784), Rodolfo, the son of Count Walter, loves the humble Luisa, daughter of the soldier Miller. Rodolfo refuses to marry Frederica, Duches of Ostheim, and is imprisoned by his father. The Count's henceman Wurm forces Luisa to write a letter saying that she loves someone else. When Rodolfo is is released, he makes Luisa to admit that she wrote the letter, and poisons both himself and her. Before the poison take effect, Luisa reveals that Wurm coerced her into writing the letter, and Rodolfo kills him.

The two children in Humperdinck's *Königskinder* (1907), the Goose-Girl and the young Prince who comes to the woods disguised as a beggar, are poisoned by the Witch with whom the Goose-Girl lives.

Cilea's *Adrianna Lecouvreur* (1902), based on Eugène Scribe's (1791-1861) and Ernest Legouvé's (1807-1904) drama (1849), recounts how the famous actress Adrienne (1692-1830), rival of Princess Bouillon for the love of Maurice de Saxe, dies from inhaling the scent of a bunch of poisoned violets sent by the Princess.

In Italo Montemezzi's *L'amore dei tre re* (1913) (words by Sem Benelli [1877-1949], after his verse tragedy, 1910) Fiora has been obliged for political reasons to marry Manfredo, son of King Archibaldo. Fiora secretly takes a lover, Avito. While Manfredo is absent fighting a campaign, the old king murders his daughter-in-law for refusing to divulge the name of her paramour. The King poisons her lips, hoping that Avito will come to see her body. Avito dies from kissing the poison, as does Manfredo, who cannot live without his wife.

Some of the most famous deaths in opera are vivid instances of the *crîme de passion*. The advent of the literary movements of Realism and Naturalism in the later 19th century saw a focus on sordid circumstances in the new *versimo* operas, with adultery, betrayal and murder as recurring themes.

The central figure of Bizet's *Carmen* (1875), based on the short story by Prosper Merimée, having ruined the life and career of the soldier Don José, becomes enamoured of the glamorous bullfighter Escamillo. Carmen spurns Don José's pleas for her return and is stabbed to death by her suffering former lover outside the arena where Escamillo is successfully performing.

In Mascagni's *Cavalleria Rusticana* (1890), after Giovanni Verga's (1840-1922) drama (1886), Santuzza, pregnant by Turiddù, is distraught when she finds out he has abandoned her for the attractive young Lola, wife of the village drover, Alfio. She appeals to Turiddù to return to her but to no avail, and in revenge tells Alfio what has happened. The wronged husband challenges Turiddù to a duel and kills him on Easter morning.

In Leoncavallo's *Pagliacci* (1892), Canio, the leader of a group of strolling players, discovers through the rejected Tonio that his wife Nedda and Silvio are having an affair and plan to run away together. During the play put on for the villagers, on the old *commedia del arte* topos of Pagliaccio cuckolded by Colombine and Harlequin, Canio finds the stage situation mirroring his own life. He stabs Nedda and then Silvio, who runs to her aid.

Eugen D'Albert's (1864-1932) *Tiefland* (1903) is a German *verismo* opera of passion and murder. The gentle Martha is seduced and betrayed by her employer Sebastiano, a rich landowner. He gives her to the ingenuous shepherd Pedro, without revealing the truth about her past (as in Donizetti's *La Favorite*). Martha eventually confesses her story to Pedro who, when Sebastiano tries to take her back, strangles him, and returns to the hills with Martha.

In Richard Strauss's *Salome* (1905), the stepdaughter of Herod Antipas develops a seething passion for the imprisoned John the Baptist. After Salome has performed the voluptuous Dance of the Seven Veils for Herod, she demands the head of Jokannan as her reward. Her necrophiliac obsession is revealed as she fondles and kisses the head until the revolted Herod orders his guards to crush her with their shields.

Berg's *Lulu* (1937), depicts the career of the *femme fatale* Lulu and her various lovers, though there is little love in any of her transactions except from the lesbian Countess Geschwitz. Lulu comes to London as a prostitute and is murdered by Jack the Ripper.

6. Death Caused by Suicide

Many deaths in opera are caused by suicide, and these are always associated with complications caused by love. Purcell's *Dido and Aeneas* (1689), one of the earliest operas still in the repertory, uses the material from Book 4 of Virgil's *Aeneid* (19 BC), recounting the Carthaginian queen's desertion by the Trojan hero who is called by destiny to found the city of Rome. The topic of Dido's betrayal has been treated operatically some 90 times, with 40 settings of Metastasio's popular libretto *Didone abbandonata* (1724). In despair Dido takes her own life, a moment famously conveyed by the lament "When I am laid in earth" in Purcell's score. Berlioz's *Les Troyens à Carthage* (1864) concludes with a vast fresco in which, among Carthagian curses, Dido mounts the funeral pyre to die as the Trojan ships sail away.

Fenella in *La Muette de Portici* and Edgardo in *Lucia di Lammermoor* both kill themselves on losing their loved ones and hence the very purpose of their lives. The queen Sélika in *L'Africaine* allows Vasco da Gama, whom she loves, to sail away with his Portuguese beloved Inès. She takes her own life, inhaling the poisonous flowers of the Upas tree, to be translated to a higher spiritual realm where the mysteries of love will be made clear.

Wagner's Senta in *Der fliegende Hollander* swears to redeem the legendary Dutchman from his curse by remaining true until death. When he too sails away, thinking she has betrayed him, Senta throws herself into the sea, enacting the reality of her promise, ending the curse and opening the gates of transcendence. When the hero Siegfried is murdered in *Götterdämmerung*, the deceived Brünnhilde, now knowing the truth, follows him into death and bids her farewell to life and the old world-order from her funeral pyre. Analogies with Dido and Sélika are strong.

Amilcare Ponchielli's (1834-1836) *La Gioconda* (1876), after Hugo's drama *Angelo, tyran de Padoue* (1836), presents a complication of the frequent love triangle. Gioconda, a street singer, adores Enzo Grimaldi, who in turn loves Laura, the wife of Don Alvise. Gioconda is the object of the spy Barnaba's passion. He falsely accuses Gioconda's mother of witchcraft but she is saved by Laura's intercession. Barnaba now informs against the love of Laura and Enzo, and Gioconda helps them to escape by promising herself to Barnaba. But she evades his clutches when he comes to claim her by stabbing herself. The work looks back to the heroic self-sacrificing actions of Sélika, and also forward to the *verismo* works of Mascagni and Leoncavallo, where murder follows on betrayal.

Otello, in Verdi's opera, takes his own life in sorrowful despair after murdering Desdemona. Hermann, the deluded hero of Tchaikovsky's *The Queen of Spades* (1890), is preoccupied with winning at cards after the Countess's ghost appears to him and reveals her secret hand (three, seven, ace). As he becomes more obsessional Lisa, his beloved, drowns herself in despair. Hermann continues gambling and loses on the third card, and as the Countess's ghost reappears to mock him, he loses his reason and kills himself.

It is interesting to note that other operatic deaths by drowning are mainly female. Stanislaw Moniuszko's (1819-1872) *Halka* (1854) and Léo Delibes's (1836-1891) *Kassya* (1893), after a short story by L. von Sacher-Masoch (1836-1895), are other examples, and relate with thematic pertinence to the despair associated with the elemental beings of the Undine and Rusalka.

Puccini's Tosca, like Fenella and Senta, throws herself to her death when confronted by the executed Caravadossi and the treachery of Scarpia. But the most heartrending suicide comes in *Madama Butterfly* (1904), after David Belasco's (1853-1931) drama (1900). Pinkerton, an American naval officer, falls in love and arranges a marriage with a young Japanese geisha, and goes thorough a nuptial ceremony with her. He departs for America, but Butterfly, with their child Trouble, waits for him to return. He does eventually come back, but with his new American wife, and the grief-stricken Butterfly kills herself.

In Puccini's *Turandot* (1926), after Carlo Gozzi's (1720-1806) drama (1762), the slave-girl Liù is tortured and kills herself rather than reveal the name of her beloved master, the Prince Calaf, to the icy Princess Turandot. She, realizing what true love really is, accepts Calaf's hand.

The *verismo* naturalism is carried further by Wolf-Ferrari (1876-1948) in *I Gioielli della Madonna* (1911), again with a complication of love intrigue and deaths ensuing from it. Raffaele is willing to go to any length to prove his love for Maliella, even to stealing the jewels from the Madonna's statue. But he is pre-empted in this deed by Gennaro who wins Maliella for himself. In remorse she confesses to the enraged Raffaelo, and rushes away to drown herself. Gennaro returns the jewels to the statue, and then stabs himself.

7. Death Caused by Cataclysm

Fenella in *La Muette de Portici* dies by leaping from the battlements into a lava-flow from the erupting Mount Vesuvius—the uproar in nature reflecting the tumult in human affairs, both public and private.

Irene and Rienzi in Wagner's opera seek to support each other in death as the fire caused by rioters engulfs the Capitol, their love a sign of value and commitment amidst the unpredictability and violence of human affairs.

Fidès and John of Leyden in Meyerbeer's *Le Prophète* join each other as the palace is overtaken by explosion and fire. The conflagration caused by the vengeful Berthe is used as punishment of the traitorous Anabaptists, the cleansing flames bringing about their transcendence and a new world order.

The fated love between the Queen of Sheba and the courtier Assad in Karl Goldmark's (1830-1915) *Die Königin von Saba* (1875) is given fatal external correlative in the huge sandstorm that causes the death of her beloved in his banishment.

Alfredo Catalani's (1854-1893) *La Wally* (1892), after Wilhelmine von Hillern's (1836-1916) novel *Die Geyer-Wally* (1875), describes how La Wally is drawn closer to her lover, Hagenbach, when her suitor, Gellner, tries to kill him. The lovers perish in an avalanche in a situation of authentic Love-Death.

8. Death Caused by Rapture/Cessation

By far the most mysterious of the variants of the *Liebestod* is the sudden and ostensibly inexplicable death ensuing in a situation where excessive emotion leads to some kind of psychosomatic intensity of unendurable pressure on a character's constitution. This results in cessation of vital autonomic function. Arguably Isolde's sudden death at the end of *Tristan und Isolde* is the most famous.

We return to this theme in Section 4 of our book, but first wish to present in more detail some other Romantic themes which we have touched on so far rather briefly: notably water, fire, light and darkness (the Elemental Being) and some masculine perspectives on Love-Death.

Note

[1] This is one of Ivor Gurney's (1890-1937) most important songs: a setting of words from John Masefield's play, *The Tragedy of Pompey the Great*, 'By a Bierside'. Gurney composed the song in 1916 whilst serving as a private soldier with the 2/5 Gloucestershire Regiment. See Harrison, N., 2016, Ibid., 46.

CHAPTER 9

WATER AND FLAMES: THE CONCEPT OF THE ELEMENTAL BEING

The mind evolved in the sea, water made it possible.

(Peter Godfrey Smith. *Other Minds: The Octopus and the Evolution of Intelligent Life*)[1]

1. The Elemental Being

The concept of the elemental being is of ancient origin. The being, almost invariably female, falls in love with a mortal and then draws him into a union that is fatal in human terms, but proposes some sort of metaphysical transference or transcendence. It is given exquisite description by Apollonius of Rhodes, in Book 1 of the *Argonautica* (3rd century BC).

> Meantime Hylas with pitcher of bronze in hand had gone apart from the throng, seeking the sacred flow of a fountain, that he might be quick in drawing water for the evening meal and actively make all things ready in due order against his lord's return. For in such ways did Heracles nurture him from his first childhood when he had carried him off from the house of his father, goodly Theiodamas, whom the hero pitilessly slew among the Dryopians because he withstood him about an ox for the plough …. And quickly Hylas came to the spring which the people who dwell thereabouts call Pegae. And the dances of the nymphs were just now being held there; for it was the care of all the nymphs that haunted that lovely headland ever to hymn Artemis in songs by night. All who held the mountain peaks or glens, all they were ranged far off guarding the woods; but one, a water-nymph was just rising from the fair-flowing spring; and the boy she perceived close at hand with the rosy flush of his beauty and sweet grace. For the full moon beaming from the sky smote him. And Cypris made her heart faint, and in her confusion she could scarcely gather her spirit back to her. But as soon as he dipped the pitcher in the stream, leaning to one side, and the brimming water rang loud as it poured against the sounding bronze, straightway she laid her left arm above upon his neck

> yearning to kiss his tender mouth; and with her right hand she drew down his elbow, and plunged him into the midst of the eddy.[2]

The water-nymph depicted is in essence a beautiful but destructive being. (Her equally ambiguous and passionately destructive counterpart later in the poem is the terrifying witch Medea.) The situation and the symbolism will be the same in all the Undine and Rusalka stories of Romanticism which in turn drew from folktales of both Western and Eastern Europe. The myth underscores all too clearly the fundamental motif of all these stories of elemental beings and humans. In whatever disguised or transposed format, the search is for some kind of healing fulfillment we might call salvation, whether in the search of the water nymph for ensoulment, or in the pathetic dilemma of the betraying/betrayed human searching for peace or redemption.

What the modern version of the myth does is to focus more on the elemental being, who is in need of a human soul, and can find this only in sacrificial love from a man. The inevitable disillusionment leads to the destruction of the human with some kind of post-mortal, elemental relocation, the implications of which seems to vary from one re-creation of the myth to another, from the tragedy of fatal vengeance to a sentimentalized concept of some sort of pagan paradise.

The Importance of Undine

Undine is a fairy-tale novella (*Erzählung*) by Friedrich de la Motte Fouqué in which Undine, a water spirit, marries a knight named Huldbrand in order to gain a soul. It is an early German romance, which has been translated into English and other languages. During the 19th century the book was very popular and was, according to *The Times* in 1843, "a book which, of all others, if you ask for it at a foreign library, you are sure to find engaged".[3] The story is descended from Melusine, the French folk-tale of a water-sprite who marries a knight on condition that he shall never see her on Saturdays, when she resumes her mermaid shape. It was also inspired by works of the occultist Paracelsus (Theophrastus Bombastus von Hohenheim, 1493-1541). This Swiss physician, alchemist and natural philosopher was prominent in the Renaissance reaction against the medical doctrines of Galen and Avicenna, combining the beginning of modern scientific spirit with a Neo-Platonic mysticism. Fouqué's story captures the essence of the myth.

"It appears," said the workmen to one another in astonishment, "as if the confined water had become a springing fountain." And the stone rose more and more, and, almost without the assistance of the work-people, rolled slowly down upon the pavement with a hollow sound. But an appearance from the opening of the fountain filled them with awe, as it rose like a white column of water; at first they imagined it really to be a fountain, until they perceived the rising form to be a pale female, veiled in white. She wept bitterly, raised her hands above her head, wringing them sadly as with slow and solemn step she moved toward the castle. The servants shrank back, and fled from the spring, while the bride, pale and motionless with horror, stood with her maidens at the window. When the figure had now come close beneath their room, it looked up to them sobbing, and Bertalda thought she recognized through the veil the pale features of Undine. But the mourning form passed on, sad, reluctant, and lingering, as if going to the place of execution. Bertalda screamed to her maids to call the knight; not one of them dared to stir from her place; and even the bride herself became again mute, as if trembling at the sound of her own voice.

While they continued standing at the window, motionless as statues, the mysterious wanderer had entered the castle, ascended the well-known stairs, and traversed the well-known halls in silent tears. Alas, how different had she once passed through these rooms!

The knight had in the meantime dismissed his attendants. Half-undressed and in deep dejection, he was standing before a large mirror, a wax taper burned dimly beside him. At this moment someone tapped at his door very, very softly. Undine had formerly tapped in this way, when she was playing some of her endearing wiles.

"It is all an illusion!" said he to himself. "I must to my nuptial bed."

"You must indeed, but to a cold one!" he heard a voice, choked with sobs, repeat from without; and then he saw in the mirror, that the door of his room was slowly, slowly opened, and the white figure entered, and gently closed it behind her.

"They have opened the spring", said she in a low tone; "and now I am here, and you must die."

He felt, in his failing breath, that this must indeed be; but covering his eyes with his hands, he cried: "Do not in my death-hour, do not make me mad with terror. If that veil conceals hideous features, do not lift it! Take my life, but let me not see you."

"Alas!" replied the pale figure, "will you not then look upon me once more? I am as fair now as when you wooed me on the island!"

> "Oh, if it indeed were so," sighed Huldbrand, "and that I might die by a kiss from you!"
>
> "Most willingly, my own love," said she. She threw back her veil; heavenly fair shone forth her pure countenance. Trembling with love and the awe of approaching death, the knight leant towards her. She kissed him with a holy kiss; but she relaxed not her hold, pressing him more closely in her arms, and weeping as if she would weep away her soul. Tears rushed into the knight's eyes, while a thrill both of bliss and agony shot through his heart, until he at last expired, sinking softly back from her fair arms upon the pillow of his couch a corpse.
>
> "I have wept him to death!" said she to some domestics, who met her in the ante-chamber; and passing through the terrified group, she went slowly out, and disappeared in the fountain.[4]

Hoffman's opera presents a pared and powerful presentation of this archetypical Romantic topos. Undine appears quietly in the party, bidding a good night. The guests are filled with disastrous premonition, and see Huldbrand doomed by this terrifying power. But Undine with detached imperturbability only smiles on lovingly, and repeats her good night. Huldbrand feels death upon him, and his touching concern is that there is 'not a single kiss'. Undine assures him that there is, that she is constrained to kiss him, because she must kiss him to death. Huldbrand's riposte captures the aspiration all have in dying;

> When such a departure beckons
> It surely means gaining salvation.

The words are beautiful and apparently reassuring, but probably promise only watery oblivion.

This version of the Undine story by Hoffmann and other adaptations are discussed in more detail in Chapter 18, but it is so very representative of our theme. Undine's endearing laughter is mentioned twice by Huldbrand as she approaches to implant the kiss of the Love-Death. It is a motif that recurs with tragic prescience also at the conclusion of Wagner's *Siegfried,* where the rapturous love celebrated by the young hero and the awakened Brünnhilde reaches its highpoint in the paradoxical celebration of *lachendes Tod* ("laughing death"). The mystical inextricability of love and death finds overwhelming expression here.

The Rusalka

In Slavic mythology, a *rusalka* (plural: *rusalki*) is something akin to the Greek sirens and Celtic mermaids. Rusalki are beautiful young women who dwell in bodies of water and enjoy enticing men. The concept of rusalki originated from a Slavic pagan tradition where the young women were symbols of fertility. These nymphs did not interfere too much with human life and mainly served to provide life-giving moisture to the fields and forests every spring when they came ashore to dance in the spring moonlight. The water spirits were believed to help crops grow plentifully and so were generally treated with respect. In the modern era, a large mythos mostly unrelated to fertility has sprung up around the beguiling young women of the water. By the 19th century, the main objective of a rusalka had transformed into harassing the human population.

While the pagan water nymph occurred naturally, the mermaid rusalka has its origins in quite varied stories. Typically, they involved young women dying violent deaths. Sometimes it is murder, sometimes it is a suicide, but usually it is a death by drowning. Stories of rusalki often revolve around women betrayed by a husband or lover. Some rusalki are young women who plunged into a lake or river because they were pregnant out of wedlock. Alternatively, other origin stories say a rusalka is any young woman who dies a virgin, regardless of whether the death was violent or natural. Still other myths say that babies who die before being baptized are reborn as water sprites. This version of the myth ties in with the idea that babies born out of wedlock were supposedly drowned. (There is also a link here to tales of witches, and to the 'reverse baptism' of trial by ducking: if the accused drowned they were seen as innocent; if they emerged from the water unscathed the 'baptism' had failed, the water had rejected them.) Finally, some myths claim that a rusalka is any unclean soul: in other words, anyone who killed himself or herself by jumping into the lake or river.

Many rusalki were said to have finite lives. These souls linger on in water until their allotted time on earth is complete (this version of events usually accompanies the violent death or suicide origin-stories). Others must remain until their death is avenged (this version of events usually accompanies the origin-stories of murder or jilted lovers). Rusalki haunt lakes, rivers, ponds, marshes, swamps, and any other body of water. They were often described as being slim with large breasts. They are pale-skinned and have long, loose hair that is either blonde, light brown, or green. Their eyes are said to not contain pupils and, if the rusalka is wicked, can be blazing green. Invariably, the women wear light, sheer robes as though made of mist. Rusalki are representations of universal beauty and even the wicked ones are greatly admired and feared in Slavic society.

The motivations of a rusalka vary depending on where she is living, a fact that may have its roots in the ancient pagan myths. In places where plant life is bountiful and crops grow well, such as in Ukraine and areas around the Dnieper River, the rusalki are charming and playful. However, in harsher climes, the rusalki are wild and wicked. These malevolent spirits would crawl out from the water in the middle of the night in order to ambush humans, especially men, whom they would then drag, alive, back into the watery depths.

These later variants have also appeared in the mythology of Western Europe in the concepts of the Wilis, the spirits of young women who die before their wedding day and become angry avenging spirits who appear in the woods at night. By way of atavistic vengeance on the male sex, they drown or alternatively dance any unwary man to death. Victor Hugo refers to the subject in a poem called "Fantômes" in *Les Orientales* (1829), and Heinrich Heine described the legend in a passage about the Wilis in *De l'Allemagne* (1833-34). The story was adapted by the librettist Henri Vernoy de Saint-Georges (1799-1875) in the scenario of the ballet *Giselle* and immortalized in the musical setting by Adolphe Adam (1803-1856) (1841).

> There is in a part of Austria a tradition which has much in common with that of the elves, although it is Slavonic in origin. This is the tradition of the nocturnal dancer who is known in Slav countries as by the name of *Wilis*. The Wilis are betrothed girls who have died before their wedding day. The poor young creatures cannot remain at peace in their graves. In their stilled hearts, in their lifeless feet, there remains that love of dancing which they could not satisfy in their lifetime, and at midnight they rise and gather in bands by the roadside, and woe betide the young man who crosses their path! He is compelled to dance with them; they surround him with frenzied desire, and he dances with them until he falls dead. Clad in their wedding dresses, with flowered wreaths on their heads and rings sparkling on their fingers, the Wilis dance like elves in the moonlight. Their faces, though white as snow, have the beauty of youth; their laughter has terrible joy, their call is seductive, and their expressions hold out such sweet prospects. These lifeless bacchantes are irresistible.
>
> Seeing betrothed girls die who are so full of youth, people should not persuade themselves that such radiance and beauty could be so utterly

> destroyed, and this gave rise to the belief that the girl continued to seek after death the joys of which she had been deprived.[5]

The legend relates significantly to Heine's famous poem about the seductive spirit of the Rhine, the Lorelei. Here the ancient Hellenic myth of the Siren is combined with more modern developments relating to the Mermaid and the Undine:

> I know not if there is a reason
> Why I am so sad at heart.
> A legend of bygone ages
> Haunts me and will not depart.
>
> The air is cool under nightfall.
> The calm Rhine courses its way.
> The peak of the mountain is sparkling
> With evening's final ray.
>
> The fairest of maidens is sitting
> So marvelous up there,
> Her golden jewels are shining,
> She's combing her golden hair.
>
> She combs with a comb also golden,
> And sings a song as well
> Whose melody binds a wondrous
> And overpowering spell.
>
> In his little boat, the boatman
> Is seized with a savage woe,
> He'd rather look up at the mountain
> Than down at the rocks below.
>
> I think that the waves will devour
> The boatman and boat as one;
> And this by her song's sheer power
> Fair Lorelei has done.[6]

Alfredo Catalani wrote an *azione drammatica* titled *Die Loreley* (Turin 1890, with words by A. Zanardini and C. D'Ormeville). The opera describes how Walther is betrothed to Anna, but fatally falls in love with the mysterious Loreley. At the wedding ceremony (as in the Undine story), the Loreley appears and Walther rushes out irresistibly to follow her. Just like the legend of the Wilis, Anna dies of grief. Walther throws himself into the Rhine, maddened by the Loreley's elusive evasions. The plot captures the myth of the famous river naiad in its many and varied nuances: the elemental hold of her attraction, and its fatal consequences for her human victim.

Rusalki feature in many Russian and Slavic artworks including paintings, operas, and novels. One of the most renowned homages to the rusalka was written by Alexander Pushkin (1799-1837). The poem, published posthumously and titled "Rusalka" by his estate, gives a good depiction of how rusalki are imagined in Slavic cultures.

In lakeside leafy groves a friar
Escaped the world; out there he passed
His summer days in constant prayer,
Deep studies and eternal fast.
Already with a humble shovel
The elder dug himself a grave,
And calling saints to bless his hovel,
Death—nothing other—did he crave.
So once upon a falling night he
Bowed down beside his drooping shack
And meekly prayed to the Almighty.
The grove was turning slowly black;
Above the lake the mist was lifting;
Through milky clouds across the sky
A ruddy moon was softly drifting,
When water drew the friar's eye—
He looks; his heart is full of trouble,
Of fear he cannot quite explain;
He sees the waves rise more than double
And suddenly grow calm again.
Then, white as first snow of the highlands,
Light-footed as nocturnal shade,
There comes ashore and sits in silence
Upon the bank a naked maid.
She looks at him and brushes gently
The hair and water off her arms.
He shakes with fear and looks intently
At her seductive, luscious charms.
With eager hand she waves and beckons,
Nods quickly, smiling from afar,
And shoots within two flashing seconds
Into still water like a star.
The glum old man slept not an instant
All night. All day not once he prayed;
Before his eyes still hung and glistened

The wondrous girl's persistent shade.
The grove puts on the gown of nightfall;
The moon walks on the cloudy floor;
And there's the maiden—young, delightful,
Reclining on the spellbound shore.
She looks at him, her hair she brushes,
Smiles, sends him kisses sweet and wild,
Plays with the waves—caresses, splashes—
Now laughs, now whimpers like a child,
Moans tenderly, calls louder, louder …
"Here, monk, here, monk! To me, to me!"
Then vanishes in limpid water,
And all is silent instantly …
On the third day the ardent hermit
Was sitting on the shore, in love,
Awaiting the voluptuous mermaid,
As shade was lying on the grove.
Night ceded to the sun's emergence;
By then the monk had disappeared.
It's said a crowd of local urchins
Saw floating there a wet gray beard.[7]

Pushkin's narrative poem *The Water Nymph* (1832) tells the tale of a daughter of a Miller who is betrayed by the Prince and drowns herself in a stream. She becomes a water-sprite who lures men to their deaths. This piece uses the Slavonic version of the rusalka. The emphasis is more definitely on the search not for a soul but for humanity, and the searing confrontation with human weakness and betrayal. In the Russian version the betrayed heroine enters the watery realm in suicide; in the Czech work elements of both the central European and Slavonic tale shape the scenario. The transmogrified Natasha of *The Water Nymph* is impelled to seek out her faithless human lover after a period of seven years, this numerological motif expressing the idea of a preordained passage of time with mystical significance. The Seven Days of Creation and Sabbath Rest suggest that this is part of the very sway of the cosmic order, a concept that has percolated down into folklore as a sacred lapse of time. The legend of the Flying Dutchman has him come ashore every seven years to seek salvation.[8]

The bond between tragic love and death is common to all the manifestations of the Undine/Rusalka myth. Water is one elemental power: another is fire. **[Figs. 9.1-6]**

2. Love and Death in Fire

The dove descending breaks the air
With flame of incandescent terror
Of which the tongues declare
The one discharge from sin and error.
The only hope, or else despair
Lies in the choice of pyre or pyre—
To be redeemed from fire by fire.

Who then devised the torment? Love.
Love is the unfamiliar Name
Behind the hands that wove
The intolerable shirt of flame
Which human power cannot remove.
We only live, only suspire
Consumed by either fire or fire.[9]

We have already discussed how the Love-Death mystery is concomitant with the symbolic interplay of fire/light and darkness. Connected with the use of dark and light as these symbolic vectors of air and cosmos is the related elemental concept of fire. This is frequently used in issues of love and death and is obviously related in the solar concepts of light, heat, power and transformation, either for healing or destruction.

The role of Greek mythology is once again foundational. Semele (sometimes called Thyone) was a daughter of Cadmus and Harmonia, at Thebes, and mother of Dionysus (Bacchus) by Zeus. Semele's liaison with the amorous and philandering Zeus enraged his jealous wife, Hera. She disguised herself as an old nurse and induced Semele to ask Zeus to visit her vested in the same magnificence with which he would normally appear before Hera. Zeus, having already promised to grant Semele her every wish, was thus forced to grant a desire that would kill her: appearing in the elemental splendour of his firebolts, as god of thunder, he inevitably destroyed her. But Zeus saved their unborn child, Dionysus, protecting him in his thigh until he was ready to be born. The story had further developments pertinent to love and death. Dionysus, after reaching maturity, and being himself immortal though Zeus, descended into Hades and brought back his mother Semele. That she too became an immortal, even a goddess, is attested in both pottery (on an Attic black-figure hydria from the Leagros Group, c. 520 BC), and also in poetry (in Pindar's 'Olympian Ode 2').[10]

Handel treated this subject in his masque (or musical drama) *Semele,* first given in London at Covent Garden on 10 February 1744, setting the version of William Congreve's (1670-1729) drama (1710). The title role was sung by Elisabeth Duparc, who created several roles in Handel's oratorios. Here the flames capture the dilemma of love and death in one symbol, and also serve as the ultimate cause and medium of Semele's divine transcendence in Act 3.

Scribe and Auber's prototypical grand opera *La Muette de Portici* (1828) contains one of the most dynamic presentations of death in flames, in a scenario where the fires involved take on an almost eschatological implication. The revolt in Naples (1647), when the fishermen rose against their Spanish oppressors, is the historical background to the touching account of the folk hero Masaniello (who leads the revolt), and his mute sister Fenella (based on the lithe and acrobatic character in Sir Walter Scott's *Peveril of the Peak*, 1822). It depicts the oppression of poor people, the rebellion, the betrayal of its noble-minded leader and his death by poisoning. These tumultuous events are given a doomsday correlative in the simultaneous eruption of Mount Vesuvius (famously depicted in Pierre-Luc-Charles Ciceri's [1782-1868] epic stage design). The abandoned Fenella, deprived of her brother's love, propels herself in despair from the parapet of the palace into the encroaching lava flow. By implication this threatens all the assembled characters, and becomes a symbol of the cleansing of society in the imagery of apocalypse.

This notion of self-sacrifice as personal penitence and punishment for sin, serving as a metonymic recompense or ransom for the whole of society, becomes one of the most striking and powerful symbols of Romantic opera. An example of this self-offering bound up with both expiation and martyrdom from the history of the saints, finds an icon in Joan of Arc (1412-1431). A virginal young woman with access to the divine in her private locutions with heaven in the voices of St Margaret and St Barbara, Joan accepts a mission to help save her people by leading the resistance to the English invasion. But she is captured and condemned as a heretic or a witch, and burned to death in the marketplace of Rouen.

Joan attracted powerful Romantic interest, as a figure of freedom in the search for human liberty, embodied most famously in Schiller's drama *Die Jungfrau von Orleans* (1801). The poet sought to elaborate her story by adding another favourite trope of Romantic opera, the consecrated virgin who compromises her calling by falling in love, and must pay the inevitable penalty of death. But in Joan's case this death is more than a personal recompense, but rather death on behalf of the French nation, and indeed for all humankind on behalf of the pure in heart. In this way she becomes a Christ-like figure (as do all the martyrs through the ages), and her death in the flames extends over time and space to heaven itself.

Nicola Vaccai (1790-1848) (1827), Michael William Balfe (1808-1870) (1837), Vesque von Püttlingen (1803-1883) (1840), J.A.A. Langert (1861) and Emil Rezniček (1860-1945) (1886) all narrate her story in opera. But it is in those of Verdi (*Giovanna d'Arco*, 1845) and Tchaikovsky (*Orleanskaya dyeva*, 1881) that became most famous. Both use Schiller's fiction of Joan's betrayal of her quasi-priestly ministry in her love for Lionel as dramatic vector. The flames of Joan's death, far from destroying any hope of resurrection in the annihilation of her body (a popular concept behind the burning of witches in the great persecutions from 1486-1782), are a medium of transcendence to her immortality.

The priestess Norma in the opera (1831) by Felice Romani (1788-1865) and Bellini is in love with the Roman consul Pollione, by whom she has two children. He betrays her noble love by his liaison with the neophyte Adalgisa. Norma tries to persuade Pollione to renounce the younger woman and return to her. He refuses, and in her disillusionment Norma publicly admits that she has betrayed the high office vested in her and her nation by intimately consorting with the enemy of her people. She is condemned to death. She mounts the flaming pyre erected to punish her, actions which so impress her erstwhile lover, that he decides to join her in this expiatory death, and together they enter the flames. Love and death are combined in a symbol of the wider cleansing of the social order.

These expiatory and redemptive topoi are found in other famous operatic scenarios. For Wagner, the denouement of *Rienzi* (1842), based on Mary Russell Mitford's (1787-1855) drama (1828) and Edward Bulwer-Lytton's (1803-1873) novel (1835), relates closely to the scenario of *La Muette de Portici*. Rienzi, the folk hero who tries to liberate the people of Rome from aristocratic and factional tyranny, is betrayed by political faction and a fickle, disloyal people. He is excommunicated, and abandoned by all except his sister Irene. She finds him at prayer, and joins him out of loving devotion as the mob set fire to the Capitol. Her lover, Adriano, rushes in to perish with them in the flames. Concepts of liberty, personal hubris, popular instability and treachery, and the issues of private sinfulness and national culpability, are all questioned in the destructive but cleansing flames.

A similar scenario is found in Scribe and Meyerbeer's *Le Prophète* (1838-49), where the story traces the Anabaptist takeover of the city of Münster (1534-35) during the earlier years of the Reformation. The idealistic desire to found a theocracy heralding a New Jerusalem is compromised by a grasping for power and personal aggrandizement. The opera ends with the hero, John of Leyden, betrayed by his followers during his coronation banquet. His abandoned love Berthe has set the palace of Münster on fire, and the stored powder magazine in the cellars blows up, killing the Anabaptist traitors. John is joined in death by his mother Fidès. United in love in a fiery death, they invoke the Holy Spirit to bring about a new age cleansed of human folly. The original sin that haunts all human history and endeavour is overcome by love and sacrificial death that alone can bring about a new world order.

In Vladimir Stassov's (1824-1906) and Mussorgsky's great Russian music drama *Khovanshchina* (1886), based on complicated political and religious affairs at the time of the accession of Peter the Great in 1682, there is strife between the Streltsy (Guards) led by Prince Ivan Khovansky, and a rival faction led by Prince Golitsyn. Andrey Khovansky is saved from committing rape and murder by the long-suffering Marfa, who is involved with the ultra-orthodox sect, the Old Believers, who refuse to accept the ecclesial reforms of 1650. Led by Dosifey, they go to their deaths in a huge fire (like the Albigensian heretics, walking into the fiery pit at Carcassonne in 1209) together with the reconciled Andrey and Marfa. The couple give witness to their love in dying, just as the Old Believers perceive themselves as martyrs.

The Dove of Pentecost, manifested in the tongues of fire on those in the Upper Room at Pentecost (The Acts of the Apostles, 2:1-4), provides the only hope from the despair of the human condition—redemption. In life, with King Lear, we are "bound upon a wheel of fire", whether in profound suffering, or consumed by the fire of desire or the fire of love. The fire of being can also be the means of becoming, with those embracing the fire of expiation and purgation enter into a trajectory of transcendence, liberated from history and memory, beyond time and space. **[Figs. 9.7-8]**

3. Light and Darkness, Obfuscation and Illumination

Richard Wagner's *Der Ring des Nibelungen* (1856-76) covers a symbolic history of the world and human effort from the origin of things until an apocalyptic final scenario. The trajectory of the *Ring* Cycle is initiated with just such a symbolic interplay, and ends in flames. In this tetralogy the great hero, Siegfried, is betrayed by deception and traitorously murdered. His noble spouse and formerly demi-divine Valkyrie, Brünnhilde, orders a funeral pyre to be built for Siegfried, and, taking the Ring of Destiny, wrought from the gold of the Rhine by the renunciation of love, she places it on her own finger. She bids farewell to the greatest of heroes, and to the reign of the gods, and plunges into the flames to join Siegfried in immolation. The great hall behind her collapses, the Rhine overflows its banks, and as Hagen tries to snatch the Ring from Brünnhilde, he is dragged below the elemental, cleansing waters by the Rhine Maidens. Valhalla rises in flames, the kingdom and era of the gods is destroyed. The Ring is returned to the Rhine and a new age of human love dawns.

In *Das Rheingold* the opening scene at the bottom of the Rhine transmutes into light as the rising sun penetrates the water and strikes the dormant treasure, filling the dark waters with a new radiance. This is the accompaniment to the revelation of the secrets of nature narrated by the Rhine daughters to Alberich. It is about the mystery of love and power: only renunciation of love, the essence of the natural order, empowers the coercion of nature. The course of action is thereafter pulled between realms of light and darkness. Wotan is confronted with his new castle in the light of dawn, but ventures fatefully into the darkness of Nibelheim to wrest the Ring and the treasure from Alberich with the help of the universal trickster, Loge, the god of fire. Eventually the gods will enter the splendour of Valhalla using the rainbow bridge summoned up by Donner, the rainbow being the colour spectrum of refracted light. Glass and light, colour, reflection and refraction all seem media of transcendence, like the rainbow, which is sunlight refracted through water (raindrops) into the constituent primary colours. This relates to nature, to creation, and to the biblical covenant, so that, as with Ann Radcliffe, William Wordsworth and Casper David Friedrich, nature becomes a moral emblem, a source of spiritual elevation (transcendence) and a stimulus to artistic creativity.

In *Die Walküre* the meeting of Siegmund and Sieglinde happens in storm and night, and their love unfolds in a movement towards light. Siegmund sees the hilt of the sword gleaming in the trunk of the World Ash Tree, and his love for Sieglinde blossoms with the light of the spring moon which bursts upon them. Siegmund will die in the night of Hunding's vengeance, and Brünnhilde will be imprisoned in sleep on her mountain top in the night. She is, however, significantly protected by Loge's fire, a Magic Fire, which Wotan directs to defend the high place until the advent of a great hero.

The whole symbolic course of *Siegfried* is plotted in terms of light and darkness, as the hero develops from an angry ignorant youth through processes of confrontation and growth to the full enlightenment of maturity realized in love. The darkness of cave and forest are lit by the fires of creativity (as he supersedes the craftsman Mime to re-forge the Sword) and the dawning of light in the woodland (where tasting the blood of the vanquished Dragon gives him a transforming insight into nature). Now in the increasing light he is able to overcome Wotan, in his guise as Wanderer, and initiate the advent a new order, from the divine to the heroic. He has the natural instinct (through the Wood Bird) to approach the light of the fire around the high rock, and entering through it, to claim his bride, in adult emotional and sexual union, amidst the full blaze of sunlight ("Heil Dir, Sonne").

In *Götterdämmerung* the night and break of day symbolize the beauty of radiance of Siegfried's union with Brünnhilde. The patterns of deception and betrayal that follow use both the physical reality of night and the mental darkness of obfuscation (in the potion administered by Hunding) in the destruction of the pure hero. The antidote restores his faculties and enlightens his memory to celebrate his love for Brünnhilde before his murder reinforces the night anew. Now Brünnhilde's great act of farewell will consecrate Siegfried's sacrifice in an act of complete self-offering and immolation in the flames of Loge's fire. The light and cleansing power of the fire symbolize a new world order, and the return of the Ring to the overflowing Rhine completes the circle of light and darkness initiated at the beginning of *Das Rheingold*.

Night and day, of course, are most famously explored in *Tristan und Isolde* as the very metier of the mystery of love and death. The day is associated with the normal expectations of life, with its structures and expectations, its control of emotion and feeling, its obedience to the norms and norms of society. For the lovers, evading the potion of death for one of love, the day represents the blockage of true feeling and the impeding of the expression of authentic feeling. Only in the obscurity of night are these structures and barriers blurred and removed, and in the realm of the changeable moon can the unconventional and forbidden find expression. The polarities of light and darkness become synonymous with life and death, and only in the latter can true love find untrammelled expression and fulfilment, can the individual be relieved of the burdens of responsibility, individuality, and the relentless pressure of yearning and sexual desire.

Notes

[1] Godfrey Smith, Peter. *Other Minds: The Octopus and the Evolution of Intelligent Life* (London: William Collins, 2018), 200.
[2] Apollonius of Rhodes, *The Voyage of the Argo: The Argonautica.* Translated with and Introduction by E. V. Rieu (London: Penguin Books, 1959), 1:1207-1239).
[3] Au, Susan. "The Shadow of Herself: Some Sources of Jules Perrot's *Ondine*". *Dance Chronicle*. Taylor & Francis, Ltd. 2:3 (1978): 160.
[4] Friedrich de la Motte-Fouqué. *Undine* [1811] (Stuttgart: Philipp Reclam Jun., 1969); 90-91; *Undine by De La Motte-Fouqué.* Adapted from the German by W. L. Courtney and Illustrated by Arthur Rackham (London: William Heinemann, 1909), 130-132.
[5] Heinrich Heine. *De l'Allemagne* (1833-34), 453.
[6] Heinrich Heine. "Die Lorelei" in 'Die Heimkehr', *Buch der Lieder* (1827). The well-known German poem (1822) has been the inspiration to many ballads, songs, and operas. Here it is given in an anonymous American translation. Also see *Heine. Introduced, Edited and Translated by Peter Branscombe* (The Penguin Poets), London: Penguin Books, 1967), 40-41.
[7]Alexander Pushkin. *Selected Works in Two Volumes. Volume One: Poetry* (Moscow: Progress Publishers, 1974), "The Water Nymph". Trans. Avril Pymon; 61-64.
[8] Magic numbers percolate through myth and religion. Seven is composed of four and three, Pythagorian special numbers: the Seven Ages of Man (Shakespeare, Jacques in *As You Like It*): there are Seven Deadly Sins and Seven Virtues; seven dwarves; the Seven Joys of Mary (Annunciation, Visitation, Nativity, Epiphany, Finding in the Temple, Resurrection and Ascension), and also the Seven Sorrows of Mary (The Prophecy of Simeon, The Flight into Egypt, The Loss of Jesus for Three Days, The Carrying of the Cross, The Crucifixion of Jesus, The Deposition from the Cross, The Burial of Jesus).
[9] Eliot, T S. *Four Quartets*: *Little Gidding* (London, Faber and Faber, 1944), 41-42.
[10] See Carl Kerényi, *The Gods of the Greeks*, ibid. 1951, 225-228.

CHAPTER 10

THE *LIEBESTOD* IN MASCULINE PERSPECTIVE

> 25 "How are the mighty fallen in the midst of the battle!'' Jonathan lies slain upon thy high places.
>
> 26 I am distressed for you, my brother Jonathan; very pleasant have you been to me; your love to me was wonderful, passing the love of women.
>
> 27 "How are the mighty fallen, and the weapons of war perished!"
>
> (2 Samuel 1: 25-27)

The *Liebestod,* which we discussed in some detail in Chapter 7, is invariably associated with the tragic fatal outcome of doomed romantic love, realized in the traditional love duet for soprano and tenor. But a subgenre and gender variant of this is manifested in the male duet for tenor and baritone that began emerging from the later-18th century. All these duets take place in situations of menace and danger where death is ever-present in recollection or in imminence: here love (very often love-hate) is expressed in the face of threats, rivalry or sacrifice that prefigures a situation where one of the pair will soon face grave danger and probably die. In all instances the concept of Love-Death, and its relation to transcendence, is realized in the rhapsodic nature of the music, often in the unison singing that is so much a part of these duets.

A prototype is provided in the duet for Orestes (baritone) and Pylades (tenor) in Act 3 of Gluck's *Iphigénie in Tauride* (1779) where the latter offers to die in the place of his friend. The exchange between them conveys a depth of commitment, with distinctly homoerotic overtones.

> **Pylades**
> O moment of happiness! My death,
> then will save my friend's life!
> **Orestes**
> And I would agree that yours be taken?
> Do you love me? Tell me?

Pylades
Ye gods! Ye dare ask that?...
Orestes
And you still claim that you love me
when despite the gods, you sacrifice your life …
Pylades and Orestes
Ye gods, move his heart!
Give me back my friend, that in pity
He may agree that all my blood
May satisfy you and suffice your severity.[1]

Neither Pylades nor Orestes dies eventually but the threat and their mutual preparedness to offer themselves for the other make this a situation of Love-Death. The recurrence of such situations increases with the emergence of Romantic opera, with the tenor-baritone combination becoming a specific topos of the era.

Scribe's and Auber's opera *Le Maçon* (*The Mason*, more accurately *Maurer und Schlosser*, 'Mason and Locksmith' in the German version) (1825), designated an *opéra français*, depicts the friendship between the artisans Roger, a mason, and Baptiste, a locksmith. Both are induced by threats to attend to a secret commission in the Turkish Embassy in Paris, where they are obliged to install chains and build walls to immure the Greek odalisque Irma and her dashing lover Léon, who has tried to help her escape. The two friends work in fearful harmony as they complete their sinister task, fearing the ominous threat of harm at the completion of their commission. Singing in unison, they try to keep each others' spirits up in their workers' song as they labour under duress, and in alternating exchanges recount their street abductions, fears, and hopes of recompense. They fear death, which they endeavour to deflect in their harmony of work and mutual encouragement.

Roger & Baptiste
Without rest or peace,
Diligently begun,
Freshly attentive,
One earns an honest wage.
Baptiste
Are you frightened?
Roger
I am never fearful! ...
Baptiste
I feel afraid!

Roger
So much the worse!
Baptiste
So you have courage?
Roger
More than enough [for us both]![2]

Roger's ebullient spirits serve to strengthen and encourage the failing Baptiste, providing resolution for them both and enabling them to complete the task in brotherly support. The strongly rhythmic music, with an interaction suggesting a shared heavy labour, returns rondo-like amidst their fearful asides to each other. After their work is completed and observing the terrifying situation into which they have been forced, they are returned home blindfolded, but are able to deduce where they were taken and alert the Watch, who are able to save the poor victims from their entombment. The duet is a celebration of simple daily labour in the midst of the dangerous circumstances into which the pair have been forced. Their demotic activity assumes an elevated and heroic role of liberation as love overcomes the threat of death for themselves and the imperilled victims.

Of even greater impact, and of very great significance for the history of opera, is Scribe's and Auber's *La Muette de Portici* (1828), which established the definitive form of a new type of grand opera and provided a tenor-baritone duet of the greatest influence. The "Amour sacrée de la patrie" takes place in Act 2. The fisherman folk-hero Masaniello is joined by his boon companion Pietro as they swear to rise up against Spanish tyranny in Naples.

Masaniello & Pietro
It is better to die than live in misery!
Is there any danger for a slave?
The yoke that chokes us all must fall.
And the foreigner perish under our blows!
Masaniello
Will you follow me?
Pietro
I will be at your side,
I want to follow you to the death…
Masaniello
To glory!
Pietro
We will be united by the same death.

Masaniello
Or crowned by the same victory.[3]

They break into a unison refrain that became a clarion call to arms against any unjust social system, famously signalling the beginning of the Belgian Revolution in August 1830. Their revolutionary aspirations are realized at the end of Act 4, but in Act 5 the friendship turns into betrayal and death. Pietro, feeling that Masaniello has abandoned their cause, poisons him, and the revolution fails just as Vesuvius erupts, an objective correlative of the turmoil among men. The noble hero dies amidst this tumult, recalling in his madness, with the typical fleeting thematic recollection of the Romantic mad scene, the past days of hope and idealism. The patriotic fervour and love shared with Pietro is ironically prefigure the topos of love betrayed in this tragic end. [**Fig. 10.1**]

The same loyalist sentiments, the pull between love and duty, are relived in Rossini's *Guillaume Tell* (1829), where the emotional heart of Act 1 is the duet between William Tell (baritone) and young Arnold (tenor). Tell is a Swiss patriot who, like Masaniello, is just waiting for the right moment to raise a rebellion against the Austrian overlords. Arnold is in love with Mathilde, a Habsburg princess, and torn between his private passion and his devotion to his country. The duet "Où va tu?" sees Tell waylaying Arnold, who seems to avoid him. Tell questions Arnold's attitudes and seeks to persuade his young compatriot to join the rebels. Arnold reveals his passionate feelings for Mathilde in his famous paean of love: but he is responsive to the call of duty, and his soaring aside captures his dilemma

Arnold
Ah! Mathilde, I love you, that is true,
but the time has come when I must leave you.
To my country, to my duty
My whole heart is dedicated.

(Tell notes his changing attitude with sympathy and hope.)

Tell
In his face I clearly read
The anguish of his heart.
Though he was disloyal to us,
He feels remorse at last,
And by repentance may expunge
The dishonourable past.

Tell tries to pull him into the patriotic cause, despite the fervour of his love. Tell calls out his father's name when the governor Gessler passes by. Arnold vows to insult him, but Tell retrains him.

> Dare nothing of the kind,
> Plan how to protect your father,
> and how to liberate your country.

It is only when Gessler vengefully murders Arnold's father, Mechthal, that Arnold has the incentive to join the conspiracy of the cantons. The themes of love and death in a sense shape the intention and direction of this duet, Arnold's distraction in personal emotion underpinning the disengagement that eventually contributes to his father's murder. The intertwining of love and death is developed further when Tell and Walter bring him the news of the murder. Arnold's lament captures the agony of love challenged by death. [**Fig. 10.2**]

> **Arnold**
> Alas! ... my heart fails me.
> His days which they dared proscribe
> I did not defend.
> My father, you should have cursed me!
> My heart is torn by remorse!
> O heaven! O heaven! I shall never see him again!

The theme of Arnold's heartbreak, already prefigured in the duet with Tell, reaches a highpoint at the beginning of Act 4, when Arnold visits the ruins of his home. His distraught aria is a threnody of love lost in death.

> **Arnold**
> Ancestral refuge
> Where my eyes first saw the light,
> Until yesterday your guardian shelter
> Protected a father for my love.
> Now I call in vain, o bitter sorrow!
> I call, and he no longer hears my voice!
> Beloved walls where my father lived,
> I come to see you for the last time.[4]

It might be surprising that there is a baritone-tenor duet in Donizetti's *Lucia di Lammermoor* (1835). At the beginning of Act 3 in the original conception of the work, Henry Ashton comes to see Edgar of Ravenswood in his decrepit castle, Wolf's Crag, to announce the marriage of Lucy to Arthur Bucklaw, and to demand satisfaction for the insults of Edgar's intrusion at the betrothal ceremony earlier on. The duet "Orida è quesa notte" depicts the venomous confrontation between these two strong characters. With a storm breaking outside the ruined castle, Edgar is shown musing. Horses are heard, and Henry Ashton bursts into the room. He had already threatened Edgar at the betrothal of Lucy to Arthur Bucklaw, an act of betrayal involving Lucy's deception. The duet assumes incremental passion as Henry informs Edgar that Lucy is now married, and demands satisfaction of honour for his insults by challenging Edgar to a duel, to be fought at dawn. Edgar agrees. The duet is the inverse of the usual encounter of friends in this genre: the passions are frustration, hatred and vengeance, as the love between Edgar and Lucy is dishonoured by the implacable Ashton. The passions depicted in the melodic fervour of the *a due* singing, and the pathetic fallacy of the natural world reflecting human anguish, is a variant of love-hate, and prefigures Edgar's solution to his sorrow in the death he seeks for himself.

Edgar & Ashton
Ah! The spirit of hell
clamouring for revenge,
will take fearful control
over our souls,
is more terrible
than the roar of the thunder,
the howling of the gale.
O sun, make ready.[5]

Death hangs in the air and is realized at dawn. Edgar is shown waiting for the encounter with Ashton when the funeral procession of Lucy passes by. He learns of her tragic fate in murdering her husband on her wedding night, and her subsequent madness and death. Reflecting initially on his desolate situation at the tomb of his ancestors, Edgar is so distressed by the news of Lucy's sad fate that he stabs himself fatally, aspiring to union with his beloved in heaven. Love and death are interwoven in a masterly manner in this tale of tragic loss.

Verdi's operas contain several instances of the male love-death duet. The grand opera *Les Vêpres Siciliennes* (1855) has two related instances. The plot depicts the attempts by the young Sicilian patriot Henri/Arrigo (tenor), provoked by the zealous Procida, to defeat the French overlords of the island. The French governor Monforte (baritone) is revealed to be Arrigo's father, presenting the young man with the dilemma of loyalties divided between patriotism and family connections familiar from *Guillaume Tell*. In Act 1 Monforte questions Arrigo, neither of them knowing of the relationship that binds them together, and draws them instinctively to each other ("Ebben? Non mi rispondi tu?"), Arrigo declares he has no father while Monforte urges him to join the French against the rebels. Although they part, threatening each other, the sense of affinity has been palpable. The troubled relationship comes to a head in Act 3 where the truth of their relationship is known and explored in the long duet "Quando al mio sen per te parlava". What follows is really a love duet between father and son, despite the furious resistance of Arrigo. The long broad melody of Monforte's address is the chief theme of the opera,

Monforte
When sincere pity for a blind error
Spoke for you to my bosom,
when in you, I saved a rebel,
Arrigo, did your heart say nothing to you?

Arrigo
At his voice I shudder,
I banish my terror in vain!
Hapless me!

(The father's pleading become more passionate.)

Monforte
As I contemplate that beloved face,
I feel my heart leap with joy.
At last on earth I am happy
That I can say: my son once more!

(But this love is spurned by the shocked Arrigo who cannot adjust to the fact that his enemy is now his parent.)

Arrigo
If it is true that you love me,
Let me flee, yes flee,
to another shore to another land….
Ah! I would like to fly to your bosom,
But I cannot.

(Monforte is desolate but goes on hoping.)

Monforte
Is your father's ardent prayer
Nothing, O Arrigo, nothing to you?
Open your bosom to a holy love,
Give way at last, O son, to me.[6]

This love will end tragically in death, as at the moment of reconciliation, when Monforte blesses Arrigo's union with Hélène/Elena, he will die in the massacre following the Vespers bell. The duet captures a tragic irony of secrets, dilemmas, foiled restoration and union.

The themes of love, vengeance and a fatefulness that swirls around human endeavour are represented with an almost superstitious adherence to fate and providence in *La forza del destino* (1862), a throwback to the German Romantic *Schicksalstragödie*. Don Alvaro (tenor) accidentally kills his beloved Leonora's father, who interrupts their planned elopement. Neither Alvaro nor Leonora is able to escape the curse which seems to be part of the very natural order. Leonora's brother Don Carlo (baritone) becomes an avenging angel, pursuing both of them with an intransigent hatred. The nexus of love and death is focused on the fateful encounters between the two male protagonists and finds particular expression in their duets. Love is seemingly found in their first duet, and death in the second.

Don Alvaro seeks out a new life as a soldier. In Act 3 the two men meet, unaware of each other's identity. Alvaro saves Carlo from a fight in an inn and the latter, overwhelmed with gratitude, joins with Alvaro in a vow of unbreakable friendship:

> **Alvaro & Carlo**
> Friends in life and death
> The world shall see us.
> both it will find united
> In life and death.
> United in life and death
> Both of us it shall see.

Alvaro is wounded in the succeeding battle and, apparently close to dying, entrusts Carlo with the key to his strong-box. Carlo's promise to destroy a secret packet of papers imbues their exchange with the tenderness and emotion of a love duet ("Solenne in quest'ora"), invested with a beneficent serenity.

> **Carlo**
> I swear, it shall be done

> **Alvaro**
> Now I die tranquil. I press you to my bosom.
> Farewell.

Carlo
My friend, put your trust
In heaven, trust in heaven.
Farewell.[7]

But Carlo breaks his solemn promise and opens the packet, whose contents reveal Alvaro's true identity to the grim satisfaction of his erstwhile friend. Later, Alvaro is alone and sad ("Nè gustare m'è dato") when Carlo reveals himself and demands satisfaction for the dishonour of his family: Alvaro and his sister must both die. Love has veered into its dark default, and the prospect of death taken its place.

Carlo
Death! Ere I fall dead
I will reach Leonora
And plunge in her this blade
Still reeking with your blood.

Alvaro
Death, yes! With my sword
I will kill a cut-throat villain;
Turn your thoughts heavenward,
Your hour has struck at last.

In Act 4 Alvaro has renounced violence and sought refuge in a monastery, Here Carlo tracks him down ("Invano Alvaro ti celasti al mundo") and, proving inexorable, provokes Alvaro into a duel and is mortally wounded. Carlo dies, but not before striking down his sister. The fraught interplay of love and death reaches its tragic and fated conclusion.

But perhaps the most famous tenor-baritone love-death occurs in Bizet's *Les Pêcheurs de perles* (1863), set in ancient Ceylon. In Act 1 Nadir (tenor), a hunter, meets his erstwhile friend Zurga (baritone), chief of the pearl-fishers, after many years. They recall how the appearance of a young and beautiful priestess in the temple caused them both to fall in love with her and destroyed their amity, turning it to hatred ("Au fond du temple saint"). Their reminiscence, however, seems to instil a grace of reconciliation, and they are brothers once more. However, their unison refrain, now one of the most famous duets in all opera, seems more like a passionate love duet. [**Fig. 10.3**]

Nadir & Zurga
O yes! We vow to remain friends!
Yes, it is she! It is the goddess,

who on this day comes to unite us.
And faithful to my promise,
I will cherish you like a brother!
It is she! It is the goddess
Who on this day comes to unite us.
Yes, let us share the same fate!
Let us be one until death![8]

This erotically tinged fervour, though, is once more put under unendurable stress when Leila returns as priestess and Nadir, disregarding his re-established friendship, enters into a sacrilegious relationship with the consecrated virgin. This is discovered, and will be punished with death. As in *La forza del destino*, any promises made to his friend are cast aside. But Zurga discovers that he has a debt of gratitude to Leila, who once saved him from death, and he causes a distraction, enabling the lovers to flee as he dies in their place. The last scene sees Nadir and Leila escaping, appropriating for themselves the earlier rapturous refrain between the companions, while Zurga dies abandoned in the valley below, offering his life in love for his friends (cf. The Gospel of John, 15:13).

Another example of an even more idealistic celebration of self-sacrificing friendship is provided in Verdi's *Don Carlos,* based on the drama by Schiller. The heir to the Spanish throne, Don Carlos (tenor), is pitted against his father, King Philip II, and the obscurantist religious establishment led by the Grand Inquisitor. The Infante is closely associated with Rodrigo, Marquis of Posa (baritone). Their desire is to procure freedom for the Protestant people of Flanders, suffering under the punitive campaign of the Duke of Alva to restore the Catholic faith. Their happiness at seeing each other again after a long separation finds expression in their celebration of friendship, inextricably bound up with the aspiration to find freedom for a people.

Rodrigo
Ah, my dear prince!
The hour has struck!
The voice of the Flemings calls you!
Help them Carlos, be their saviour!

The aspiring but weak and lonely, prince looks to his friend for support at every level of their relationship.

Carlos
My saviour, my friend, my brother,
Let me weep in your arms!

Rodrigo
To the voice of a sincere friend
Let your heart not be shut!

The climax of the colloquy is a unison celebration of their friendship, which once again has all the fervour of a love duet, couched in their common idealism and affection.

Don Carlos & Rodrigo
God, you sow in our spirits
A ray of the same fire,
the same exalted love,
the love of freedom!
Go, who has made our honest hearts,
the hearts of two brothers,
accept our vow!
We will die loving one another!

Their duet is interrupted, with portentous symbolic implications, by the arrival of the royal party, and a chorus of monks is heard off-stage. Their loving colloquy is resumed, shadowed by the thought of death.

Let us be united in life and in death!
God receives our vow,
To die loving one another!
Let us be united in life and in death!

It will be the intervention of the Inquisition that curtails both Posa's work for freedom and his very life when he is later assassinated by a sniper, the bullet intended for Carlos. He dies calling on the prince to be strong in their work for freedom and love, a real love-death.

Rodrigo
Ah! I die joyful at heart,
for you live, saved by me.
Ah! I see Spain happy!
Farewell! Carlos, ah! Remember![9]

Verdi's penultimate opera *Otello* (1887) uses Shakespeare's play depicting how the noble love between Othello (tenor) and Desdemona is turned into jealous hatred and murder by the scheming Iago (baritone), anxious to bring forth "the monstrous birth" of his evil instincts and his nihilistic beliefs ("Credo in un Dio crudel"). He sows doubt in Othello's mind by insinuating that Desdemona is having an affair with the young officer Cassio, who has spoken of her in his sleep and has her handkerchief, a present from Othello, in his possession. The radiant conjugal love of Othello is turned into murderous vengeance, with the tenor-baritone interchange finding unison resolve only as love is turned into a torrential desire for death.

> **Otello & Iago**
> Yes, I swear by the marble heaven!
> By the forked lightning!
> By death and by the dark destroying sea!
> Let this hand which I raise
> And stretch forth
> Soon blaze in wild transport of rage!
> God of vengeance![10]

Iago's fiendish actions destroy love and generate death, with Desdemona strangled and the tragic Othello dying by his own hand.

The male variant on the Love-Death presents an intriguing tangential perspective on the trope, casting the key presuppositions into an effective series of contrasts with mainstream Romantic concerns. These powerful duets are not positioned at the end of operas but reflect on key emotional and dramaturgical issues during the course of the action. This reflective dimension is underpinned by the types of character depicted in these colloquies: fathers and sons, mentors and pupils, tempters and victims, working colleagues, political allies or sworn enemies, but also friends (sometimes estranged and reconciled, sometimes confronting potentially destructive situations). Whatever the case, these duets always essentially consider in whatever form the issues of love and the challenges they pose in confronting alienation, separation, loss. This type of duet, in intense emotional rapture, generates processes of discovery, forgiveness or alienation that either intensifies the central concerns of romantic love or helps to precipitate the denouement.

Notes

[1] Gluck, *Iphigénie en Tauride* (words by Nicolas-François Guillard). The duet comes across powerfully as sung by John Aler and Thomas Allen (Philips, 416 148-2).

[2] Auber, *Le Maçon* (words by Scribe). The duet is sung in German by Walter Anton Dotzer and Franz Fuchs in a recording from Austrian Radio ORF (14 Dec. 1950) (Orfeo C 985 191).

[3] Auber, *La Muette de Portici* (words by Scribe), sung by Alfredo Kraus and Jean-Philippe Lafont (EMI 7492842).

[4] Rossini, *Guillaume Tell* (words by Étienne de Jouy & Hippolyte Bis), with Nicolai Gedda as Arnold and Gabriel Bacquier as William Tell (EMI 769951 2).
[5] Donizetti, *Lucia di Lammermoor* (words by Salvatore Cammerano), sung by Luciano Pavarotti and Sherrill Milnes (Decca 410 193-2).
[6] Verdi, *Les Vêpres siciliennes* (words by Scribe), sung in Italian by Placido Domingo and Sherrill Milnes (RCA 09026 63492 3).
[7] Verdi, *La forza del destino* (words by Francesco Maria Piave), sung by Mario del Monaco and Ettore Bastianini (Decca 421598-2), also by Jussi Björling and Robert Merrill (Prima Voce NI 7945) (rec. 1949-51).
[8] Bizet, *Les Pêcheurs de perles* (words by Eugène Cormon & Michel Carré). The performance by Jussi Björling and Robert Merrill is legendary (Prima Voce NI 7945) (rec. 1949-51).
[9] Verdi, *Don Carlos* (words by Joseph Méry & Camille du Locle, after Schiller). The performance by Carlo Bergonzi and Dietrich Fischer-Diskau is very good (Decca 455 437-2). It is also famously sung by Jussi Björling and Robert Merrill (Prima Voce NI 7945) (rec. 1949-51).
[10] Verdi, *Otello* (words Arrigo Boito after Shakespeare), sung by Jon Vickers and Tito Gobbi (RCA GD81969 [2]); Jussi Björling and Robert Merrill (Prima Voce NI 7945) (rec. 1949-51).

PART 3:

MUSICAL CONSIDERATIONS

CHAPTER 11

MUSIC AND OPERA FROM THE LATE 18TH CENTURY TO THE LATE 19TH CENTURY PART 1

Vissi d'arte, vissi d'amore.
I lived for art, I lived for love.

(Luigi Illica, *Tosca*, Act 2)

1. Rescue Opera

From its very inception, opera has incorporated the concept of release, of liberation, of setting free. In Chapter 6 we briefly touched on the theme of Rescue Opera, which is incipient to the myth of Orpheus and Eurydice. The dramatic spark of the story is the resolution and courage of Orpheus in descending to the Underworld to bring back Eurydice from the realm of death. During the early centuries of opera the chosen mythology often returns to this question of liberation through sacrifice. Handel's *Alcina* (1735) entails freeing her many captives from her enchantment through the power of pure love. All Gluck's reform operas also focus on this idea, with rehearsal of the Orpheus myth in *Orfeo ed Euridice*. In his *Alceste* (1767) the heroine, the wife of Admetus, braves the Underworld to give up her own life to save her husband. Hercules, moved by love, brings her back from Hades. The *Iphigenia* operas provide several variants of rescue. The daughter of Agamemnon and Clytemnestra, Iphigenia is rescued from being sacrificed by her father in Aulis by the goddess Artemis/Diana, and is taken to Tauris where she becomes the priestess of the goddess (*Iphigénie en Aulide*, 1773). Her brother Orestes is wrecked on the island in a storm. Iphigenia recognizes him when he is brought to her for sacrifice, and is able to escape with him (*Iphigénie en Tauride*, 1779).[1]

The changing political scene at the end of the 18th century, especially the onset of the French Revolution in 1789, led to an unprecedented explosion of fear, terror and death. The escape of aristocrats and ideological enemies of the movement meant that rescue and liberation from imprisonment and death became one of the dominating issues and symbols of the age. As if intuitively pre-empting these matters, André Gretry's *opéra-comique Richard Coeur de Lion* (1782) presents a nigh-perfect blueprint for the rescue scenario in operatic literature, via the route from mythology to medieval history. Blondel, Richard's minstrel, disguised as a blind singer, wanders in search of the abducted king. Helped by an English knight and by Marguerite of Flanders he manages to make contact with his imprisoned master by means of the romance, "Une fièvre brûlant", and is able to rescue him. The romance appears nine times in the opera in various melodic and rhythmic transformations and became the harbinger of the systematic use of the *Leitmotif.* [2]

The operas from the revolutionary period that capture the quintessence of this tumultuous period are both by the Italian Luigi Cherubini (1760-1842). His *Médée* (1797), returning to opera's ancient mythological roots, depicts a tale of passion, betrayal, vengeance and infanticide. Medea kills her own children by the perfidious Jason and escapes in her dragon chariot, just as France was murdering its own offspring in its ideological frenzy. *Les Deux journées* (*Le Porteur d'eau*) (*The Two Days*, or *The Water-Carrier*) (1800), with the Revolution more in perspective, is a classic rescue tale of human kindness overcoming death, as the folk-hero of the story, Mikéli, rescues some fleeing aristocrats by secreting them out of Paris in the great barrels used for his trade.

Beethoven was deeply aware of Cherubini's example, and his *Fidelio* (1805/1814) has become the archetypal rescue opera. The brave self-sacrificing Leonora disguises herself as a man in order to penetrate the prison where her husband, Florestan, has been unjustly incarcerated for years by the evil Don Pizarro. She finds him and protects him from murder, just as the visiting minister inspector is announced, the famous trumpet-call signalling his arrival becoming an emblem for the age and a cipher for this rescue topos in opera. Love triumphs over death.

The Revolution and the subsequent Napoleonic Wars saw Europe plunged into tumult and widespread conflict for a quarter of a century (1789-1815). The impact on music and on opera in particular was potent, not least in the emergence of the incendiary anthem of the Revolution, *La Marseillaise,* by Rouget de Lisle (1760-1836). He wrote both the tune and the words for the work, associated with the body of Marseilles Volunteers who sang it on entering Paris in 1792. This marching song, extolling the overthrow of tyranny and the sacrifice demanded by liberty, was spread all over Europe by the French armies which were always accompanied by their military bands. Marches, which had epitomised the *gloire* of Louis XIV, now came to signify the triumph of Napoleon and the very spectacle of his military prowess. Marches now became inseparable from the language of freedom, rescue and aspiration in opera, just as the lament had characterised its first centuries.[3]

Beethoven's *Fidelio* contains a low-key march for the entry of Pizarro, but more especially a triumphant one heralding the advent of the liberating figure of Don Fernando. It is a hallmark of the style of Gasparo Spontini (1774-1851) whose operas *La Vestale* (1807) and *Fernand Cortez* (1811) reflect the grandiose pretensions of the First Empire: both are dominated by marches and concepts of liberation. The latter work was specially commissioned by the Emperor Napoleon as a piece of propaganda to further his ambitions in the Peninsular War. The passage where the Spanish troops, beleaguered by the Aztecs, are liberated sounds rather like the French revolutionary anthem.[4] Auber's opera *La Muette de Portici* (1828), describing the bid for freedom from Spain by the fisherfolk of Naples, featured the duet equally incendiary for the age: "Amour sacré de la patrie" in which the hero and his friend espouse the cause for liberation. Two years later, on 25 August 1830, it did indeed do just that, when a performance in Brussels signalled the commencement of the Belgian Revolution against the Netherlands.[5] Henceforth, the occurrence of conspiracies became a recurring motif in operatic scenarios, none more so than in Meyerbeer's *Les Huguenots* (1836), where the machinations on St Bartholomew's Eve reflecting the division of faith and polity led to the notorious massacre, and where the main melody, rousing the partisan Catholics, clearly identifies with the *La Marseillaise.* The pattern would continue into the 20th century with the conspiracy scene in Prokofiev's *Semyon Kotko* (1940): here the fight for the motherland is central.[6]

2. The Rise of the Soprano and the Tenor

The rise of *opera seria* in the 18th century is inextricably associated with the form and poetry provided by the great librettist Pietro Metastasio (1698-1782), who dominated dramatic music in this century. The highly formal construction, with the set scene types, special categories of arias and artificially stratified procedures of this genre, were underscored by the universal hegemony of the *castrati* singers who became cardinal features of this period. Boy sopranos, through surgical procedures, were able to grow into adulthood without the usual lengthening of their vocal chords, something which made them less flexible and raised the tone they could produce. Combining the register and agility of a soprano, but with the power and projection of an adult man, these singers were capable of a kind of sound and virtuosity that has become legendary. Composers were encouraged to develop the techniques of *bel canto* to new levels of unparalleled difficulty and brilliance in writing for individuals with the castrato voice. Their unique abilities quite eclipsed the other singing registers, and despite the fame of many female sopranos, the castrati were the supreme stars of opera.[7]

In London between 1720 and 1730, for example, Handel worked closely with the male mezzo-soprano Senesino (Francesco Bernardi, c.1680-c.1750), for whom he created roles in fifteen of his operas; and the male contralto Antonio Mario Bernacchi (1785-1756) (two roles). Farinelli (Carlo Broschi, 1705-1782), the male soprano, sang in Vienna and then London where he was the star of Nicola Porpora's (1686-1768) rival establishment at Lincoln's Inn Fields. He became the focal point of a cult, with women fainting from excitement at his performances, and for 25 years was the personal musician to the troubled Philip V of Spain, singing the same two arias for him every night for years.[8]

But by the mid 18th century the castrati had lost appeal in Paris, and when Gluck took his Viennese reform operas there, he transposed the role of Orfeo for a tenor. Indeed, Gluck's demand for dramatic truth and the growing sense of humanity in the age of liberty, equality, and fraternity, began a process of decline, eventually leading to the demise of this type of singer. The situation changed gradually. Mozart, with a natural sense and aptitude for the Classical distinction of genres, wrote for castrati in seven of his operas including *Mitridate, re di Ponto* (1770), *Lucio Silla* (1772), *La finta gardiniera* (*The Feigned Gardener*) (1775), *Il re pastore* (1775), *Idomeneo* (1781) and *La Clemenza di Tito* (1791), as well as his joyful motet *Exsultate Jubilate* (K.165), composed for the celebrated male soprano Venanzio Rauzzini (1746-1810). There were also arias for castrati by Mozart's near-contemporaries Thomas Arne (1710-1778) and J.C. Bach (1735-1782).

By the turn of the century a sea-change in social attitudes and expectations had arrived. The last great castrato, Giovanni Battista Velluti (1780-1861), was active from 1800-30, with Rossini and Meyerbeer the last significant composers to write for this voice type (*Aureliano in Palmira*, 1813; *Il Crociato in Egitto,* 1824). The last castrato to sing at the King's Theatre in London was in 1800. When roles composed for these singers have been revived, they are now assumed either by a counter-tenor (who possesses a different timbre and capacity), a female soprano (where the gender difference generates a certain awkwardness), or by a tenor (which distorts the balance of the music as originally conceived).[9]

The renowned vocal prowess of the great castrati stars cast an indelible memory. This was reflected in the anomalous assumption of heroic masculine roles by mezzo-sopranos and contraltos in Italian opera of the 1800s (like Arsace in Rossini's *Semiramide*, 1823; Orsini in Donizetti's *Lucrezia Borgia*, 1833) and acknowledged by Wagner in the role of Adriano in *Rienzi* (1842). As well there was the emergence of the *travesti* part, especially favoured for the depiction of adolescent men (usually pages, such as Cherubino in Mozart's *Le nozze di Figaro*, Oscar in Auber's *Gustave III* (1833), Urbain in Meyerbeer's *Les Huguenots*, Oscar in Verdi's *Un ballo in maschera*, Tebaldo in *Don Carlos*), with Richard Strauss providing an historic retrospective in the major role of Octavian in *Der Rosenkavalier* (1911).

The other voice types were of course always valued and utilized in the developing genres of opera, but none carried the kudos of the castrato until the turn of the 18th century. Mozart's series of concert arias gives an idea of the types of singers he worked with, like his sisters-in-law, the brilliant Josepha Weber (1758-1819) and Aloysia Weber (1761-1839), and the famous tenor Anton Raaff (1714-1797).

Figures who emerged in the early 19th century began to generate the star quality formerly enjoyed by the castrati. The soprano Angelica Catalani (1780-1849), who appeared all over Europe between 1797 and 1821, could perhaps be seen as the forerunner of these superstars. She was followed by others like Henriette Sontag (1806-1854) who was so famous as to have operas written about her artistic life (Auber's *L'Ambassadrice*, 1836). She sang with exquisite taste and charm, in clear bright soprano, with an unsurpassable execution, said even to have excelled Catalani's. She created the role of Euryanthe for Weber (1824).[10]

In the efflorescence of the *bel canto* school of Rossini, Donizetti and Bellini (the period1812-45), the transition from castrati to what we may call the triumph of the soprano and tenor clearly emerges. Rossini's first wife, Isabella Colbran (1785-1845), created the leading roles in six of his operas (*Elisabetta*, 1814; *Otello*, 1816; *Armida*, 1817; *La donna del lago*, 1819; *Mosè in Egitto*, 1818; *Maometto II*, 1820; *Semiramide*, 1823) and was regarded as the finest dramatic coloratura soprano in Europe from

1801-22. Wilhelmine Schröder-Devrient (1804-1860) is considered the first great singing actress, the supreme Fidelio of her age (1822), who inspired many composers, among them Wagner. He wrote three roles for her (Adriano in *Rienzi*, Senta in *Der fliegende Holländer*, Venus in *Tannhäuser*) despite a decline in her vocal power. Giuditta Pasta (1797-1865) became an almost legendary figure, with her dramatic interpretations and poignant singing. Donizetti and Bellini both wrote for her (*Anna Bolena*, *Norma*, *La Sonnambula*, *Beatrice di Tenda*).

The first half of the 19th century thus heralded a period of change in the operatic world. It manifested itself in the art of composition, in singing and in acting techniques, in theatre practice, as well as in the audience itself and their expectations. It was a period when the female singer moved into the centre of attention, gaining a remarkable power over the whole of theatre practice of the time, and influencing not only the way music was composed, but also acting on stage and the whole process of opera production. Theatrical practice of this period challenged female singers. They were required to be their own stage directors, costume designers and make-up artists. The style of *bel canto* was an essential part of their musical education and central to the outstanding performing skills of the two most famous singers of the time—Giuditta Pasta and Maria Malibran.[11]

The première of *Lucia di Lammermoor* (26 September 1835) is of great importance for focusing on the superlative soprano and tenor leads in this opera. Donizetti wrote these exceptionally vital title roles for the brilliant soprano Fanny Persiani (1812-1867) and the tenor Gilbert Duprez (1806-1896). Both seemed to epitomize the power and beauty of the human voice. Earlier in the same year Bellini had created *I Puritani* in Paris (25 January 1835), using artists who became known as 'the Puritani Quartet': the soprano Giulia Grisi (1811-1869), the tenor Giovanni Battista Rubini (1794-1854), the baritone Antonio Tamburini (1800-1876) and the bass Luigi Lablache (1794-1858), all of whom assumed a quasi-mythical status in operatic history.

The figures of Rubini (with his sweet timbre, huge power and volume, and extended register into falsetto) and the highly intelligent Duprez (who sang in Italy from 1828-35 and in Paris from 1837-49), shifted the gaze from the high rhapsodic soprano voice to the visceral thrill of the tenor, especially when reaching for the high notes. The role and importance of the tenor had been increasing from the 1820s. Singers like Domenico Donzelli (1790-1873), the creator of Pollione in *Norma*, added to the growing prominence of the voice type encouraged by Rubini's great presence and style. But it was in Paris that this development found definitive shape. Rossini moved to the French capital in 1824 and, working in both the Théâtre Italien and the Paris Opéra, had access to a range of superlative singers from both Italy and France: the Italians Pasta, Donzelli, Franceso Graziani (baritone, 1828-1901) and Giulio Marco Bordogni (tenor, 1789-1856); the French sopranos Laure Cinti-Damoreau (1801-1863) and Julie Dorus-Gras (1805-1896), the tenor Adolphe Nourrit (1802-1839), and the bass Nicolas-Prosper Levasseur (1791-1871).[12]

Adolphe Nourrit was an artist of the highest integrity and imagination, a tenor of supreme intelligence, who helped to shape the history of opera in the period 1827-37, a source of inspiration to Rossini, Auber, Meyerbeer, and Halévy, all of whom wrote roles expressly tailored for his unique vocal gifts and poetic insights in the some of the most famous operas of the age: *La Muette de Portici*, *Guillaume Tell*, *Robert le Diable*, *Le Philtre*, *Gustave III*, *La Juive*, *Les Huguenots*. Nourrit was famous for his beautiful tone and dramatic understanding, and used the French technique of the *voix mixte* (a blend of chest and falsetto registers) to achieve his high notes. But the arrival back in Paris of Duprez, with his technique of singing high C from the chest, brought about a revolution in the apprehension of the tenor voice and audience expectations. This development, which led to the fraught departure of Nourrit from Paris and a visit to Italy to try to learn the new technique, tragically resulted in his suicide. From then on, the type of robust vocalism typical of the *spinto* style has dominated tenor singing. Neither Rossini nor Meyerbeer favoured the new approach, but tenors straining for high C have become the subjects of a universal adulation of this voice type that has prevailed ever since.[13]

Nourrit and Duprez were superseded by the phenomenon of Giovanni Mario (Cavaliere di Candia, 1810-1883), who studied in Paris and made his debut there in 1838, but found his favourite platform in London, where he appeared from 1838-67, usually in partnership with the soprano Giulia Grisi, whom he married in 1844. He sang with her, Tamburini and Lablache in the premiere of Donizetti's *Don Pasquale* (1843), taking the place of Rubini in the famous Puritani Quartet. His voice was considered as among the most beautiful ever heard. This, when combined with the elegance and style of his delivery and his good looks, made him the operatic idol of the age. His association with Grisi left a model of such tenor-soprano partnering that was particularly noticeable in the Golden Age of late 20th-century opera during the 1950s and 1960s: Mario del Monaco (1915-1982) and Renata Tebaldi (1922-2004); Maria Callas

(1923-1977) and Giuseppe di Stefano (1921-2008); Wolfgang Windgassen (1914-1974) and Birgit Nilsson (1918-2005); Nicolai Gedda (1925-2017) and Beverley Sills (1929-2007); Joan Sutherland (1926-2010) and Luciano Pavarotti (1935-2007). The mythical implications and access to a rapture or transcendence provided by the combination of music and the human voice seems to approach the Olympian ideal here.[14]

The adulation of the tenor continued through the rest of the 19th century, and indeed began to assume even greater prominence at the turn of the century. The lyrical Polish singer Jean de Reszke (1850-1925) became a living legend at the Metropolitan Opera in New York, as did Francesco Tamagno (1850-1905), the creator of Verdi's Otello, who is regarded by many as the greatest *tenore di forza* of all time. The talents of both were seemingly combined in the artistry of Enrico Caruso (1873-1921), who appeared over 600 times in 40 operas in New York. He has generally been thought of as possessing one of the most beautiful tenor voices ever known, and is also significant as being the first great singer to use the gramophone extensively.

Other famous voices followed, like Beniamino Gigli (1890-1957), who was regarded as Caruso's successor, and Jussi Björling (1911-1960), with his warm and appealing lyric talent. The mystique continued into the last years of the 20th century with the tremendous interest and popularity of the famous 'Three Tenors' phenomenon. In the wake of the 1990 World Cup, the three most famous tenors of the time were brought together in the Baths of Caracalla in Rome to sing a concert that has come to share in the legendary mystique of the voice type: Luciano Pavarotti, Plácido Domingo (1941-) and José Carreras (1946-). The thrill generated by the tenor voice remains undiminished, and central to the power of opera to lift the human spirit.

3. The Development of Scenery

The earliest known forms of stage presentation and lighting derive from the early Grecian and later Roman theatres, built facing east to west so that plays could be performed in the afternoon, with natural sunlight striking the actors but behind the audience seated in the arena. Natural light continued to be used when playhouses were built, with a large circular opening at the top of the theatre. Early Modern English theatres (like the famous Globe in Southwark) were roofless, allowing natural light to illuminate the stage. When theatres began moving indoors, artificial lighting became necessary, and evolved with the advance in theatre construction and technology. It is not known when candlelight was introduced, a revolutionary step in the development of theatrical lighting across Europe.[15] The first operas appeared not to have scenery. It was only in the period 1630–50, with the spectacular performances in Cardinal Barberini's palace in Rome, and the subsequent establishment of court theatres, that stage spectacle, so characteristic of the Baroque, became fashionable. These theatres, with tiers of boxes, and the audience separated from the stage action by the proscenium arch, lent themselves to the development of two-dimensional flat scenery, with the perspective painting of much Renaissance art now adapted by theatre designers. The first opera house, San Cassiano, opened in Venice in 1637. Monteverdi's *L'incoronazione di Poppea* was first performed at the Teatro Santi Giovanni e Paolo as part of the 1642-43 carnival season. The theatre, established in 1639, had earlier staged the première of Monteverdi's opera *Le Nozze d'Enea in Lavinia*, and a revival of the composer's *Il ritorno d'Ulisse in patria*. The opera house was later described by an observer: "... marvellous scene changes, majestic and grand appearances [of the performers] ... and a magnificent flying machine; you see, as if commonplace, glorious heavens, deities, seas, royal palaces, woods, forests". The theatre held about 900 people, and the stage was much bigger than the auditorium.

A milestone was reached in the production of Cesti's *Il pomo d'oro* (*The Golden Apple*) (Vienna 1668) for the marriage of the Emperor Leopold I to Margaret of Spain, where there were 23 different sets designed by Lodovico Burnacini (1637-1707). The Bibiena family, descended from Giovanni Maria Galli (1619-1665), produced several generations of theatrical architects of great importance for the history of stage design.

In England, stage production was suspended in 1642 under the Puritan Commonwealth, but great advances were made in theatres on the Continent. During his exile, Charles II witnessed Italian theatrical methods, and brought them back to England at the Restoration of the monarchy in 1660. New, much larger playhouses were built in England, a development that called for more elaborate lighting. Candlelight became the main source of light in Restoration theatres. This was concentrated towards the front of the house, over the forestage, with dipped candles used to light chandeliers and sconces (cylindrical candles were made by repeatedly dipping a wick into hot wax). The serious disadvantage was that candles needed frequent trimming and relighting regardless of what was happening on stage, as they were constantly dripping hot grease on actors and audience alike, while the chandeliers themselves blocked the view of some of the audience. Commercial theatres tended, for economic reasons, to be more conservative with their lighting, while Court theatres (like Drury Lane and Covent Garden) could afford to follow Continental innovations, and were lit by a great central chandelier, with varying numbers of stage chandeliers and candle sconces on the walls of the auditorium. By the 1670s, the new Hall Theatre began using footlights, with candles or lamps. Not much had changed by the middle of the 18th century.

Gas lighting was adopted by the English stage in the early 1800s, beginning with the Drury Lane and Covent Garden theatres. The 1820s saw the development of a new type of artificial illumination. A gas flame was used to heat a cylinder of quicklime (calcium oxide) which would begin to incandesce on reaching a certain temperature. The ensuing light was then redirected by reflectors and lenses. After a slow start, this new method found its way into theatrical use around 1837. It became popular in the 1860s and beyond until displaced by electrical lighting. The advances made in illuminating theatres during this time paved the way for the many innovations in the modern theatrical world.

Giuseppe Galli-Bibiena (1696-1757), working in Vienna, and Ferdinando Galli-Bibiena (1657-1743), working with Bernadino Galliari (1707-1794) and Fabrizio Galliari (1709-1790), the principal designers of the old Ducal Theatre in Milan, set the pattern for Italian stage scenery. The Galliari brothers were the first scene painters at La Scala, followed by Pietro Gonzaga (1751-1831), who, influenced by the paintings of Canaletto (1697-1768), introduced more colour and contrasts of light and darkness, an important development at a time of poor stage lighting. These significant trends reached a highpoint in the work of Alessandro Sanquirico (1777-1849), whose stage settings for Rossini and Bellini have endured as models of their kind. He also produced the spectacular designs for Meyerbeer's *Il Crociato in Egitto* (1824) at La Fenice in Venice. Carlo Ferrario (1833-1907, Verdi's favourite designer) and Antonio Rovescelli continued in this tradition of creating splendid scenery for La Scala and other Italian houses. Scenery had to be moved on and off stage in grooves in the absence of stage lifting-towers, introduced to La Scala only in 1919-21.

Italian models were widely emulated in Germany and elsewhere in central Europe. There were developments with the use of perspective in creating three-dimensional sets in Frankfurt by the Italian Giorgio Fuentes (1756-1821), as exemplified in his famous sets for Mozart's *La Clemenza di Tito*

(1799). Other landmark achievements were in Berlin, with Karl Friedrich Schinkel's (1781-1841) famous designs for Hoffmann's *Undine* (1816) and Mozart's *Die Zauberflöte* (1816), and Karl von Gropius (1793-1870) for *Der Freischütz* (1821), where the spatial emphasis contributed to conjuring up the Romantic atmosphere.

Landmarks in the further development of German stage designs came in the work of Simon Quaglio (1796-1878), who in the mid-century provided splendid scenery for Mozart and Wagner productions in Munich. These were highly realistic and anticipated the style of scenery at Bayreuth before the First World War. Angelo Quaglio (1829-1890) also worked in Munich and designed the first Wagner productions there, showing a special gift for epic architecture. Johann Kautsky (1827-1896) designed Wagner sets for Vienna (1880-90). Josef Hoffmann's (1831-1904) designs for Valhalla were "a truly inspired stroke of genius" and his heroic landscapes suited Wagner's ideals above others of the time: he provided the scenery designs for the first *Ring* in Bayreuth (1876).[16]

Development of a modern German style began with the advent Alfred Roller (1864-1935) in Vienna (during Mahler's directorship), Leo Pasetti (1889-1937) in Munich in the 1920s and 1930s, and Adolph Mahnke (1891-1945) and Hans Strohback (1891-1949) in Dresden during the same period. These modern trends were influenced by the writings and designs of Edward Gordon Craig (1872-1966) and Adolphe Appia (1862-1928). Appia was unhappy with the visual aspect of productions at Bayreuth. He saw two-dimensional flat scenery as detrimental to theatre, and regarded light as the most important element in production. The ideas of Craig and Appia in the early part of the century were adopted especially after the Second World War in the work of the producers Wieland (1917-1966) and Wolfgang Wagner (1919-2010), where there was a conscious move away from older ideas in an attempt to de-Nazify the reputation of Bayreuth. The sets were reduced and replaced by often geometrically abstract landscapes of the mind, with brilliantly imaginative coloured lighting.

On the other hand, theatre after the Second World War, with advanced modern technology at its disposal, seemed in many respects to revert to the ancient primitive techniques of stage production and [non-] lighting. Modern producers have been overwhelmed by the *Verfremdungs-Technik* of Vsevolod Meyerhold (1874-1940) and Bertold Brecht (1898-1956), whereby stage effects and imaginative transformation were replaced by a deliberate attempt to debunk any form of illusion, be it emotional empathy or recreated verisimilitude. The increased resources of even greater computerized technology are largely neglected as producers have turned the techniques of alienation to usually bizarre re-interpretations of dramaturgy in the fashion for *Regietheater*. This invariably destroys the mystique of past times, exotic space, historical concern, representation of nature and ancient symbolism, often leaving audiences bewildered, puzzled or distracted. Re-inventions of dramatic scenarios, often by highly elaborate but dysfunctional stage structures and procedures, distort the original and destroy the aim of drama for an audience to re-live the dramatic experience as a bridge to emotional transcendence or even spiritual catharsis. The powerful traditions of stage production and lighting, developed over four centuries, are largely ignored, or less done today, and certainly given less prominence.

The place of light and darkness in the history of operatic production and dramaturgy has bearings on the polarities of Apollo and Dionysus, one of the fundamental concepts we have emphasized, underlying the innate drama and tensions in opera from its Florentine inception. Whether overt or tacit, dramaturgical or symbolic, illumination and obfuscation are key to a deeper understanding or perception of the presented world. Light and clarity (the object) opposes shade and diffusion (what might the object actually be?). What lies behind the object? The clear rational ideas of Descartes and Locke seem to oppose the emergence of post-Kantian phenomena. This *flourish of realism* has a teasing temporal dimension, implications of time standing still as opposing the time of becoming (an issue tangentially raised in Wagner's *Parsifal*).

In France advances were made over the old Italian system of stage flats by the early photographer Louis-Jacques-Mandé Daguerre (1787-1851) and the artist Pierre-Luc-Charles Cicéri. Their work had antecedents in the theories of Lavoisier, who had experimented with the redirection of light from stage sets by means of reflectors (1781), while Colonel Grober published a treatise (1809) proposing the abolition of painted scenery and footlights, the dimming of house lights during performances, and the use of relief in the construction of décor. Joseph Antoine Borgnis (1718-1863), an engineer and professor of applied mathematics, proposed a more sophisticated use of lighting and of stage machinery (1820). At the Paris Opéra, rue Le Pelletier, during the 1820s and 1830s, these theories were realized artistically by Daguerre and Cicéri with revolutionary success. Using extensive resources, they developed powerful three-dimensional scenery with vivid lighting effects, corresponding to the dramatic and historical scenarios of the works being composed for the house. The result was to influence and even determine a whole style of opera presentation, with effects even on composition throughout the 19th century and even into the 20th century.

The emphasis on historical and nature verisimilitude, based on close observation and scholarly research, resulted in an astonishing series of *mise en scène* that became almost legendary: the final scene of Auber's *La Muette de Portici* (1828), where the panorama from the gardens and balustrades of a palace in the foreground led the eye across a rising landscape into the sublime vista of Vesuvius erupting; the monastery in Act 3 of Meyerbeer's *Robert le Diable* (1831), where the ruins of the monastery of Monfort l'Amaury were faithfully copied and recreated in several layers of pillared elevation and also of recession, with two angles of the cloister looking into the central garden with cypress trees, the tombs of the perfidious nuns to the fore, with gas lighting to represent the moonlight, and dimmed houselights to sustain the eerie mood of the supernatural ballet. The last scene in Auber's *Gustave III* (1833) depicted a ballroom in the Royal Palace at Stockholm, with a staircase descending onto the stage, great columns, and hanging chandeliers that created a space across which the fabulous succession of dances was enacted, culminating in the celebrated galop for the whole company. The first act of Halevy's *La Juive* (1835) displayed a square in the city of Constance, with looming Gothic tenements, the steps of the cathedral to the left, space for the Emperor's procession to cross the stage, soldiers in real armour, and horses borrowed from Franconi's Circus, in a scene of unparalleled pomp. In Act 3 of the same opera, the palace of the Emperor, with beautiful gardens looking across a parapet to an exquisite landscape beyond, was regarded as a masterpiece of loveliness. Developing technology helped portray Romantic illusion. [**Figs. 11.1-8**]

Notes

[1] Robinson, Michael F. *Opera before Mozart* (London: Hutchinson University Library, 1966).
[2] Charlton, David. *Grétry and the Rise of Opéra-Comique* (Cambridge: Cambridge University Press, 1986), 226-250.
[3] Dent, Edward J. *The Rise of Romantic Opera* (Cambridge: Cambridge University Press, 1976).
[4] Everett, Andrew. *Josephine's Composer: The Life, Times and Works of Gaspare Pacifico Luigi Spontini (1774-1851)* (Bloomington In: Author House, 213).
[5] Letellier, Robert Ignatius. *Daniel-François-Esprit Auber: The Man and His Music* (Newcastle: Cambridge Scholars Publishing, 2010), 164-186.

[6] Letellier, Robert Ignatius. *The Operas of Giacomo Meyerbeer* (Madison: Associated University Presses, 2006).
[7] Barbier, Partick. *The World of the Castrati: The History of an Extraordinary Operatic Phenomenon* (Souvenir Press, 1998).
[8] D. F. E. Auber's opera *La Part du Diable* (Paris, 1843), also known as *Carlo Broschi*, is built around these events. The play *Farinelli and the King,* with music by Claire van Kampen, premiered in London in 2015 and on Broadway in 2017. Also see Barbier, Patrick, *Farinelli: Le castrat des Lumières* (B. Grasset, 1994).
[9] Baker, Theodore. *Baker's Biographical Dictionary of Musicians* (8th ed.) (London: Macmillan Publishing Company, 1995).
[10] Pleasants, Henry. *The Great Singers: From the Dawn of Opera to Our Own Time* [1966] (London: Macmillan, 1981).
[11] See Sofia Livotov, "In the Centre of Attention" (The Challenge of being a Prima Donna in the First Half of the 19th Century) (diss., The Conservatorium van Amsterdam, 2017). Artists like Giuditta Pasta and Maria Malibran questioned their own performances in directing their own stage appearances and producing opera scenes in the style of the historical setting.
[12] Celletti, Rodolfo, *A History of Bel Canto,* new ed. (Clarendon Paperbacks) (Oxford University Press, 1996).
[13] Pleasants, Henry. *The Great Tenor Tragedy: The Last Days of Adolphe Nourrit As Told (Mostly) by Himself* (Portland OR: Amadeus Press, 1995); Zucker, Stefan, *Franco Corelli and a Revolution in Singing: Fifty-Four Tenors Spanning 200 Years* (New York: Bel Canto Society, 2015).
[14] Steane, J. B. *The Grand Tradition: Seventy Years of Singing on Record, 1900 to 1970* (London: Duckworth, 1974, 1978).
[15] Hartnoll, Phyllis. *The Theatre. A Concise History*. rev. ed. (London: Thames & Hudson, 1968, 1985). The illustrations and Select Bibliography are a rich resource, 271-284; Sadie, Stanley (ed.) [1992]. *The New Grove Dictionary of Opera* (London: Macmillan, 1997), 4: 491-518. The comprehensive Bibliography covers general studies of the history of staging and particular studies in the history of opera century by century, 4:516-518. Individual entries in Larue, Steven C (ed.), *International Dictionary of Opera* (2 vols) (Detroit MI: St James Press, 1993) have extensive bibliographies.
[16] Carnegy, P. *Wagner and the Art of the Theatre* (London: Yale University Press 2006), 76-81.

CHAPTER 12

MUSIC AND OPERA FROM THE LATE 18TH CENTURY TO THE LATE 19TH CENTURY PART 2

À cet amour qui est le battement de coeur
De le univres entire,
Mysterieux, sublime,
Croix et delices pour le coeur.

(Alexandre Dumas *fils*, *La Dame aux camélias*, 1852)

A quell'amor ch'e palpito
Dell'universo intero,
Misterioso, altero,
Croce e delizia al cor.

(Francesco Maria Piave, *La Traviata,* 1853)

Yes, for a year,with that love which is the very breath
of the universe itself,
mysterious and noble,
both cross and ecstasy of the heart.

Music, Myth and Social Change: Alteration of Music Style

1. The Unbroken Musical Line, the Librettist, and the Aria

Opera began as a rhapsodic musical setting of the poetic text in an unbroken declamatory sequence, with a continuous accompaniment (for organ, lute) interrupted by special orchestral effects (often on the trombones), representing the changing drama of the text, with the chorus commenting on the action from time to time. Monteverdi's *La favola d'Orfeo* (1607) provided the prototype for the whole operatic genre. His *L'incoronazione di Poppea* (1642) finishes with a love duet of alternating parts. This style of an unbroken musical line continued through much of the 17th century, with the works of Pier Francesco Cavalli (1602-1676). In the emergence of *tragédie lyrique*, however, the declamatory parts begin to separate into more concentrated lyrical sequences that become arias, sometimes extended formal sections. The composer Jean-Baptiste Lully (1632-1687), working directly under King Louis XIV (reg. 1643-1715), was helped by the librettist Philippe Quinault (1634-1688). He wrote formal scenes in verse expressing stereotypical noble sentiments whose stateliness reflected the expression of Royal splendour, or *gloire*. The inseparability of song and dance in this concept of glory at Versailles (1669) was a great stimulus in the evolution of specific generic forms, with each major event, and the finale, marked by ballet.[1]

The melodic setting of the words generated its own specific harmonic support in the accompaniment. Dido's Lament in Purcell's *Dido and Aeneas* (1689), with its ground-bass, is a classic example of the elaboration of form. The traditions of *tragédie lyrique* were later sustained and developed by Jean-Philippe Rameau (1683-1764), whose fascination with harmony often saw melody emerging from an elaborated accompaniment and a more diverse use of instrumentation. His development of Lully's recitative, with far greater flexibility of treatment, was his notable contribution to the genre. The *da capo* aria was restrained, with more emphasis in arioso and recitative, with the role of the orchestra expanded in illustrative colour and dramatic commentary. Debates in French opera about these developments saw the emergence of the factions of *Lullyistes* and *Rameauneurs*.

The expansion of *opera seria* in the 18th century saw the increasing development of fundamental incipient structures, with growing emphasis on the librettist. Metastasio's texts were often set many times by different composers, reflecting both the grandiose nature of the Baroque and the tremendous concern with patterns and order. The whole of opera became subject to a series of formalized expectations and conventions, with set types of characters and arias appropriate to certain emotions and events; the French and Italian style of overture and the universal *da capo* aria, adapted to every occasion; and, following sequences of presentation, often with short independent interludes or *intermezzi* played between acts.

The rigidity of the format dominated opera, and was challenged firstly by the emergence of the comic genre in John Gay's (1685-1732) *The Beggar's Opera* (London, 1728) and Pergolesi's (1710-1736) *La serva padrone* (Naples, 1733). The first used folk songs and ballads, and the latter using the *intermezzo* to initiate a whole modal shift in operatic genre, with simpler stories and shorter forms serving a humorous, often satirical subject. We have commented on these ideas and the development of the *opéra-comique* deriving from the simple popular entertainments at the great winter and summer fairs of Paris. The Paris production of Pergolesi's work in 1748 led to the battle of the genres in the famous *Guerre des buffons*, with controversy between those supporting the traditions of French serious opera and those in favour of the new simplicity of Italian *opera buffa*.

The second invention of opera came with the famous reform operas of Gluck, first in Italian in Vienna, 1762-69, then in French in Paris, 1773-79. He stressed the importance of subordinating the music to the dramatic exigencies of the plot, removing the formal stratification of the genre, simplifying the nature of the aria, and dispensing with the dry recitative (with harpsichord continuo) which dominated *opera seria*, replacing it with dramatic accompanied recitative (for the full orchestra). All these ideas were expounded in his famous preface to his opera *Alceste* (1767).

Another great controversy ensued between the supporters of Niccolò Piccinni (1728-1800), composer of the most successful *opera buffa* of the age, *La buona figliola* (*The Good Daughter*, 1760), upholding the traditions of Italian opera, and those of Gluck, with his severe but often sublime reforms. It would take a genius like Mozart to feel at home in all these operatic traditions, and write music in most of the generic forms of music drama that had evolved to date. His most admired achievements in the three operas he wrote with the librettist Da Ponte, especially *Don Giovanni* (1787), reveal a complex amalgam of convention with innovatory ideas that look forward to a new age. This was also true of the German Singspiel *Die Zauberflöte* (1791), which, in its folksy spirit and sustained use of archetypal imagery, fuses many musical traditions into an enduring comment on the human condition.

It fell to another decisive genius, Rossini, to shape the operatic inheritance of the early 19th century into a set of formal and structural arrangements that would characterize the musical and dramatic profile of opera for the next 40 years—the so-called Code Rossini. The hegemony of this approach (characteristic of Donizetti, Bellini and most of Verdi's work) was challenged only by the emergence of French Grand Opera (1828-1870) as the logical successor to the traditions of *tragédie lyrique*. This portrayed a drama in five acts that dealt with the struggle of individuals caught up in the clash of classes and peoples. Developments included noble declamation, great crowd scenes, a more pliable melodic structure, a renewed perception of orchestral colour, acknowledgement of the tradition of ballet, and a revolutionary stage presentation where spectacle shaped by pictorial and historical factors became a dramaturgical principle. This second generic pattern, to which Rossini himself contributed (in 1827, 1829), is epitomized in the work of Meyerbeer, working with the librettist Scribe. This genre influenced the whole development of opera for the rest of the century, in Germany, Italy and especially Russia.

Grand Opera would be challenged by the third generic revolution initiated by Wagner, writing his own libretti and creating the concept of music drama, where the text is served by an unbroken flow of orchestral commentary that assumes an increasingly psychological function in elaboration of the *Leitmotif*-system and the expanded capacity of the symphonic tradition. The continuous texture, almost a reversion to opera's first concern to rediscover Greek tragedy, became the new way forward for opera, *Zukunftsmusik*, the music of the future, in the rest of the 19th and into the 20th century.

As with French Grand Opera, there was a concern for a fusion of artistic media in pursuit of a unified dramatic experience (the total art work, the *Gesamtkunstwerk*). Imitation of Wagner, without his musical genius, has remained a besetting challenge for opera, adapting to the new aesthetic systems of the modern age, where only a few composers have achieved enduring success, among them Puccini, Richard Strauss, Berg, Janáček and Britten.

2. Gods and Priestesses and Myths

The origins of opera are bound up with the myths of Ancient Greece and the desire to rediscover Antiquity and reinvent the approach to drama. So often we have recourse to mention Orpheus, the obvious patron of the new genre (1597) with its emphasis on music and song, and his self-sacrificing actions in seeking to liberate his lost wife Eurydice from the Underworld (Peri, 1600). The power of song and the search for freedom are integral to each other and to the genre itself. The lament for the loss, the trip to the dark regions of fear and death (*katabasis*), and the act of rescue are features that recur in opera in one form or another throughout its history. The foundational myth is re-enacted in Monteverdi (1607) and in Gluck (1762). There is the famous spoof in the operetta of Offenbach (1858), and modern handling of the story by Hans Werner Henze (*Orpheus behind the *wire,* 1966), Harrison Birtwistle (*Mask of Orpheus,* 1986) and Phillip Glass's *Orphée* (1993). Love and death are intertwined, and the myth presents the polarity of the gods Apollo and Dionysus.

The reinvention of opera under Gluck not only re-presents this myth, but introduces significant developments in the utilization of other stories from Antiquity that begin to move the scenario from the realms of the divine to the more human world of experience. Alceste is not a child of the gods like Orpheus, but an ordinary woman who seeks to offer herself in place of her husband, and is in her turn rescued by the hero Hercules. The Iphigenia stories also present scenarios of family love and sacrifice. The goddess Artemis (Diana) makes two rather cursory theophanies both in Aulis and Tauris, but the emphasis of the stories is the family relationship between Iphigenia and her father, Agamemnon, and then her brother, Orestes. Iphigenia is saved from sacrifice by Diana and Orestes is saved from a similar fate by his sister. On Tauris Iphigenia is not demi-divine, but a priestess in the service of the goddess. The power of the stories lies in the very real human emotions presented in the context of extreme challenges within society and the family. In the Tauris opera the family drama is still about sacrifice, but enacted in the context of cultural tensions between the Greeks and the Scythians, and for the heroine in the strain between her priestly office and that of her personal engagement with her family.

After Iphigenia, conflict between the demands of religious office and the demands of love become a very significant theme in opera, with the issues of a divine commitment vying with the strength of human passion. Medea, the priestess in Cherubini's opera (1797), espouses the dark divinities and murders her children to be avenged of Jason's betrayal of her. Spontini's Vestal virgin Julia (1807) allows the strength of her human passion for the Roman hero Licinius to overwhelm her sacred duty in safeguarding Vesta's fire. She must die as a consequence, but is vindicated by lightning from heaven and so rescued from death.

The demands of sanctity and sacred duty have always been couched in the language of sexual restraint, with continence, celibacy and chastity providing a hierarchy of detachment, engagement and value on the part of the sacred votary. Each concept represents a higher rung in perceiving the mystery of commitment and volition, with chastity embodying the paradox of total passion in self-giving, be it of virginal commitment or loving sexual union. For Bellini's Norma, high priestess of the druids (1831), there seems to be no existential conflict between her celibate public persona and her secret passion for the Roman consul Pollione which has produced two children. Unlike Medea, she is not prepared to sacrifice them to vengeance, but accepts and assumes a public purification of her perceived unfaithfulness in loving union with her lover, ironically an act of chastity. This will also characterize the behaviour of Lelia in Bizet's *Les Pêcheurs de perles* (1862) where, as with the Vestale and Norma, she does not hesitate to place her personal love for Nadir before her celibate sacred duty. [**Fig. 12.1** Norma]

Other variants of this very powerful trope are found in Halévy's *La Juive* in the figure of the Jewess Rachel (1835), who is prepared to elope with her passionate lover Samuel until his religion and status in life as the Christian Prince Leopold are revealed. She will die rather than recant her love or her faith, despite his betrayal. The queens Dido in *Les Troyens* (1863) and Sélika in *L'Africaine* (1865), invested with regal responsibilities, are also prepared to brook divine and social displeasure in loving the man of their choice, be it the Trojan hero Aeneas, called to found Rome, or the explorer Vasco da Gama, called to found the Portuguese empire in India. Like Norma, both heroines sacrifice themselves rather than be separated from the man they love. Both die at their own hands, caught up either in prophetic visions of future Roman glory or in the transition in death to a celestial paradise of untrammelled love.

The passion of Tristan and Isolde (1865) also disregards sacred knightly and family codes in almost blindly pursuing their mutual love, and death must come, either through self-induced destruction or a mystical relinquishment of life in this world. The same scenario applies to Nikia, the priestess heroine of Ludwig Minkus's (1826-1917) grand ballet *La Bayadère* (1877) who with single-minded defiance gives herself in love to the hero Solor even though this will result in her death. She will find her transcendent vindication in the mystical visions of the Kingdom of the Shades where she can be united in a spiritual paradise with the one she loves. This is also the scenario for Delibes's *Lakmé* (1883), in which the daughter of the Brahmin High-Priest Nilakantha is in love with Gerald, a British officer. Despite the difficulties posed by her personal office, her race and her family, she is immovable in her private love and prefers to die by suicide. This will be echoed in the fate of Cio-Cio-San in Puccini's *Madama Butterfly* (1904).

3. Politics and Revolutionary Periods in European History

During the minority of Louis XIV there was an attempt by the regent, Cardinal Mazarin, to introduce Italian opera to Paris; Cavalli himself visited the French capital in 1660. The first establishment of a national opera came in 1669 with the foundation of the Académie de Musique by Pierre Perrin (c. 1620-1675) and Robert Camber (1628-1677). They were succeeded by Lully, who having devised music for Molière's plays, in 1672 composed the first French opera, *Les Fêtes de l'amour et de Bacchus*. Lully set a standard for French recitative, and after the model of the *ballet de cour*, gave dance a place it would always retain in French serious opera. He also initiated a tradition of spectacle in French music drama, with an emphasis on formal manners, all of which led to a certain rigidity in the *tragédie lyrique*, reflecting the glory of *le roi soleil* (the Sun King). Only the eventual appearance of a composer of Rameau's imposing qualities (1733) could really build on Lully's achievement. The flimsy gaiety of popular works emerging in contradistinction, and called *opéra-comique* in 1715, established what was to become a polarization of genres and taste in French operatic tradition.

Despite the various factions that emerged in the 18th century around the appearance of Italian *opera buffa*, and *the* rivalry between Niccolò Piccinni (1728-1800) and Gluck, it was the humble *opéra-comique* that proved the more influential genre before, during and after the huge challenges of the French Revolution in 1789. This popular type of opera served as a medium for social comment and fed into the emergence of Romantic opera. New ideas were incorporated by librettists like Jean-François Marmontel (1723-1799) and Michel-Jean Sedaine (1719-1797). They introduced contemporary philosophical trends from various different ages and countries, from different social strata and milieux, from fairy tales, from the supernatural, and from Rousseau's ideas about the dignity of the natural man.

The Terror developed the already extant topos of the Rescue opera (Grétry, *Richard Coeur de Lion* 1782; Berton, *Les Rigueurs du cloître* 1790). Under the influence of Cherubini, an Italian, and Étienne-Nicolas Méhul (1763-1817), passionate and tragic themes were promoted. These gave attention to the clarity of French declamation, a new emphasis on developing action during musical numbers, combined with a deep concern for freedom and human worth (Cherubini in *Lodoïska* 1792, Jean-François Le Seuer [1760-1837] in *La Caverne* 1793, and Méhul in *Joseph* 1807).

With the establishment of the Consulate (1799) and then the Empire (1804), a growing sense of stability, order and French hegemony once more needed the traditions of *tragédie lyrique* to express the grandiose ambitions and Neo-Classical aspiration of the times. This was embodied in Empress Josephine's favourite composer, the Italian Spontini (*La Vestale*, 1807), who wrote also to promote Napoleonic military propaganda for the Peninusla War (*Fernand Cortez*, 1811). The defeat of the First Empire and the Bourbon Restoration (1815), and the ultramontane ambitions of King Charles X, saw Rossini, another Italian immigrant, invited to Paris, where he was instrumental in bringing about dynamic new developments in the capital's operatic life. With Auber's *La Muette de Portici* (1828) and Rossini's own *Guillaume Tell* (1829), popular aspiration to freedom found prescient voice for the next Revolution in July 1830 (August 1830 in Belgium). This saw the end of the Restoration and a new plutocracy shape political and social aspiration under the bourgeois monarchy of the citizen king Louis-Philippe. French Grand Opera, already formed, now took on its definitive expression in the works of Meyerbeer and Halévy, where issues of freedom were now focused on personal liberty (*Robert le Diable*, 1831) and religious toleration (*La Juive*, 1835; *Les Huguenots*, 1836).

These social forces found a deflected, gentler expression in the renewed flourishing of *opéra-comique* which now reached its apogee in the works of François-Adrien Boieldieu (1775-1834), Ferdinand Hérold (1791-1833), Auber, Adolphe Adam, and Halévy.

The next social upheaval would be the Revolution of 1848 which shook all Europe. Already the cry for more popular freedom had been expressed in the appearance of the first really vital works of Verdi in Italy (*Nabucco*) and Wagner in Saxony (*Rienzi*) (both 1842). The oppression of peoples and their desire for liberty are voiced, be they in the guise of the ancient Hebrew people in the Babylonian Exile of 587-535 BC (the chorus "Va pensiero") or the citizens of 14th-century Rome, uniting in 1347 under a popular folk hero against an oppressive nobility (the chorus "Santo Spirito, cavaliere"). Verdi became an *ersatz* hero of the Risorgimento for a united Italy, while Wagner, much influenced by the radical ideas of Mikhail Bakunin (1814-1876), actually participated in the 1848-49 Revolution in Dresden and was obliged to flee, and to live in exile in Switzerland for 16 years.

The Revolution in Paris (March 1848) was witnessed first-hand by Meyerbeer who was then writing the incendiary first act of *Le Prophète* (1849). This depicted the Anabaptist sect, with its preachers inciting social insurrection against their oppressors, just as the movement during the Reformation had secured the German city of Münster for their millenarian movement, to initiate the beginning of the 1000 years of Jesus' return in the Endtimes (1534-35). In France the revolution lead directly to the Second Republic which in turn provided Louis Napoleon with opportunity to stage his *coup d'état* (1851) and secure the plebiscite leading to the proclamation of the Second Empire (1852).

This began a period of great prosperity and social dynamism for France, although at the cost of certain political liberties. Attention was focused on a belligerent foreign policy against Russia (the Crimean War) and against Austria (with French influence in Italy). The policy of the Opéra, still at the centre of the operatic world, was to tone down the politico-revolutionary nature of grand opera scenarios, and deliberately to encourage and intensify further the manifestation of spectacle in its productions. The operas of Halévy during the 1850s reflect this move away from the confrontational scenarios of the 1820s-30s and a return to fantastic or exotic subjects. The two ancient operatic genres of France were slowly merging into the synthesis known as *opéra-lyrique*, with key works like Charles Gounod's (1818-1893) *Faust* (1859) and Ambroise Thomas's (1811-1896) *Mignon* (1866) providing models for this fusion of the tragic and comic elements of the French musical heritage.

The burgeoning spread of colonial enterprises at the time was reflected in operas of conquest and destiny like *Les Troyens* (1863) and *L'Africaine* (1865). This period of opulence and expansion was brought to a close by the Franco-Prussian War of 1870, another example of Bismarck's provocative aggression like the staged conflicts against Denmark (1864) and Austria (1866), designed to secure Prussian hegemony and the unification of Germany. An opera like Wagner's *Die Meistersinger von Nürnberg* (1868) reflects a search for national pride in Germany's history and culture. After 1870, and the trauma of the Paris Commune (1871), the French nation began to disregard its own ancient operatic traditions, espousing more abstract forms of music (e.g. César Franck 1822-1890, and Gabriel Fauré 1845-1924), seeking new directions in the Impressionist movement in art and music (Monet, Debussy), and in adulation and imitation of Wagner (Vincent d'Indy 1851-1931; Ernest Chausson 1855-1899).

Note

[1] See Smith, Patrick J. *The Tenth Muse: A historical study of the opera libretto* (London: Victor Gollancz Ltd., 1971).

PART 4:

UNEXPECTED DEATH EXPLORED

CHAPTER 13

THE HEART OF THE MATTER

As he dressed for dinner his thoughts hung effortlessly on the moment when he would see her again, just as the gymnast already touches the still-distant trapeze towards which he is flying, or just as a musical phrase seems to reach the chord that will resolve it and is already pulling it towards itself, by virtue of the very distance between them, with all the strength of the desire that promises that phrase and summons it into being.

(Proust, *End of Jealousy: Pleasure and Days*)[1]

Introduction

Siegfried (*with calm astonishment*).
What is this about fear?...
(*forcefully*)
If it is an art
Then why do I not know it? –
Out with it? What is this fear?

Mime
Have you never felt
in the gloomy forest
as darkness spreads
to some twilit spot
and, from afar, a rustling,
humming, roaring sound draws near,
a furious booming
crashes closer.
a whirling flicker
flits around you
and, swelling and whirring,
floats towards you,
(*trembling*)
have you not felt the terror then
that, creepingly, seizes hold of your limbs?
(*quivering*)
Searing shuddering

shakes your frame,
(*with quavering voice*)
quaking and quivering in your breast,
your hammering heart is bursting?
If you've never felt all this,
then fear remains unknown to you.[2]

One of the sustaining tropes of Wagner's Ring cycle stems from Wotan's grand idea, expressed by the Sword Motif, heard for the first time at the end of Act 1 of *Das Rheingold*. In Act 1 of *Siegfried*, Wotan has informed Mime that only one who has never known fear will forge the sword from the shattered pieces salvaged from the encounter between Siegmund and Wotan in *Die Walküre* Act 3, and unify its strength. Mime realises that he has never taught Siegfried such an emotion, but he can lead him to where he will experience fear, at the dragon Fafer's lair, *Neid-Höhle.*[3]

Later, when Siegfried finds Brünnhilde, removes her breast plate, and for the first time encounters a woman ("This is no man!"—*Das ist kein Mann!*), he experiences terror, fiery magic in his heart, he staggers, swoons and trembles.[4] He imagines his mother, whom he has never seen, and he calls to her for help. Then after Brünnhilde awakes:

Around me everything floats
and sways and swims;
searing desire
consumes my senses:
on my quaking heart
my hand is trembling!
What is this, coward, that I feel?
Is this what it is to fear?[5]

Seeing Brünnhilde "incarnates for him all the love for which he has longed" and has taught him fear.[6] [**Fig. 13.1** Frau Minna]

1. Sudden Unexpected Death

By far the most mysterious of the variants of Love-Death is the sudden and ostensibly inexplicable death ensuing in a situation where excessive emotion leads to some kind of psychosomatic intensity of unendurable stress, resulting in destabilisation and ultimately cessation of vital autonomic functions. We introduced these circumstances in Chapters 1 and 5, noting origins from biblical times through to modern opera. The association with the heart is there from the beginning.

The great Greek tragedies which flourished in the 5th century BC may be referred to as the true beginnings of Western culture as we know it today. Euripides (480-406 BC), whom Aristotle considered the most tragic of the poets, wrote in the century when what can be called the ancient art of Tragedy evolved into a dramatic spectacle, uniting myth, religion and music, probing relationships between gods and men, and establishing a form of art which later evolved into medieval church music, secular dances and concerts, opera and the theatre and the cinema of today. It was a *Gesamtkunstwerk*, a unified work of art.

Euripides was credited with changing the style of Greek drama, bridging the world of the gods and humans. He portrayed his characters as speaking an everyday language but with philosophical and psychological introspection. He was a realist and an observer of things around him, living at a time of great intellectual ferment in Athens. This must have included medical knowledge and the ideas of Hippocrates (460-370 BC). His plays sift the mental registers of Tragedy, revealing an awareness of mental states, including madness, and disorders such as 'the sacred disease' (epilepsy). For Euripides the madness of epilepsy could provide a part of the tragedy for the Tragedy, the meaning behind the play being a mirror to human life. Madness was close to the surface of existence, even in a society bent on proclaiming the importance of philosophical debate and logic. In his plays, and in the Hippocratic writings, the gods are set in the background, as issues of causality shifted ground. As George Steiner put it: "Tragic drama tells us that the spheres of reason, order and justice are terribly limited".[7] [**Fig. 13.2** Melodrama]

Euripides also knew about the effects of stress on the body. No better examples are found than in his plays *Hecuba* and *The Trojan Women*. In the aftermath of the Trojan War, when the triumphant Greeks are finally sacking Troy, the terrible consequences for the defeated are portrayed. Hecuba, with the total collapse of her social structure, the death of so many of her kith and kin, and the knowledge that she, the queen and mother to the great hero Hector will be a slave, cries out: "Never before now has my heart shuddered or quailed so incessantly … Oh, my sorrow! I am fainting and my limbs are giving way", She collapses. In *The Trojan Women* she asks to be taken away, harrowed by tears, considering that no one has good fortune until dead ...

Hecuba

Yet, if god had not turned the world upside down, we would vanish into obscurity.

> We would never have given men to come the inspiration to sing of us in their song.[8]

2. Tragedy, Death and Opera

From the time of the story of Orpheus, death and bereavement have woven the thread of opera's tragic stories; we have discussed many examples in previous chapters. In this section of our book we develop our interrogation those apparently *Unexpected Deaths*, linked not only with social and political circumstances, but also with tragic incarnation.

A good starting point is with Hugo in *Undine;* in fact the hero in all the Rusalka operas provides a prototype in which the mysterious kiss is a metaphor for death, where an elemental attraction initiates a process of mortal transcendence. Of great importance to the theme of our book are the recurrent instances of such deaths in Wagner's operas, where Tannhäuser, Elsa, Isolde, and Kundry all die of such apparent inexplicable compulsion.

There are hints as to what may be happening. In Bellini's *I Capuletti e I Montecchi* (*The Capulets and the Montagues,* 1830), the sleeping potion which gives the appearance of death is given to Giulietta by Lorenzo, the Capulets' doctor. News of her 'death' arrives as Romeo and Tebaldo are about to fight a duel. Romeo takes poison in Giulietta's crypt. On awakening, Giulietta finds her lover dying, and she dies—perhaps of grief?

Mefistofele (1868), which Boito based on both parts of Goethe's *Faust,* portrays Margherita as delirious in prison after poisoning her mother and murdering her baby she has had by Faust. Mefistofele offers to save her but she rejects his help. She regains her sanity briefly and dies—perhaps of guilt?

Bluebeard, as set out by Béla Bartók in *Duke Bluebeard's Castle* (1918), is not the monster Gilles de Rais, but a sorrowing, idealistic man who takes Judith, his latest bride, home to his murky castle. She makes him unlock his secret doors one by one, and when she has penetrated the innermost secret, she takes her place behind the last door, another failure among the other wives, and dies leaving Bluebeard in his loneliness—fate or necessity?

In Janáček's *The Makropoulos Affair* (1922), Emilia Marty is the victim of a process invented 300 years before for prolonging life, and is unable to die until she finds the formula. Life has become an intolerable burden for her, with every pleasure grown stale. In the end the formula is burned, and she is able to die—perhaps of *Weltschmerz*?[9]

Daphne (1938), the opera by Richard Strauss, based on the classical legend, finds Daphne, the daughter of Peneios and Gaeafalls, in love with Apollo. He comes down to earth disguised as a shepherd, but Daphne refuses to yield to him. Leukippos, a real shepherd who is in love with Daphne, also tries to win her love. He is killed by Apollo, who reveals himself as a god, and turns Daphne into a laurel tree—metamorphosis, transfiguration or transcendence?

There are many other examples of sudden, unexpected operatic deaths, and we have listed the ones we will shortly consider in List 1 (page 169). First, we want to return to the heart, literally and metaphorically, and discuss what is referred to as the autonomic nervous system—ANS.

3. Autonomic Nervous Activity

This is not the place to digress on complicated matters of neuroanatomy and neurophysiology, but some understanding of what and who rules the heart, our hearts, is essential to our enquiry.

Neuroscience as we know it today was extremely rudimentary until the 19th century. Then it became possible for dedicated researchers, with newer technologies such as microscopes and techniques of electrical stimulation, to explore what was connected to what in the nervous system, how the connections were made, and in what way nervous impulses travelled and their energy dispersed within the apparatus of the brain and spinal cord.

The ANS as we now understand it is closely tied to our emotional states. In the past the concept of *sympathy* was used to designate an idea of a 'consensus' or an affinity between bodily organs that are not directly connected anatomically. How might action in, say, the stomach influence the heart and vice versa? The physician often described as the first neurologist, who coined the word *neurology*, was Thomas Willis (1621-1675). He also gave us the term *solar* plexus: the celiac plexus (from Greek *koilia* meaning belly), a confluence of nerves in the abdomen, being like the sun, with rays of fibres emerging from it. This emphasized our 'gut' feelings and the widespread nature of their influence. [**Fig. 13.3** *Solar Plexus*]

The expression *sympathetic response* was used well into the 19th century. Other terms for these nervous actions included *visceral* (*vegetative Nervensystem*), and in 1898 the physiologist Walter Gaskell (1847-1914) suggested the term *autonomic*. This implied a degree of independent action, which by this time included the activity of one of the major nerves leaving the skull, the vagus nerve (wandering), with direct connections of the brain to the heart.

An influential text book by Smithy Ely Jelliffe (1866-1945) and William White (1870-1937), *Diseases of the Nervous System: A Text-Book of Neurology and Psychiatry* (1915), was an attempt to unite more closely what was at the time becoming two independent disciplines, namely neurology and psychiatry. This was driven by the success of finding localized lesions in the brains of people with motor and sensory abnormalities (neurology), and, from the other side, the Freudian onslaught on our understanding of human behaviour, normal and pathological from an exclusively psychological perspective. Id and unconscious forces commanding the ego, "Reason is, and ought only to be the slave of the passions, and can never pretend to any other office than to serve and obey them", as the philosopher David Hume (1711-1776) put it.[10] Jelliffe and White united what they referred to as 'sensory-motor neurology' with knowledge of the "historically oldest portion of the nervous system, the sympathetic and autonomic (vegetative) and … the increase in our knowledge of the mechanisms that operate at the psychic or mental levels." The individual was acknowledged as a biological unit, with broadly defined goals, the nervous system being a part of the larger whole.

> Man is not only a metabolic apparatus … nor do his sensori-motor functions make him a feeling, moving animal ... nor yet is he exclusively a psychical machine … he is all three, and a neurology of today that fails to interpret nervous disturbances in terms of all three of these levels, takes too narrow a view of the function of that master spirit in evolution, the nervous system.[11]

Later investigators were able to identify some subcortical structures, such as the hypothalamus, that were integrative areas of autonomic and somatic functions, allowing for theories in which transitions from physiological to pathological states could arise.

The idea that cortical areas of the brain, usually associated with intelligence and higher cognitive functions, and subcortical systems were distinctly separated, keeping the 'animal passions' away from contaminating the *logos,* feared much by Plato and so many thinkers since, has been quite undermined by advances in neuroscience, especially from the mid-19th century onwards.

The physiologist Walter Cannon (1871-1945) stimulated the hypothalamus in animals and observed intense emotional output. He famously coined the phrase 'flight or fight' to describe an animal's response to threats. The idea that a small electrical stimulus to a part of the subcortical brain could dramatically elicit emotions was an anathema to psychiatry and of little interest to neurology, yet it provided a gateway to our understanding of the power of earlier nervous elements of our evolutionary past to command our actions. It is now known that nearly the entire cortex (the outer sixlayered mantle of the brain) is connected to brain areas such as the basal ganglia (large sub-cortical structures related to movement and emotional action), which themselves are also closely allied to what is referred to as the limbic system. The latter forms an extensive network of evolutionary very old structures. In these, integration of information coming from inside the body is united with that which comes to us through our sense from our environment, a kernel wherein 'I am' is embodied. The limbic influences on our motor system form the basis of what makes us tick, what drives us on, emotion united anatomically with motion, as it is in many languages and in life: in English the word 'motion' is six-sevenths of 'emotion'.[12]

The psychiatrist Stephen Porges (1945-) associates the emergence of empathy in primates with the developing evolutionary complexity of the ANS and an increased sophistication of what he refers to as the Social Engagement System. He describes an important mammalian anatomical feature, namely, the shift to audio-vocal communication with the evolution of the mammalian middle ear, which developed from the jawbones of earlier reptiles.[13] The human middle ear carries sound at only specific frequencies. It is naturally attuned to the sound of the human voice, although it has a range greater than that required for speech to move us. Further, the frequency band that mothers use to sing to their babies, so-called 'motherese', or child-directed speech, with exaggerated intonation and rhythm, corresponds to that which composers have traditionally used to compose melodies. Our ear-ways have direct communication with the limbic structures and basal ganglia, of much relevance for our love of music and the power of the human voice.[14]

The human brain has the capacity to transform the mechanical energy of sound into the electromagnetic firing of neurones. The human auditory system has the important capacity to detect harmonicity, and discriminates vowels, the sounds of early chant-like proto-European music. The 'vagal effect', in which the parasympathetic system dominates over the sympathetic, tones down the ANS. The use of prolonged vowel sounds of early Christian music and chanting, altered sympathetic-parasympathetic balance which enhanced reflection and contemplation, an effect which seems universal, found in all cultures. The hidden mysteries of early musical symbolism were based on the harmonies of the human voice.

Communicating musically is a part not only of our biological and cultural heritage but a feature of every human infant's experience. We are mammals, all mammals have mothers, and our mothers sing lullabies. These are sung in the higher vocal frequencies, with slower tempos allowing for the unfolding of the harmonies bringing emotional interplay between mother and baby, well before the development of words, and vocal semantic structures of language. Increasing parasympathetic tone decreases heart and breathing rates, increases social engagement, and alters facial expression. We sing because 'singing' was of selective value to our pre-linguistic, stone age human ancestors, a legacy handed down to us.

Entrainment is when two oscillating systems assume the same period ratio: synchronisation of endogenous rhythms (heart and lungs) with an exogenous (external) one is what music does to and through our bodies. Humans are the only species to entrain to music over long periods of time: future events are predicted based on past regularities. New-borns perceive and anticipate beats, we spontaneously synchronise to the beat of music— best at 50-100 beats per minute, the heart rate. Beat perception is predictive. Between-person oscillatory couplings of ANS activity among singers in a choir leads to phase synchronization both in respiration and heart rate, also recorded using mantras, in comparison to resting recordings.

Delay of musical goal, motion in the music, a rising of tension, increased attention, expectation and ambiguity, intentionality, can be trance inducing: these are fundamental to the enjoyment of music and relate to the reactions via entrainment of our physiological responses.[15]

In summary, the ANS is made up of those nerves that co-ordinate the functions of the internal body organs, essential for survival. The limbic connections directly influence autonomic activity, and the regulation of heart, lungs and internal abdominal organs are their principal demesne. These same organs transmit, via the same vagus nerve, information back to the brain. The physiology provides our interoception (a sixth sense), the central core of our 'feelings', essential to our human embodiment of the musical experience.

4. Love is a Miracle of Civilisation—Stendhal

Henri-Marie Beyle (1783-1842), a man of more than two hundred pseudonyms, and best known as Stendhal, wrote several novels, the most famous being *Le Rouge et le Noir* (*The Red and the Black,* 1830) and *La Chartreuse de Parme* (*The Charterhouse of Parma,* 1839). He wrote his own epitaph: "Here lies Arrigo Beyle, from Milan. He lived, he wrote, he loved.1783-18-", and predicted he would be famous around 1880. All was, and became, true. [**Fig. 13.4** Stendhal]

A proverbial lover of love, a lover of women, who had many love affairs, and was deeply affected by the death of his mother (1790), Stendhal wrote a rather rambling text *De l'amour* (*On Love*, 1822). This describes the delights and trembles of love that arise as the lover's imagination takes over, such that there is in nature nothing that does not bring *the* woman into attention, a developing and accreting process which he referred to as 'crystallization'.

He went every evening to the opera, such that "music, when perfect, lifts the heart exactly as when you delight in the presence of your beloved. This means that music gives what must be the most profound happiness available on this earth". The meeting of eyes, presumably often from the theatre boxes, in which "a certain lively sparkle caught flashing from her eyes, are more diamonds of crystallization for [his] encrusted branch" was captivating. "It is because we will never understand the *reason* for our feelings that the wisest men become obsessive lovers of music." "Beauty is a quality which nourishes the soul, and offered *probabilities* about a woman." He described four different types of love: Passionate; Mannered; Physical and Vain; and has a syndrome named after him.[16]

The Stendhal Syndrome and Other Variants

On 22 January 1817, Stendhal described the following after visiting Florence and viewing art in the Santa Croce Cathedral:

> I was in a sort of ecstasy, from the idea of being in Florence, close to the great men whose tombs I had seen. Absorbed in the contemplation of sublime beauty … I reached the point where one encounters celestial sensations … Everything spoke so vividly to my soul. Ah, if I could only forget. I had palpitations of the heart, what in Berlin they call 'nerves'. Life was drained from me. I walked with the fear of falling … I had reached that point of emotion that meets the heavenly sensations given by the Fine Arts and passionate feelings. Leaving Santa Croce, I had an irregular heartbeat, life was ebbing out of me, I walked with the fear of falling … [I] felt close to heaven …

The eponym Stendhal Syndrome was coined by a Florentine psychiatrist, Graziella Magherini (1989), but it has also variously been referred to as the Florence Syndrome, hyperkulturemia, or as Artattack.[17] She described 106 cases that experienced autonomic disturbances after viewing paintings and sculptures, from chest pains, sweating, anxiety, dizziness, depersonalization and palpitations to occasional episodes of paranoia and psychosis. It is suggested that Freud (at the Acropolis in Athens), Jung (at Pompeii), Dostoevsky and Proust may have had episodes of the syndrome. Dr. Bar-El and his co-workers described a variant referred to as the 'Jerusalem Syndrome' with the addition of intense religious ideas, and delusions of grandeur.[18]

Siegfried, with the flames that raged around Brünnhilde firing up feelings in his breast, sings "All hail to the mother" (*O Heil der Mutter*):

Siegfried
The light of your eye
is clear to see;
the sigh of your breath
is warm to feel;
the sound of your singing
is sweet to hear:
but what you say to me singing,
stunned, I cannot understand.
With my senses I cannot
grasp far-away things,
since all these senses
can see and feel only you.
You bind me in fetters
of anxious fear;
You alone have taught
me to dread it.[19]

Stendhal and Siegfried, captured by the mother, the female, love, fear and beauty, are overwhelmed by the enrichment of feelings. Such is the way the ANS affects us. As Schiller observed of the sublime: "It is a composition of melancholy which at its utmost is manifest in a shudder, and of joyousness which can mount to rapture ... this combination of two contradictory perceptions in a single feeling demonstrates our moral independence in an irrefutable manner."[20]

> A daze had come over his mind, he had another centre of consciousness. In his breast, or in his bowels, somewhere in his body, there had started another activity. It was as if a strong light were burning there, and he was blind within it, unable to know anything, except that this transfiguration burned between him and her, connecting them, like a secret power.[21]

Another syndrome of significance for our theme is Takotsubo cardiomyopathy, also called the 'Broken Heart Syndrome'. It is a transient dysfunction of the left ventricle of the heart, accompanied by electrocardiographic alterations associated with extreme emotional stress. It is found with higher risk in those with a background of anxiety or depression. It occurs almost exclusively in women.[22]

A variant of autonomic sensitivity recently identified is 'Autonomous Sensory Meridian Response', a tingling feeling that that begins at the crown of the head and then descends through the rest of the body. It is triggered by a gentle stimulus, which includes music.

Then there is the congenital central hypoventilation syndrome, eponymously called 'Ondine's Curse'. This is a rare condition in which sufferers lack autonomic control of their breathing and are hence at risk of suffocation while sleeping. Ondine is a variation of Undine.[23]

5. The Heart and Sudden Death

There are no reported deaths from the Stendhal Syndrome, but descriptions of sudden unexplained death, as we have noted, stem forward from biblical times. Shakespeare poses the question, "Tell me where is fancy bred/ Or in the heart or in the head?"[24] For Plato it was from the head, the part of the body nearest to heaven; for Aristotle in *De Anima* (*On the Soul*) the heart was the prime organ of the soul.[25] The Greek physician Galen (130-210), in his *On the Usefulness of the Parts of the Body*, also asserted that the heart was the organ most closely related to the soul, and the source of the body's innate heat (likened to the sun); it was easily upset if the balance between the bodily humours was disturbed.[26]

Michel de Montaigne in his *Essays* cites that St Augustine referred to someone who, when he heard a sad cry, would fall into a sudden swoon in which he would appear unconscious, later reporting hallucinations, but during this time he had neither pulse nor breath. Montaigne also commented on the women of Scythia, who if aroused and angry with a man, could kill him with a single glance.[27]

6. Our Interest in the Unexpected Deaths in Opera

From a medical perspective we are aware of sudden death, from *expected* causes. For example, this may be due to cerebral events such as a brain haemorrhage, or respiratory arrest from inhalation difficulties; sometimes drug ingestion may be a factor (notably such as cocaine. This is sometimes attributable to prescribed psychotropic agents). But in many the final event is a cardiac arrest. Akin to the *Unexpected* operatic deaths is our experience with epilepsy and the syndrome referred to as SUDEP (Sudden Unexpected Death in Epilepsy). This tragedy is still a very active area of our research, and just as attempts are being made to discover the causes of such events, we wanted to try to understand more about the *unexplained* operatic deaths which we had witnessed so often. We set about to explore the contexts of them, in the nexus of the cultural, historical and aesthetic periods when such events became more prominent. One aspect of this was the Love-Death theme, and its philosophical and musical implications.[28]

7. Opera and the Death of Women

A well-known opera director, when asked by one of us about his ending for a forthcoming production of Wagner's *Lohengrin*, merely replied "I can't stand the deaths of all those women in opera", and moved to the next question. However, this is no light matter and is lies at the heart of our attempts to unravel the apparent puzzles regarding the *Unexpected Deaths*. As must be coming clear, there is an excess number of female as opposed to male deaths: one of the few analyses of operatic deaths is that of Michel Poizat in his book *The Angel's Cry: Beyond the Pleasure Principle in Opera*.[29]

Quite dependent on the psychoanalytic works of Jacques Lacan (1901-1981), Poizat reintroduces us to the word *jouissance*. This has no adequate English translation. He equates it with 'bliss' or 'ecstasy' and it is beyond the pleasure principle.[30] It is associated with a feeling "known to opera lovers, the thrill or shiver that courses through the body in those supreme moments of musical ecstasy … tears of joy". These tears are as common or more common in men as compared to women.

Poizat describes in some detail the experiences that he and several friends had one evening on the steps of the Opéra Populaire de la Bastille in Paris and draws us into a discussion of the pre-eminence of the vocal and musical aspects of the performances of several operas. These drive the desire of the itinerant opera devotee, more than the visual moments, and the female voice "presents itself as singing, as pure music free of all ties of speech". The recitatives never evoke the *jouissance* that is the glory of the aria.

Tracing the history of the latter he writes of the "progression towards the high notes… [in opera] … uninterrupted, as witness the works of Wagner, Richard Strauss, and Berg" among others:[31]

> The strange power of singing resides essentially in the emotion it sets off. And if we look more closely, we can see that this emotion itself has its origin in the 'strongest of feelings', in those of sexual arousal.[32]

In a chapter titled "Angels, Woman, and God", Poizat reminds us of the Angel Musicians in medieval iconography, those messengers from the celestial hierarchy that sing such sublime hymns they are not interpretable, yet descending to us below on earth, their sounds are heard in churches of worship (as in the *Angel of the Smile* in Reims cathedral): high voices, transmitted from on high, from high in the larynx, as sound moves from the chest to the head, the sensation rising upwards. [**Fig. 13.5** Smiling Angel Musicians]

In his opera necropolis Poizat analyses the tragic denouements of 257 operas from 1597 to 1973. Admitting that there were many more operas composed over that period of time, and many unknown ones simply lost to time, he observed certain themes which emerged from his sample.

1. There is a very similar number of male and female deaths (circa 85), and actually more males (if dwarfs, dragons and giants are included. (Even more so if the Helen's suitors killed by Ulysses are added).
2. When considering 'characters in torment', including those afflicted by madness, unjust condemnation, and self-sacrifice in response to unrequited love, then females begin to predominate.
3. Of 85 female deaths, more than 60 occur at the end of the opera, as opposed to 40 in males.

4. Other male deaths occur at various stages, including those due to murder etc.
5. There are a dozen deaths in which the male and female deaths take place together.
6. Taken all together, 60 female deaths, but only 37 male deaths occurred in operas composed between 1800 and 1900. Of the males, 30 were 'heroes'.
7. 'Modern times' are more dangerous for males: of 40 operas that conclude with a male death, 22 were written after 1900.
8. The male characters who perish have high voices (heroes), with exceptions (Simon Boccanegra, Don Giovanni and Boris Godunov).

Wisely suggesting caution with these data, which sadly for us are rather incomplete, and recognising that the numbers do not tally very well, he is confident that they

> support the argument for the absolute pre-eminence of the high voice over all other factors that have shaped the evolution of opera, whether sociohistorical, ideological, or anything else … Perhaps there is a lesson to be drawn from these findings to certain contemporary compositions that seem to have lost the intuition that it is the processes set in motion by this pre-eminence that allow … *jouissance* to emerge.[33]

References to *The Queen's Throat* reverberate through Wayne Koestenbaum's book of that title.[34] This is his scrapbook created for his divas, diva, from the Italian *dive*, divine, a goddess; it is an extended *billet-doux* to a life in and of opera. The book is filled with his feelings and experiences as an opera queen of listening to the female voice, bringing to us the lives of his stars, their perfections and imperfections, startling us with his personal thrills and ecstasies; how he experiences onanism through their voices, via the ear.

> The diva, when she sings, exposes *interiority*, the inside of a body and the inside of a self…those operatic moments when suddenly interiority upstages exteriority, when an inner and oblique vision supplants external verity. Voice accords presence—a myth that remains compelling … the physiology of opera singing is a set of metaphors; when we hear an opera, we are listening not only to the libretto, and to the music, but to a story about the body, and the story of a journey; the voyage of voice, travelling out from hiddenness into the world

The singer's throat is queen … the 'Ur-throat'. He apologises for ending his book with death, discussing Isolde's "Liebestod … and yet opera offers nothing less".[35]

Poizat notes that the word *diva* first appeared in French in *Trésor de la langue Français* (1832), derived from the Italian, and there is no mention of the masculine *divo.* The diva appeared at that moment in operatic history with Romanticism when "The Woman [is presented] not as such but as she is fantasised according to a logic that condemns her to death".[36]

There has been quite some criticism of this perspective, notably from Clément in her denunciation of opera as the undoing of women. A feminist criticism, blaming the genre on the middle-class with elitist urges, she examines opera from the women's point of view: "The emotion is never more poignant than at the moment when the voice is lifted to die … they suffer, they cry, they die … Yet it is … life-love-death, familiar and forgettable."[37]

8. Reprise

Peter Conrad's book on love, death and the meaning of opera, and other texts we have cited, delve in one way or another into these compelling aspects of an art form that, in spite of its detractors and naysayers, continues to thrive. In the earlier chapters of this book we have laid out several operatic themes that are central to the music and narratives we are exploring, the special variant of *Unexpected Death*—surely not forgettable.

We have already presented themes of many operatic deaths, emerging from within the plots: themes of jealousy, illness, suicide and the like. We note the historical and mythological antecedents for them, but their emergence in opera, from the late 18th century to the present, has attracted very little exegesis. Females who died at the end of the drama were unusual before Bellini's *Norma.* Poizat touched on some important epidemiological facts, related to gender, time of death, and era of composition which we believe require much more consideration than has been the case to date. To causality we add another factor, the possible autonomic instability of the characters, which, as we go on to explore, could be important in determining the fatal events.

The rise of Romanticism in part coincided with a greater understanding of neurophysiological and psychological underpinnings of our emotions, providing fertile territory for composers to explore the inner as opposed to the outer reflections of the mind and the expressions of the body. In the next section, we will list the operas we have identified in which *Unexpected Death* occurs, examining the librettos and music in some detail.

List 1: Operas with *Unexpected Deaths*

Opera	Character	Composer	Year
Females (N=31)			
Undine	Undine	Hoffmann/Lortzing	1816/1845 transcendence/ redemption
La Muette de Portici	Fenella	Auber	1828 transcendence
I Capuleti e i Montecchi	Giulietta	Bellini	1830 transc in love
Anna Bolena	Anna	Donizetti	1830 transc in ecstacy of revenge
Lucia di Lammermoor	Lucia	Donizetti	1839/1840 transc /redemp
L'Ange de Nisida/La Favorite	Countess Sylvia de Linares /Lenore de Gusman	Donizetti	1839/1840 transc/redemp
Tannhäuser	Elisabeth	Wagner	1845 transc/redemp
Lohengrin	Elsa	Wagner	1850 transc/ redemp in love
La Nonne Sanglante	Agnes	Gounod	1854 music suggests transc/redemp
Rusalka	Natasha/ Rusalka	Dargomyzhsky/Dvořák	1856/1901 trans/redemp
Faust/ Mefistolfele	Marguerite/ Margareta	Gounod/Boito	186/1859/1868 transc/redemp
Tristan and Isolde	Isolde	Wagner	1865 transcendence
Hamlet	Ophelia	Thomas	1868 transc/redemp

Manon Lescaut	Manon	Auber/Massenet/Puccini	1856/1884/1893 Transcendence
The Demon	Tamara	Rubinstein	1871 transc/redemp
Götterdämmerung	Gutrune	Wagner	1876 None
Les Contes D'Hoffman	Antonia	Offenbach	1881 transc/redemp
The Snow Maiden	The Snow Maiden	Rimsky-Korsakoff	1880/81 transc/redemp
Parsifal	Kundry	Wagner	1882 transc/redemp
Le Villi	Anna	Puccini	1884 Transc
The Queen of Spades	Countess	Tchaikovsky	1890 None
Thaïs	Thaïs	Massenet	1894 transc/redemp
Pelléas et Mélisande	Mélisande	Debussy	1902 Transc
The Wreckers	Thirza	Smythe	1906 transc/redemp
A Village Romeo and Juliet	Vrenchen	Delius	1907 Transc
Elektra	Elektra	Richard Strauss	1909 transc/redemp
La Vida Breva	Salud	De Falla	1913 Trans
Die Gezeichneten ("The stigmatized")	Carlotta	Schreker	1918 None
The Makropulos Case	Emilia Marty	Janáček	1926 transc/redemp
Das Wunder der Heliane	Heliane	Korngold	1927 transc/redemp
Daphne	Daphne	Strauss	1938 transc/redemp
Males (N=19)			
Undine	Hulbrand	Hoffman/Lorzing	1816/1845 transc/redemp
Tannhäuser	Tannhäuser	Wagner	1845 transc/redemp
Rusalka	The Prince	Dargomyzhsky/Dvořák	1856/1901 Transc

Bánk Bán	Bánk Bán	Erkel	1861 transc
Hamlet	Hamlet	Thomas	1868 transc/redemp
Die Walküre	Hunding	Wagner	1870 None
Boris Godunov	Boris	Mussorgsky	1869/1874 transc
Queen of Sheba	Assad	Korngold	1883 transc/redemp
Le Villi	Roberto	Puccini	1884 transc/redemp
Pelléas et Mélisande	Prince Golaud	Debussy	1902 None
The Wreckers	Mark	Smythe	1906 transc/redemp
A Village Romeo and Juliet	Sali	Delius	1907 transc
Osud	Žîvny	Janáček	1907 transc/redemp
Der ferne Klang	Fritz	Schreker	1912 Transc/redemp
Das Wunder der Heliane	The Stranger	Korngold	1927 transc/redemp
Lulu	Dr. Goll	Berg	1937 None
The Turn of the Screw	Miles	Britten	1954 None
Death in Venice	Aschenbach	Britten	1973 transc
St Francis	Francis	Messiaen	1983 transc/redemp

Transc = transcendence
Redemp = redemption

Notes

[1] Proust, Marcel. *The Complete Short Stories of Proust.* Trans: Joachim Neugroschel (New York: Cooper Square Press, 2001).
[2] See *Siegfried*, Act 1. *Wagner's Ring of the Nibelung: A Companion.* Trans. S. Spencer and B. Millington (London: Thames and Hudson, 2000), 217-218.
[3] A full description of the changes and exchanged between Wotan, Mime and Siegfried in relation to the development of the fear theme are given in Newman, E.

The Life of Richard Wagner Vol 2 1848-1860 (Cambridge: Cambridge Library Collection, 1933/2014), 2: 333-339.

[4]*... feurige Angst, brennender Zauber zücht mir in's Herz...mir schwankt und schwindelt der Sinn!*

[5] *Siegfried,* Act 3. Spencer and Millington Ibid., 266.

[6] Newman, E. *The Wagner Operas* [*Wagner Nights*] (London: Putnam), 1961), 586.

[7] Steiner, G. *Tragedy re-considered. In Rethinking Tragedy.* Ed. R. Felski (Baltimore: Johns Hopkins University Press, 2008), 8.

[8] Euripides, *Hecuba* in *Euripides, The Trojan Women and other plays*. Trans. J. Morwood (Oxford: Oxford World Classics, 2001), 86, 87, 438. Ibid., *The Trojan Women*, 1242-1244.

[9] *Weltschmerz*, coined by Jean Paul (1763-1825), expressing a worldly weariness, an alienation of the individual from the world's travails. With things and events which at one time seemed so vibrant now worn dull, empty of all novelty; there is simply repetitive ennui. Beneath lies an awareness of suffering and tragedy. A Romantic concept encountered in the Byronic hero, in the poetry of Heine and Baudelaire, in the novels of Hermann Hesse, and the philosophy of Nietzsche and Walter Benjamin. It is often heard in music, a sentimental melancholic sadness.

[10] Jelliffe, S. E. and White, W. A. *Diseases of the Nervous System: A Text-Book of Neurology and Psychiatry.* 2nd ed. (London: H K Lewis, 1929); Hume, D. *A Treatise of Human Nature (1739). Reprinted from the Original Edition in three volumes and edited, with an analytical index, by Lewis Amhurst Selby-Bigge, M.A.* (Oxford: Clarendon Press, 1896), 217.

[11] Jelliffe, S. E. and White, W. A. Ibid., vi vii.

[12] Many languages have double meanings of equivalents to the English 'moved' and 'being moved'. For example, the German term *Bewegung* translates as both 'motion' and 'emotion'.

[13] Porges, S. W. *The Polyvagal Theory* (New York: W.W. Norton, 2011).

[14] It has to be noted that only humans respond emotionally to music (except in a different way perhaps as found in birds and some cetaceans such as dolphins). In us music can evoke tears, and lead us to dance all night. Our emotional entrainment to external rhythm is not found in any other mammal. See more on this and related subjects in Trimble, M. R. *Why Humans Like to Cry: Tragedy, Evolution and the Brain* (Oxford: Oxford University Press, 2012).

[15] Vickhoff, B., Malmgren, H., Aström, R. et al. "Music Structure Determines Heart Rate Variability of Singers". *Front Psychol.* 4 (2013): 334. doi: 10.3389/fpsyg.2013.00334.

[16] Stendhal. *On Love*. Forward A. C. Grayling. Trans. S. Lewis (London: Hesperus Press Ltd., 2009), 11, 15, 30, 34, 52, 52. Italics in original.

[17] Stendhal. *Naples and Florence: A Journey from Milan to Reggio*. Quoted in Magherini, G. *La Sindrome di Stendhal* (Florence: Ponte Alle Grazie, 1989).

[18] Bar-el, Y., Durst, R., Katz, G., Zislin, J., Strauss Z, and Knobler, H. Y. "Jerusalem Syndrom". *British Journal of Psychiatry*, 176 (2000): 86-90.

[19] Spencer and Millington, Ibid, 269.

[20] Schiller, F. von. *On the Sublime,* in *Two Essays.* Trans. J.A. Elias (New York: 1966), 198.

[21] Lawrence, D. H., 2007. Ibid, 38.
[22] In Japanese, *tako-tsubo* refers to a fishing pot used for trapping octopus: the left ventricle takes this shape in a patient so diagnosed.
[23] This is very rarely congenital and usually secondary to some developing brain pathology.
[24] Shakespeare, W. *The Merchant of Venice,* Act 3, Scene 2, line 63.
[25] Much on the relationship between the brain and religious experiences is in Trimble, M. R. *The Soul in the Brain: The Cerebral Basis of Language, Art and Belief* (Baltimore: Johns Hopkins University Press, 2007).
[26] Humoral theories dominated medical thought for over a thousand years: Blood, phlegm, and yellow and black bile were the cornerstones of the prevailing physiologies.
[27] Montaigne, M. de. *Essays*. Trans. J. M. Cohen (London: Penguin Books, 1958), 39, 45.
[28] Thurman, D. J., Hesdorffer, D. C., French, J. A., et al. "Sudden unexpected death in epilepsy: assessing the public health burden". *Epilepsia*. 55(10):1479-85; doi: 10.1111/epi.12666. Epub 2014. Rather as with the unexpected deaths in opera, those in people with epilepsy were known about for years but evaded much attention until fairly recently.
[29] Poizat, M. *The Angel's Cry: Beyond the Pleasure Principle in Opera.* Trans. Arthur Denner (Ithaca: Cornell University Press, 1992).
[30] There is a sexual connotation: *jouir* is slang for 'to come' (M. Poizat, 1992, xiii – translator's note).
[31] Poizat, M., 3, 37, 41. "In any case, between the so-called baroque pitch of around 410 Hz and the current pitch at 440 Hz there is hardly more than a half-tone difference, which in no way can account for the difference between Donizetti's stratospheric high note and Eurydice's modest plaint when Orpheus sends her back to the dead in Monteverdi's *Orfeo*". Ibid.,43.
[32] Andréossy, V. *L'Esprit du chant.* (Plan de la Tour: Éditions d' Aujourd 'hui, 1979), 122. Quoted in Poizat, Ibid., 105. Poizat makes the important point that this is not specifically gender related, noting how the deep voice of Marlene Dietrich or Kathleen Ferrier, and the high voice of the castrato or tenor alike hold fascination for the listener.
[33] Poizat, Ibid., 136.
[34] Koestenbaum, W. *The Queen's Throat: Opera, Homosexuality and the Mystery of Desire* (Boston: Da Capo Press, 2001).
[35] Koestenbaum, Ibid, 30,103,155, 156, 239. Italics in original.
[36] Poizat, Ibid., 179.
[37] Clément, C. Ibid. 1988, 2, 10, 11.

PART 5:

LIBRETTI

CHAPTER 14

SUDDEN DEATH: LIBRETTI

Ah, even now, in spite of myself I'm weeping,
Weeping for dreams that could not last.

(Henri Meilhac & Philippe Gille, *Manon*, Act 2)

Introduction

List 1 shows a list of operas which offer a perspective of the main themes of this book. Given in order of date of first performances, the operas are divided into those concerning females and males who have suffered on stage what we have called *Unexpected Death*, sudden and often for apparently unexplained reasons. In the following chapters, we look at the libretti, having divided the operas into those in which there are suggestions of disturbances of the ANS, separating males and females. We pay special attention to examples of heart arrhythmias and references to stress or strain of the heart within the context of the drama. As we explored operas possibly related to our themes, we were continually surprised at how often the relevant motifs and undercurrents were nuanced within the texts.

We must emphasise that our list cannot be complete, as there are literally hundreds of operas that have either been lost to the ravages of passing time and changing tastes, or which are still awaiting to be returned to the stage in probably a revived or rejuvenated form.

Our survey of librettos went well beyond those of List 1. For example, we did not include borderline cases. But related themes of interest are noted in the comments which follow the brief extracts, notably Redemption, transcendence, and examples of *Liebestode.* These themes are discussed in more details in Lists 2 and 3 and in Chapters 20 to 24.

There are some 'special' operas which are included in Chapters 18 to 19, which look at the same themes but either form a group, or on account of their very nature and complexities are presented in different format. These are the *Faust* operas, the *Rusalka* operas, *La Muette de Portici* and *Pelléas et Melisande*, and then the operas of Wagner.

In what follows we provide a brief synopsis of each opera with selections from libretti, notes regarding special features (ANS activity, possible other explanations for the *Unexpected Death*), comments on relevant aspects of the music and references to *Liebestode*. With each opera we have recommended a recording which we hope to be a guide to further listening. [**Figs.14.1-14**]

Females with Autonomic References

It is love's old eternal song!
(Jules Barbier & Michel Carré, *Les Contes d'Hoffmann*, Act 3)

I CAPULETI ED I MONTECCHI (1830)

Opera by Vincenzo Bellini (1801-1835). Librettist: Felice Romani (1788-1865). Translation: Alfredo Maggioni, printed by T. Brettell, Rupert Street, Haymarket. London. First performance: La Fenice, Venice, 11 March 1830.[1]

The scene is Verona, in the 13th century.

Act 1

Scene 1. The Palace

Romeo has killed Capellio's son. Tebaldo will avenge the killing to celebrate his marriage to Giulietta. Capellio wants the marriage to take place immediately, although Giulietta is ill with fever.

Lawrence
Ah, my Lord, with ardent fever—
Heavy, afflicted, and desponding,
She—thou knowest—could onlv
By violent efforts reach the sacred altar.

Romeo, in the guise of a Montecchi envoy, suggests peace with the Capuleti by the marriage of Giulietta to Romeo. The offer is refused.

Scene 2

Giulietta longs for Romeo, who comes to her, but in the name of family, law and honour she resists his requests to escape with him, declaring that she would prefer to die of a broken heart.

Giulietta
Here, with the robe of gladness clad,
I stand—adorned—
As a victim for the altar.
Oh! at the altar's feet
Could I as a victim fall.
nuptial torches,
Abhorred and fatal,
Light ye me to my bed of death.
I burn—a flame—a fire,
Destroys my life.
'Tis the power—the power of duty,
Of law, and honour.

Romeo
Ah! cruel! thou talkest of honour,
When from me thou art taken away?
This law, that thou approvest,
Is by thy heart denied.

Scene 3

The lovers are separated by their two factions, finally proclaiming:

If all hope of ever seeing each other again in life
this will not be the last farewell.

Act 2

Scene 1

Lorenzo persuades Giulietta to take a sleeping draught that will make it seem that she has died. He will arrange for Romeo (and himself) to be present when she awakes. Taking the bottle, she declares that "only death can wrest me from my cruel father".

Giulietta
Ne'er again to meet in life,
This farewell, the last shall not be,
At least, again in Heaven we'll meet.

There united we shall live,
Far from fears and human woes.
On this earth we shall leave
Every care and every grief.

Romeo retires by the secret door. Giulietta withdraws, trembling. She begs her father's forgiveness before she 'dies'.

Scene 2

Romeo and Tebaldo fight but are interrupted by a funeral procession. It is Giulietta's. Both are united in remorse, asking each other for death.

Scene 3

The tombs of the Capuleti: Romeo enters and his companions open Giulietta's tomb. He asks to be left alone, bids her farewell and swallows poison.

Alas! You, fair soul
Rising up to heaven
turn to me,
bear me with you.

Giulietta awakes, finding Romeo, unaware of Lorenzo's plan. Romeo tells her that he has already taken the poison, and dies. Giulietta, unable to live on without him, falls onto his body dead.

Romeo
Repose with her.
One grave will both receive.
Such is, O cruel fate, the nuptial bed
thou hast prepared for us.
Oh sight! 'Tis she, whom I adored.
Ah! death itself upon thy face is beautiful:
Thy lips still seem to smile,
And sweetness shed around,
In sleep thou seemest to lie.
Ah! if thou sleepest, awake,
Arise, my love, my hope,
Come, let us fly together,
Love will conduct our steps.
But thou hearest not! alas!
I only raved and dreamt.
Her eyes are closed for ever,
She never more will wake.

Now cease to flow
My bitter tears,
'Tis vain to weep,
I now must die.
No more the day
Must dawn for me,
Be this the last
Of all my sighs.
And you receive my ashes and my woes,
O tombs of my forefathers.

Giulietta
O let me die with thee.
Ah do not leave me yet.

Romeo
Do not forget our love.

Giulietta
Repose upon my heart.

Romeo
Juliet!

Giulietta
Ah! wait for me.

Romeo.
Alas! I faint, farewell!

Giulietta
He dies, Romeo, oh heaven!

(*Romeo dies, and Juliet falls upon him.)*

COMMENT

This tale has been told many times, and Bellini's opera is fairly close to that of Shakespeare's play. Giulietta is bound to be married to another, and will die with her beloved Romeo, after a plot that goes wrong. There is a hint of a *Liebestod,* but a transcendental death of Giulietta, who perhaps, even before the sleeping draft, is vulnerable to Sudden Death (mainly referenced to the heart and tremors). Bound by the power of duty, the law and honour she begs her father's forgiveness, but anticipates her death with knowing that at least she would be united with Romeo in Heaven.

Romeo's death is by suicide. In this story there are two potions (one a magic sleeping draught, the other a poison).

The cello introduction to the last act in the grounds of the palace sets the appropriately dark mood, as does the orchestral prelude that precedes Romeo's entrance and bitter recitative, "Deserto è il loco" ("This place is abandoned"), in which he laments Lorenzo's apparent forgetfulness in failing to meet him as planned. He then hears the noise of someone entering. The confrontational duel duet for Tebaldo and Romeo generates further tension. The two men begin an angry duet (Tebaldo: "With one cry a thousand men will arrive"); Romeo: "I scorn you. You will wish the Alps and the sea stood between us"). As they are about to begin fighting, the moribund mood is underscored by the sound of a funeral procession "Pace alla tua bell'anima" ("Peace to your beloved soul"). They stop and listen, only then realizing that it is a cortège for Giulietta. In a cabaletta finale, the rivals are united in remorse, asking each other for death as they continue to fight. This mood intensifies in Scene 3. Along with his Montecchi followers, Romeo enters the tomb of the Capuleti. There is an elegiac chorus as the followers mourn Giulietta's death. At her tomb and in order to bid her farewell, Romeo asks for it to be opened. He also asks that the Montecchi leave him alone with Giulietta to sing his *romanza*: "Deh! tu, bell'anima" ("Alas! You, fair soul / Rising up to heaven / turn to me, bear me with you"). Realising his only course of action will be death, he swallows poison and, lying down beside her, he hears a sigh, then the sound of her voice. Giulietta wakes, Romeo knew nothing of her simulated death and had been unaware of Lorenzo's plan. The ensuing duet takes on the quality of a joint *Liebestod.* Urging him to leave with her, Giulietta staggers up but Romeo states that he must remain there forever, explaining that he has already acted to end his life. In a final cabaletta, the couple cling to each other. Then he dies and Giulietta, unable to live on without him, falls dead onto his body. The Capuleti and Montecchi rush in to discover the dead lovers, with Capellio demanding who is responsible: "You, ruthless man", they all proclaim.

ANNA BOLENA (1830)

Opera by Gaetano Donizetti (1797-1848). Librettist: Felice Romani, based on Tudor history. First perforamnce: Milan, Teatro Carcano, 26 December 1830. [2]

Act 1

Scene 1. Night. Windsor Castle, Queen's apartments

Courtiers comment that the queen's star is setting, because the king's fickle heart burns with another love. Jane Seymour enters to attend a call by the Queen. At the Queen's request, her page Smeaton plays the harp and sings to cheer the people present. The queen asks him to stop. She laments that the ashes of her first love are still burning. Henry VIII enters and tells Jane that soon she will have no rival, she will have husband, sceptre, and throne.

Scene 1

Chorus of Knights
Is the King coming?
And she?
Her heart groans but she covers it up.
Oh, how swiftly the lightening
descends on her head!
Perhaps, poor woman,
there is stored up for her
greater shame and pain…

Jane
And who could show themselves
to be serene when they see their Queen afflicted…

Anne
Afflicted I truly am…
Nor do I know why…
an unknown, uneasy restlessness
has stolen my peace for a few days.

Smeaton
My Queen! Oh heavens!

Anna
(How that innocent boy,
how he has shaken my heart!
The ashes of my first love
are still warm!
Ah! if my heart was not open
to another attachment,

I would not be so unhappy
in this vain splendour of mine.)
Could you but read it within me!
To no gaze has it been granted
to penetrate this sad heart;
An incomprehensible
cruel fate condemns me
to sigh in sorrow.

Scene 2. Windsor Great Park

Day, around Windsor Castle Lord Richard Percy, who has been called back to England from exile by Henry VIII. Hunters enter. Percy is agitated at the prospect of possibly seeing Anna, who was his first love. Henry and Anna enter and express surprise at seeing Percy. Henry does not allow Percy to kiss his hand, but says that Anna has given him assurances of Percy's innocence. Anna still has feelings for Percy. Henry VIII tells Hervey, an officer of the king, to be the spy of every step and every word of Anna.

Rochfort
(to Percy)
(Ah! What are you doing?
Control yourself, madman.
Every eye is turned on you,
you've turned pale,
on your face is written
the disorder of your heart.)

Scene 3: *Windsor Castle, close to the Queen's apartments.*

Percy enters. Percy says that he sees that Anna is unhappy. She tells him that the King now loathes her. Percy says that he still loves her. Anna tells him not to speak to her of love. Before leaving, Percy asks whether he can see Anna again. She says 'No'. He draws his sword to stab himself, and Anna screams and faints.

A room in the castle which leads into the rooms of Anna

Anna
I was weak … I should have firmly denied him...
Never to see him … Alas! in vain
does reason advise me:
the craven heart does not listen to its voice. ...

There he is! I'm trembling! I'm cold!
Ah! you do not know that my bonds
are as sacred as they are dreadful,
that beside me on the throne
are seated suspicion and terror.

Percy
I will leave, but first tell me,
will I see you? Promise … swear.

Anna
No: never again.

Percy
Never again! Let this
be my reply to your oath.

(*unsheathes his sword to stab himself)*

Anna
Alas! Stop…I am lost.
Someone is coming … I can bear no more.

(*she falls onto a seat*)

Smeaton
She has fainted …

(*Henry enters*)

Anna
Where am I! Oh, my lord!
I see your suspicion;
but I ask for mercy,
do not condemn me, O King.
Let this oppressed heart
recover itself for a bit.

Chorus
Oh! wherever have the sycophantic crowds gone,
Behold her … afflicted and pale, she drags her feet wearily...

Anna
Then the King is firm in his resolution?
He will require so much
from my wounded heart.

God who sees within my heart,
I turn to you … judge you
if I deserve this shame.
(*sits and weeps*)
—I feel like I'm dying

Act 2

Scene 2

Henry VIII declares that Anna has made love to the page Smeaton, and that there are witnesses. He says that both Anna and Percy will die. Percy says that it is written in heaven that he and Anna are married. They are led away by guards. *Scene 3. The Tower of London*

In Anna's cell, a chorus of ladies comment on her madness and grief. Anna enters, she imagines that it is her wedding day to the king. Then she imagines in delirium that she sees Percy, and she asks him to take her back to her childhood home. She faints.

Chorus
Who can see her dry eyed
in such anguish …
Now mute and motionless like cold stone;
now at length and suddenly studying the passage;
now sad now pale with a shadow over her face;
now composing her face into a smile:
her appearance changes as often
as thoughts and sentiments are aroused in her
in her frenzy, in her grief.

(*Anna appears in disordered dress, her head uncovered sunk in deep thought. She thinks it is her wedding day*)

The sound of drums can be heard.

Anna
(*shaking herself she awakes from her trance*)

Smeaton! Come here.
Get up, what are you doing?
Why do you not tune your harp?
Who cut its cords?

Ladies
She returns to her delirium.

Anna
They convey a low sound
like the groan cut short
of a heart that dies … it is my broken heart
which sighs its last prayer to Heaven.
Hear it, all of you.

Chorus
She's dreaming.

Anna
Heaven: grant repose at last
to my long pangs
and at least let these last heartbeats
be ones of hope.

All
Let her final delirium
be prolonged, merciful Heaven;
let her beautiful spirit
rise up to your bosom.

(*Cannon shots are heard in the distance and the ringing of bells. Anna comes to, little by little.*)

All
Heaven! Spare her wounded heart this blow
which she cannot bear.

Anna
I go down into the open grave which awaits me
with pardon on my lips,
May they obtain mercy and favour for me
in the presence of a God of pity.

(She swoons)

All
Unfortunate woman … she faints … She is dying!
The victim is already sacrificed.

COMMENTS

Of all the operas we have chosen to note the autonomic system disturbances of the heroine, none is more detailed than this one. Anna displays them to others, and calls to herself for her stricken heart. Often trembling, mute, afflicted and pale, feeling cold, surrounded by suspicion and terror, weeping, disordered in dress, she faints and swoons. Anna's end is heralded by a delirium with hallucinations, delusions and swooning: she is dead before any final execution.

Even music is dead as she implores Smeaton to tune his harp, "Who cut its cords?" They convey a low groan cut short by her broken heart, which sighs its last prayer to Heaven. "Hear it, all of you".

Anna's social entanglements, her obligations not only to the King but also to Percy (at least in heaven they were betrothed), the betrayal by Jane (who too has much rending of her heart as her guilt is revealed) are too much for her weakened heart, too much for her soul, all a part of her sudden death.

The theme of loss, betrayal and imminent death is developed throughout the second act of this opera. Donizetti especially, and uniquely in his oeuvre, infuses the proceedings with an elegiac mood sustained in four great choruses that comment on the action and the developing tragedy of the Queen's fate.

During the ongoing examination of those accused, the guards note that even Jane Seymour has stayed away from Anna. Anna enters with a retinue of ladies, who tell her to place her trust in heaven: "Oh! Dove mai ne andarano le turbe adulatrici?" ("Where are now the fawning mobs?").

In Anna's cell, a chorus of ladies comment on her madness and grief: "Chi può verderla a ciclio asciutto" ("Who can watch her dry-eyed?"). Anna enters, she imagines that it is her wedding day to the king: "Piangete voi?" ("Are you weeping?"). Then she imagines that she sees Percy, and she asks him to take her back to her childhood home: "Al dolce guidami castel nation" ("Take me back to the lovely castle where I was born"). Percy, Rochefort and Smeaton are brought in. Smeaton throws himself at Anna's feet and says that he accused her in the belief that he was saving her life. In her delirium, Anna asks him why he is not playing his lute. The others join Anna in a quartet: "Cielo, a'miei lunghi spasimi concedi alfin riposo" ("Heaven grant me respite at last from my long suffering"). Donizetti used the theme from the English folk song *Home Sweet Home* as part of Anna's mad scene to underscore her nostalgic longing. The sound of cannon is heard. Anna comes to her senses. She is told that Jane and Henry VIII are being acclaimed by the populace on their wedding day. Anna says that she does not invoke vengeance on the wicked couple. She faints. Guards enter to lead the prisoners to the block. Smeaton, Percy and Rochefort say that one victim has already been sacrificed.

LUCIA DI LAMMERMOOR (1835)

Opera by Gaetano Donizetti. Librettist: Salvatore Cammarano (1801-1852), based on the novel *The Bride of Lammermoor* by Sir Walter Scott (1819); English version by Glen Sauls. Frist performance: Naples, Teatro San Carlo, 26 December 1835.[3]

Act 1

Scene 1: *The gardens of Lammermoor Castle*

Norman [Normanno], captain of the castle guard, and other retainers are searching for an intruder. He tells Henry Ashton [Enrico] that he believes that the man is Edgar [Edgardo] of Ravenswood, and that he comes to the castle to meet Enrico's sister, Lucy [Lucia]. Enrico reaffirms his hatred for the Ravenswood family and his determination to end the relationship.

Enrico Ashton
The girl is grieving.
For who, while still mourning the ashes
of her dear mother, could turn her thoughts
to a bridal bed. Let us respect a heart
cleft with sorrow and afraid to love.

Normanno
Afraid to love?
Lucia is aflame with love!

Scene 2: By a fountain at the entrance to the park, beside the castle

Lucia waits for Edgardo. In her famous aria "Regnava nel silenzio", Lucia tells her maid Alisa that she has seen the ghost of a girl killed on the very same spot by a jealous Ravenswood ancestor. Alisa tells Lucia that the apparition is a warning and that she must give up her love for Edgardo. Edgardo enters; for political reasons, he must leave immediately for France. He hopes to make his peace with Enrico and marry Lucia. Lucia tells him this is impossible, and instead they take a sworn vow of marriage and exchange rings.

It is early evening. Lucia, in great agitation, enters with Alisa.

Alisa
Lucia, you must give up
this dangerous love affair!

Lucia
He is the very light of all my days,
the comfort and the solace of my pain.
When seized in the rapture
of his burning love,
he speaks from his inmost heart
and pledges eternal faith;
I then forget my sorrows,
my anguish turns to joy …
When I am at his side it is as if
Heaven itself had opened wide for me.

Act 2

Scene 1: *Lord Ashton's apartments in Lammermoor Castle*

Preparations have been made for the wedding of Lucia to Arthur Bucklaw [Arturo]. Enrico shows his sister a forged letter seemingly proving that Edgardo has forgotten her and taken a new lover.

(Lucia enters. She looks pale and distraught, evidence of her approaching insanity. She hesitates on the threshold.)

Lucia
If only this misery and wretchedness
would end my life of suffering—
But Death will not pity me,
and I live on in grief.
A shroud has fallen on my happiness.
Both earth and heaven have abandoned me.
I want to weep, but cannot.
Even my tears have left me comfortless.

Enrico
Enough
This paper will quickly show you
What kind of cruel wretch you think you love!

(He gives her the forged letter.)

Lucia
(She reads the letter. She almost recoils with horror and dismay.)
My heart is pounding!

Enrico
You are trembling.

Lucia
God help me!
Oh, oh … this thunderstroke!
All hope extinguished. Can I live
Through this anguish? I gave him my heart …
I put my trust in him. Oh, let me die!
Now he loves someone else!

Enrico
Your bridegroom has arrived.

Lucia
I feel coldness
coursing through my veins!

Enrico
Your marriage hour is near

Lucia
My grave is near!
There is a haze before my eyes!
Ah, God … I am terrified!
Oh, heaven help me!

O God above, who listens to my cries,
who reads this anguish written in my heart,
soften my woe! Heaven be merciful.
O, gracious God, deprive me,
deprive me of this miserable life …
Ah, in this wretched plight
how welcome death would be!
(Lucia swoons)

Scene 2: *A hall in the castle*

Arturo arrives for the marriage. Lucia acts strangely, but Enrico explains that this is due to the death of her mother. Arturo signs the marriage contract, followed reluctantly by Lucia. Edgardo suddenly appears and is shown Lucia's signature on the marriage certificate. He curses her, demanding that they return their rings to each other. Lucia, whom he still loves, swoons.

Arturo
But where is Lucia?

Enrico
She will soon be here …
But do not wonder if she seems
smitten with melancholy,
for she is still overcome with grief
at the death of her mother.

(Lucia enters, supported by Alisa and Raimondo. She is very dejected.)

Edgardo
You are forsworn before both Heaven and Love.
I curse the moment
that I fell in love with you.
May you be damned forever!

Act 3

Scene 2: *A Hall in Lammermoor Castle*

Raimondo interrupts the marriage celebrations to tell the guests that Lucia has gone mad and killed her bridegroom Arturo.

Raimondo
Lucia had still the dagger in her hand
with which she had stabbed him.
She fixed me with her gaze
and said "Where is my husband?"
Then upon her pale face
here dawned a smile!
Unhappy girl!
She has lost her reason.

Lucia enters. In the famous Mad Scene beginning with the aria "Il dolce suono" she imagines being with Edgardo, soon to be happily married. Enrico enters and at first threatens Lucia but later softens when he realizes her condition. Lucia collapses.

Scene 3: The graveyard of the Ravenswood family

Edgardo is waiting for the duel with Ashton, resolved to kill himself on Enrico's sword. He learns that Lucia is dying and then Raimondo comes to tell him that she has already died. Edgardo stabs himself with a dagger, hoping to be reunited with Lucia in heaven.

Edgardo
This brief and angry struggle shall rage no more.
I'll throw myself upon my enemy's blade.
My life is a hateful burden.
Without Lucia, the Universe itself
is nothing but a desert to me!

Chorus
Her wedding was her funeral.
Her love has bereft her of reason.
This hour may be her last.
She moans and calls for you.
Love has bereft her of reason! …
She is near to her last hour …
She will never see the close
Of this dawning day!
The death-knell is tolling already.

Edgardo
Oh, now you unfold your wing
in the sight of God, my beloved soul!
Turn your calm smile on me,
Your faithful love will come to you.
Though the wrath of our enemies
has made us cruelly at odds,
has kept us apart on earth,
we shall be united in heaven!
O, my beloved spirit,
God will unite us in heaven.
I come to you!
(He draws his dagger. The others try to disarm him.)
Ah, my beloved—I come to you—
Turn to me, ah! To your faithful love—
In spite of our enemies; wrath—
And our cruel strife—Oh, beloved,
God will unite us both—
Though on earth we were held apart
In heaven we shall be one!

COMMENT

This opera, with more than one death (a murder, and two lovers departing with a *Liebestod*) portrays the most famous of operatic mad heroines. The story has light and joy embalmed by darkness, trapped as Lucia is in a political situation, although how aware she is of its complexity is not revealed. There is trickery and deception, and no sense of a path that would allow the characters to save themselves. In contrast to her mother, the recently deceased formidable Lady Ashton, Lucia seems vulnerable, and her madness does not suddenly break-out at the end of the opera. It is revealed that she may be grieving for her mother. In Act 1 she sees the ghost—the dead murdered woman drowned by a Ravenswood: as she disappeared, the water in the fountain turned red. Lucia is odd and pale at the beginning of Act 2. The references to her autonomic instability are clear.

The whole of *Lucia di Lammermoor* sustains a mood of imminent tragedy, from the opening of the very solemn and serious prelude to the final bars marking Edgardo's suicide. The register of disturbing and brutal imagery follows closely the literary model of Sir Walter Scott's novel, where a series of events and concomitant symbolism underscores the fated nature of the love interest. In the first scene the chorus of Normanno and his retainers tells the story of Lucia's rescue from the attack of a wild bull by Edgardo's timely intervention. Lucia's entry at the ruined fountain unfolds the balladesque narrative of the tragic events that happened in the past at this Mermaid's Well, how its water that turns to blood, and how the ghost of her murdered predecessor has appeared to her at this very spot. Act 1 ends with the love duet between Lucia and Edgardo which is fraught with deeply sad intimations of farewell and loss.

The Second Act begins with another doom-laden prelude, suitably initiating the duet between Enrico and Lucia, in which the terrible act of deception is played on Lucia. Her heartbroken aside ("Souffrir nel pianto") ("Suffering amidst my tears"), with its deeply affecting horns, provides a vignette of Lucia's anguished soul, the full extent of which is revealed in Act 3, with the tragedy of her wedding night. This, like the elegiac choruses in *Anna Bolena*, is narrated by the chaplain Raimondo [Bide-the-Bent] who leads the chorus in a threnody of shock and grief that prepares the way for Lucia's entry and the beginning of her great Mad Scene, where the high clear flute (or the ethereal, even eerie, glass-harmonica) provides an obbligato that serves as a correlative for Lucia's otherworldly, mortal state. Her tragic fate is echoed in the last scene in Edgardo's great aria of loss (which uses the horns again to great melancholic effect), Lucia's funeral procession, and his suicide over her bier, providing one of the greatest sustained scenes for the lyric tenor.

L'ANGE DE NISIDA (1839) / *LA FAVORITE* (1840)

Opera by Gaetano Donizetti, libretto by Alphonse Royer & Gustave Vaëz, after Baculard d'Arnaud's drama *Le Comte de Comminges* (1764) and other material; adapted and expanded into a grand opera by Eugène Scribe as *La Favorite*. First performed: Paris Opéra, 2 December 1840. [4]

Leone has fled to Nisida from Naples for fighting a duel. He is in love with Sylvia, mistress of the King of Naples: they have met before, and they admit their love. Sylvia tells Leone that they have no future. Don Gaspar, chamberlain to the King, promises to help Leone but the latter is recognized by the King and arrested. The King affirms that Sylvia will be his Queen. Leone is freed at her request, but a Papal Bull, issued for fear that the King will make Sylvia his Queen, announces that if she is not banished she will be sent to a convent. Don Caspar has a plan for Leone to escort a young orphan woman (i.e. Sylvia) to Naples, where their marriage can go ahead. This is foiled when the courtiers reveal to Leone that Sylvia is the King's mistress. Leone renounces earthly passions and becomes a monk. Sylvia, disguised as a novice, comes to beg Leone's forgiveness. After initially rejecting her, he relents and decides to renounce his vows, but Sylvia collapses and dies.

Act 1 scene 4

> **Sylvia** (*on Leone leaving her*)
> My heart, bleeding
> With countless wounds
> Breaks, yet scorns
> Laments and tears.

Act 3 scene 8

> **Sylvia**
> Go, I love you! your absence
> means death to me.

Act 4 scene 4

> **Sylvia** (*in the guise of a novice*)
> Drained of my strength by sorrow, I am going to die

Take my broken soul, but let me win forgiveness from Leone.
What do I hear? It's a vow ascending from the altar,
a soul that heaven is taking from this world.
I shall flee this somber cloister,
But … I … cannot … death chills my blood
(*She falls exhausted at the foot of the cross*)

COMMENT

L'Ange de Nisida was unperformed, and the more familiar adaptation as *La Favorite* describes the unhappy love between Fernand, a novice in the Monastery of St James, who has been smitten by the mysterious and beautiful Leonore de Gusman, mistress of the King of Castile, who returns Fernand's love. Fernand is married off to Leonore without knowing the truth of the situation, and returns disillusioned to the monastery when he learns the facts. Leonore follows Fernand to beg his forgiveness. They acknowledge their love and are reconciled, but she dies of exhaustion in his arms.

Donizetti included some brief references to the heart and the autonomic effects of leaving a loved one, but tensions in the relationship occasioned by the dominance of the King, the Church's clasp of authority and the loss of Leone are enough to suggest the reasons for Sylvia's sudden death. The issues of continence, celibacy, chastity (as in *Lohengrin*), purity and taint, deception and truth, suffering and forgiveness, love in death, are all present in this scenario. A major theme is purity of heart over and above sexual purity, and hence the nature of true love and forgiveness. Jesus' attitude to the prostitute (Mary Magdalene?) who washed his feet in the house of the righteous Pharisee (Luke 7:36-50; John 12:1-8) and to the woman caught in adultery (John 8:1-11) illustrate the moral conundrum.

The fated love of the couple is reflected especially in the music for the tenor Fernand [Leone] in his tragic betrayal. Already in Act I when he confides in Dom Baldassare, the Superior at the convent of Santiago di Compostella, his high-flown aria ("Une vierge, un'ange de Dieu"), with its high tessitura and iterative quality, suggest a tenseness in this rapture. This becomes a cry from the heart in the ensemble of Act 3, when the scorn of the courtiers leads to the revelation of the truth and the outburst of betrayal. The famous aria of Act 4 ("Ange si pure") is an elegy for a lost dream and a tacit assertion of the purity of Leonore's [Sylvia's] heart. The following duet is of fraught intensity combining recrimination, sadness, reconciliation, a vain clutching at an illusory happiness, and the slow debilitation of the dying Leonore, before Fernand's final cry of grief.

MANON LESCAUT (1856, 1884, 1893)

L'Histoire du chevalier des Grieux et de Manon Lescaut is a novel by Antoine François Prévost (1697-1763), published in 1731. From at least eight operas based on Abbé Prévost's novel, there are three we consider here: by Daniel-François-Esprit Auber (1782-1871) (librettist: Eugène Scribe, 1856); Jules Massenet (1842-1912) (librettist: Henri Meilhac & Philippe Gill, 1884); and Giacomo Puccini (1858-1924) (librettists: Giuseppe Giacosa, Luigi Illica, Marco Praga, & Giulio Ricordi, 1893). The themes are very similar.

Synopses

AUBER'S *MANON LESCAUT* (1856)[5]

First performance: Paris, Opéra-Comique, 23 February 1856.

Set in France and Louisiana in the early 18th century, it recounts the story of the Chevalier des Grieux and his lover, Manon Lescaut. In Paris, the lovers find a blissful life together, while Des Grieux tries to satisfy Manon's taste for luxury by gambling. On several occasions Des Grieux's wealth evaporates (by theft, in a house fire, etc.), prompting Manon to leave him. He is forced to join a regiment. He deserts, and both are arrested. In Act 3 Manon has been deported as a prostitute to Louisiana and is in prison. Des Grieux contrives to see her, and they escape. They flee New Orleans and venture into the wilderness of Louisiana, hoping to reach an English settlement. Manon collapses and dies, but with an overwhelming sense of forgiveness, Des Grieux realizes the depth of her devotion.

Scene 5

(*The scene is set in a desert in Louisiana, near a wild forest, the first trees of which can be seen to one side. Des Grieux comes in, pale and wounded, helping Manon, who can hardly walk*).

Des Grieux
Since yesterday, we have been
Wandering through this barren waste.
And we have lost track
Of the road.

Manon
The heat of these brazen skies

Is consuming me!

Des Grieux (*going towards the forest*)
Come! Let us seek, in the shade
Of these great forests ...
Only a few more steps!
Let's go!

Des Grieux
All around us, these vast solitudes
Spread as far as the eye can see alas! (*aside*)
I must hide from her my terrible pain!
Let her live and I shall die!

Manon
That fiery sky, shining down on our brows,
Kills me.
In spite of myself ...
I can feel
That I am dying.
No, no, forgive me,
My love, forgive.
1 am fine, I am near you.

Des Grieux
If you can, let's walk
My beloved.
Can you?

Manon
Why ... yes … I shall lean on you.

Des Grieux (*taking the flask of water*)
Here! First let this water
Quench your thirst! (*terrified)*
There is none left ... none!

Manon (*smiling*)
What's the use?
I'm not thirsty … believe me!
I am no longer in pain,
I breathe freely,
And, with a joyful heart,
I feel once more
A smile flourishing on my lips
And hope flourishing in my soul.

Des Grieux
Her breast is breathing more easily
And with a joyful heart,
I see once more
A smile flourishing on her lips
and hope flourishing in her soul.
(*Manon collapses; Des Grieux rushes to her*).
Ah! What is wrong with you?

Manon
My strength is spent, I am dying!
Go away, go away!
(*dying*)
This is the place where my life
Must flicker and end.
Go away, I beseech you,
And let me die.

Des Grieux (*forcefully*)
Through life we were united,
And death shall unite us.
At your side, my love,
I shall remain and die! (*sobbing*)
My heart is breaking!

Manon
Come, dry these tears;
I am happy, my love,
For I die in your arms,
And my soul would be at peace
... if only I could die your wife!

Des Grieux (*with great agitation*)
You shall! by the God
Who can read in our hearts!

Manon
But how? Here we are, alone
In this vast desert! …

Des Grieux
Where everything is a proof
Of the existence of a living God!
In these deep forests,
Whose thick foliage
Will be our temple.

In the face of heaven
And in front of the Lord,
Who can see us both,
Let us kneel, let us kneel!
My God, cast on us
A merciful glance!

Manon (*praying in a low voice*)
Forgive us, forgive us.
Des Grieux
You made repentance
The virtue of the guilty!

Manon
Forgive us, forgive us!

Des Grieux
Wretchedness was our punishment,
For our unlawful love.
Des Grieux
Oh God, accept our troths
And let her be my wife!

Manon
His wife! I am his wife!
This day ends
In a sweet dream!
My heart is rising
Up to the Lord!
I am his wife,
I can feel my soul,
A ray of fire,
Go up to heaven.
Yes I, your wife, shall wait
For you in heaven.
(*Her voice dies away; her head falls on her breast*).

Des Grieux
(*in despair, he throws himself over Manon's body*).
Manon! Manon!
There I am
Calling you and loving you.
Listen to me! ... answer me! Nothing!
(*grief-stricken*)
She is no longer alive,
Death has closed her eyes

Manon I must follow you
Follow you to heaven!
Here I come, here I come!
Alas, what shall I do,
Now that death has closed your eyes!
I shall follow you to heaven!
(*Enter Marguerite, Gervais, and a band of black slaves.*)

Gervais
Here they are!

Marguerite
Here they are!

Gervais (*rushing to them*)
It is our friends!
Come!

Des Grieux (*showing them Manon*)
Dead!

Marguerite and Gervais (*with awe*)
Dead!

The Chorus
Dead!
Love is ending
In a sweet dream,
And her heart is rising
Up to the Lord!

MASSENET'S *MANON* (1884)[6]

First performance: Paris, Opéra-Comique, 19 Janauary 1884.

Act 1

Set in Amiens in 1721, Guillot has ordered dinner with his friend for three actresses. The officer Lescaut comes to wait for his young cousin, Manon: Guillot makes advances to Manon but is repulsed. Chevalier des Grieux arrives, he and Manon are smitten with each other, and they abscond to Paris in a coach.

Des Grieux
O Manon, do not be afraid; Rely on me!
I am the one who is to blame.

Oh my dear love, you're trembling so!

Act 2

Des Grieux and Manon are in Paris. An unknown suitor sends her flowers. Des Grieux goes off to post a letter asking his father's permission to marry Manon. He returns, dreaming of an idyllic life with Manon. She hears some news from the elderly Count des Grieux, telling her that his son will enter holy orders. Des Grieux is abducted by his father's agents.

Des Grieux
Oh my dear love, you're trembling so!

Manon
I'm filled with doubt, I'm so afraid,
And my heart must be breaking!
(*very troubled*)
Ah, even now, in spite of myself I'm weeping,
Weeping for dreams that could not last.

Des Grieux
Your hand … it is trembling

Act 3

Cours la Reine, Paris. The promenade celebrates a holiday. Des Grieux is due to take holy orders, Manon does not believe he has forgotten her. She seeks him out at the seminary of Saint-Sulpice, and exerts her charms on Des Grieux, pressing his hand, and they rush off together.

Manon
Tell me now, Des Grieux, am I still your beloved?

Des Grieux
Manon, siren and sphinx calling man to destruction,
Many women in one, how I love you and hate you!
Gold is your heart's desire, and pleasure is your god!
But yet, in spite of all, oh, how I love you!

Act 4

Manon and Des Grieux arrive in Paris seeking to improve their fortunes gambling. The Chevalier plays against Guillot. Losing every hand, Guillot accuses Des Grieux of cheating. He is arrested, and Manon is threatened with deportation.

Act 5

The road to Le Havre. Manon is to be deported to Louisiana. An attempt is made to rescue her. Manon falls exhausted into Des Grieux' arms. She asks for pardon, murmuring that now she can die in peace. Her lover tries to rouse her, assuring her of his forgiveness and love, but Manon dies in his arms.

Manon
Oh! I feel the light
Of a pure flame illumines me.
I see days of happiness to come!

Des Grieux
Yes, Heaven itself
Forgives you. ... I love you!

Manon
Oh! I can die then!

Des Grieux
Die! ... .no, live!
And together, free from danger,
We both shall travel this happy road!

Manon
(*as in a dream, leaning on Des Grieux*)
Yes. ... I can still be happy!
We will speak of the past ...

Des Grieux
Yes, a charming dream,

Manon
(*her hand at her heart*).
I stifle! ... I am fainting!

Des Grieux
Be brave! ... see the night is falling!
There's the first star! ...

Manon (*looking at the sky*).
Ah! the pretty diamond!
(*smiling*).
You see, I am still a coquette!

Des Grieux
Come! All is ready
For our liberty!

Manon
No! Sleep overcomes one!
A sleep without awakening!

Des Grieux
Let us go!

Manon
(*With infinite tenderness*)
I love you!

Des Grieux
(*trying to lead her off*)
Manon!

Manon
No!... No!...
This kiss is a last farewell.

Des Grieux
No! ... you shall not die! ... Hear me!
Remember!
Is it not my hand that presses your hand?

Manon
(*falling asleep*).
Do not wake me!

Des Grieux
Has it no longer a caress for me?

Manon
Cradle me in your arms!

Des Grieux
Know my voice through my tears!

Manon
Let us forget the past! ...

Des Grieux
I have forgiven you!
The future smiles for us!

Manon
Ah! Can I forget those unhappy days;
Those sad loves!

Des Grieux
They have flown from my memory!

Manon
No ... I am dying! ... I must!
It is pardon ...

Des Grieux
No! I will not believe it ...

Manon
(*as if falling asleep*).
And this is the story
Of Manon Lescaut.

(*She dies*).

PUCCINI'S *MANON LESCAUT* (1893)[7]

First performance: Turin, Teatro Regio, 1 February 1893

Act 1

Des Grieux is in Amiens, occupied with preparations for his departure but captivated by Manon, who is distressed, on her way to a convent. He is warned that his new-found love will soon leave him. Edmond (a student) notices his concern and agrees to help him rescue Manon by abducting her himself. The two lovers flee in the carriage hired for Geronte (enamored of Manon). Lescaut (Manon's brother) tries to calm Geronte's anger by offering to find Manon and take her to him.

Des Grieux
One day she will appear,
Divine and haunting
And her beauty will enchant me.
Then love will find me, forever bind me …
May I ask you to hear me …

I was strangely enchanted
when I first saw you near me …
I even thought that I had seen you before.

Manon
I surrender.
when the darkness will conceal us I shall come…

Act 2

Manon, high-strung and irritable, is now Geronte's mistress. Manon remembers her happiness with Des Grieux. Lescaut tells her that Des Grieux is gambling and hopes to win her back. Lescaut warns them that Geronte has summoned the police to arrest Manon.

Manon
Ah, entice me no longer!
Let me dream till the end of time!
There's a silence deeper than the tomb,
There's a silence, cold and grey and hopeless!
He, with his arms around me in love had bound me
And love had found me,
And in his burning kisses I forgot the world,
But now this is over!
Once more to hold you,
To drink your burning kisses,
As united in rapture we were!

Des Grieux
Manon, I am close to dying …
I, your slave now,
sliding down into Hell and perdition.

Act 3

Manon is awaiting deportation to America in a prison near Le Havre. Lescaut and Des Grieux are ready to free her: Lescaut has bribed a prison guard but the plan fails. Manon and Des Grieux say sad farewells, but then he is allowed to go with her.

Manon
Frightened, I tremble for you …
Ah, I am afraid that something will happen.
Evil foreboding has gripped my heart.

Act 4

A vast plain in Louisiana. Manon and Des Grieux are alone, having escaped from prison. Manon, pale and exhausted, leans on Des Grieux who can barely hold her up. Manon delirious. She faints, and as the sun sinks over the horizon, dies his arms.

Manon (*her voice becoming weaker*)
I cannot … forgive me!
You are strong and will live
But I … can bear it … no longer …

Des Grieux
You're suff'ring!

Manon
Forgive me: All I need is a moment …
one moment only …
Oh stay beside me …
stay close ... stay near me, my love!
(*She faints*).

Des Grieux
Manon! ... Hear me, my love!
Why are you silent, my darling? ...
Manon … oh answer! ... oh speak! ...
Nothing! Oh God, my God!
Manon, will you not answer me?
(*Manon looks at Des Grieux almost without recognizing him. He gently takes her into his arms*)

Manon
I'm thirsty, I am dying …
Bring me some water … some water!
Lonely, forlorn, my life is ending!
And darkness is descending …

Ah, It's all over!
Asil of peace now the grave invokes …
Oh God, I do not want to die! My love,
come help me!

(*Des Grieux rushes back from getting water, just in time to catch the falling Manon in his arms*)

Ah, in your arms, my darling, once more
Enfold me!

Manon
I'm dying! Night of eternity,
The darkness has descended.
(*her voice becoming weaker, transfigured by delirium*)
The time if fleeting … kiss me!

Des Grieux
Oh Angel come down from Heaven,
Oh star that will guide me forever! ...

Manon
Here ... here … come close to me …
Ah, let me touch you …
Come here … like this … kiss me! …
Darling, come near me! Now I can feel you!
Alas!
(*she collapses*)
Stay near me! ... I cannot see you …
I shudder …
(*smiling, her eyes closing*)
Ah, how she loved you, your Manon!
Remember …
When first we met I was young and lovely …
And now I have to die ….
All my songs will shortly be forgotten …
But my love … shall never die …
(*Manon dies. Des Grieux, crazed by grief, collapses over her dead body*)

COMMENT

Manon is a young woman (a poor girl) who becomes the mistress of several men, although in Auber's opera they are amalgamated into one man, the Marquis d'Héringy. In the versions of Auber and Puccini, the end is in Louisiana, with Manon dying in the desert. Massenet sets the end in Le Havre. Whether a whore or an angel, she is for Massenet 'Manon' (giving a name akin to Carmen, or Aida, Emma Bovary or Nana). She is given much more to desperate passion by Puccini. Des Grieux is a man under the spell of a woman, and both are driven to sensual passions, material gain and their downfall (*cortese damigella*): Des Grieux cannot resist her. Puccini also raises the issue of a premonition of Manon by Des Grieux, and their duet in Act 2 shows the harmonic influence of the love music in *Tristan und Isolde* of 30 years earlier. Manon seems to have been too fragile a person to have withstood the times she was born to, blown about, eventually going west literally and metaphorically. The references to autonomic and physiological instability are present in all three librettos, especially Massenet's and Puccini's. Forgiveness and resolution in Auber's version would seem at least to hint at a redemptive finale.

Manon's death scene is of an unusually serious nature for Auber, with a dignity and tenderness that hardly finds its parallel in the scores by Massenet and Puccini. It was felt in the 19th century that one could only regret that such talent, grace, and skill had been employed in such an ephemeral manner, and with such prodigality on such "an unworthy subject".[8] The last scene, devoted to the death of Manon and the despair of Des Grieux, is a unique passage in Auber's opera, and provided the composer with the opportunity of writing a type of dramatic symphony, powerfully expressive in its simple grandeur and real emotion.[9] The approach of death is reflected in the melodic contours of this scene. Melodies with non-triadic leaps, beginning at a low point and ascending, alert one to something unusual in Auber—as in one of his most moving outbursts "Dans ces déserts la terreur" ("The terror in this wilderness"). The scene is devoid of any sentimentality and tackles the sentient target of death with fixed intensity.

Massenet's score is rich in melodic invention, with a profusion of motifs, and an emotional flow of real power. The music he wrote for Manon appeals deeply to the heart, and sustains a consistent musical portrait of youthful vulnerability and impending dissolution. This is present in her first appearance with its affecting motif, in her farewell to the little table where she and Des Grieux have shared their meals, in the erotic power of her renewed blandishment of Des Grieux in St Sulpice (where she squeezes his wrist and hand), and in the death at Le Havre. All her old emotional tropes emerge in dying exhaustion as she once again presses Des Grieux's hand, and the reminiscences of earlier times intensify the pathos of her exhaustion and death.

Puccini's characterization of Manon presents a more adult and worldly-wise figure, as is also the portrait of Des Grieux. Manon's music has a sumptuous erotic vividness and veristic power, as revealed in the love duet which has a kind of maturity that is rather Wagnerian. Manon's music does not depict the vulnerability of Massenet's nor yet the emotional clarity of Auber, and does not stir the emotions in the same way. The death scene in the wilderness of Louisiana is extended in the extreme. The physical dissolution becomes a prolonged metaphor of Manon's fated moral disintegration.

THE DEMON (1871)

An opera by Anton Rubenstein (1830-1894). Libretto: Pavel Viskovatov, based on a poem by Mikhail Lermontov (1814-1841). Translation by Juan Gherzi, Montevideo, Uruguay. First performance: St Petersburg, Maryinsky, 25 January 1875.[10]

Prologue

Scene 1

During a storm in the Caucasian mountains a chorus of evil spirits call upon the Demon to destroy the beauty of God's creation. The Demon sings of his hatred for the universe and rejects an Angel's plea for him to reconcile with heaven.

Chorus of Infernal Spirits
The demon is flying,
The storm, to us, from the darkness
Seems to be calling.
We are waiting impatiently
From above, the order
To roam wildly the earth
In a whirlwind of destruction.

Act 1

Scene 2

Tamara, awaiting her wedding with Prince Sinodal, is by a river. The Demon, ranging through the world, sees her and falls in love with her. He promises her that all the world will worship her if she returns his love.

Demon
What do I see?!
I am confused
By long forgotten dreams.
With angelic beauty radiate
Her celestial traits.

Tamara
Something strange suddenly
oppresses my breast,
I cannot breathe,
As if some danger threatened me!

(*Tamara, deeply shaken, looks around her, trying to find out who has addressed her; she finally detects the Demon, who immediately vanishes.*)

Nanny
What is the matter, the poor little one is trembling,
She has grown pale, and has a wild look.

Tamara
My dears, there he was, right there he stood,
Enchanting songs he sang to me,
Passionate words he directed to me.
Divine joys he promised me!

Scene 3

Prince Sinodal is on his way to Prince Gudal's court for his marriage to Tamara, but is delayed by a landslide. The Demon appears and avows that the Prince will never see Tamara again. The Prince and his attendants are attacked by Tatars, and he is mortally wounded. Before he dies, he tells his servant to take his body to Tamara.

Act 2

Scene 4

At the festivities, a messenger announces that Prince Sinodal has been delayed. Tamara senses the presence of the Demon and when the Prince's body is brought into the castle she is overwhelmed by grief and keeps hearing the voice of the Demon and his promises. She urges her father to let her enter a convent.

Tamara (*softly to her nanny*)
Ah, nanny, nanny, I am afraid, I am afraid!
All night I did not sleep, and the spectre again
Was standing before me.
Passionate whisperings I kept hearing! ...
(*looking at her dead bridegroom*)
My prince, wake up, Wake up for an instant!
Tamara is close to you, get up!
Ah, get up for this day so full with joy! …
But you are cold… Your lips are silent …
Then bury me with him, yes, bury me!
I had loved him, for so many years—since my childhood.
Ah, it is the end of everything, the end of love's dreams,
The end of the heart luminous raptures!

They ceased like a fleeting dream my hopes, my joys.
What could I now hope for! Only the grave,
Only the grave can bring me together with him!

Demon (*to Tamara*)
He is far away, he will never know,
He will not value your anguish;
Heavenly light now caresses
The ethereal look of his eyes;
He listens to songs from paradise …
What are life's petty dreams,
What the sighs and tears of a young maiden
To a guest of the celestial regions?
But you, I, the free son of the ether,
Shall take to the space beyond the stars;
And you will be queen of the universe,
My eternal friend!

Act 3

Scene 5 In front of the monastery. Night with moon.

The Demon, believing that his love for Tamara has opened his spirit to goodness, intends to enter the convent where she is now living. An Angel tries in vain to stop him.

Scene 6

Tamara is constantly troubled by thoughts of the Demon, who comes to her in her dreams. The Demon now appears in reality, declares his love for her and begs her to love him in return. Tamara tries to resist her attraction for him but fails. The Demon kisses her, and the Angel suddenly appears and shows her the ghost of Prince Sinodal. Tamara, horrified, struggles from the Demon's embrace and falls down dead.

Tamara's cell. It is night.

Demon
I am the one, whom no one loves
And all living souls curse;
I am the king of knowledge and of freedom,
The enemy of heavens, the evil of nature …
But, you see, I am at your feet!
I brought to you as token of affection
The pure prayer of love,

To goodness and heaven
You could return me with a single word.
Dressed with the holy cover of your love,
I would present myself there
As a new angel under a new splendor …
Oh, only listen to me, I beg,
I am your slave, I love you!

Tamara
Tell me, why, why do you love me!
Ah! By a fatal poison
My weakening mind is seized!
Your words are fire and venom,

Demon
I love you with a passion not of this earth,
In a way you cannot love:
With all the ecstasy, all the passion
Of an immortal thought and dream.
In my soul, since the beginning of the world,
Your image has been impressed,
In front of me it has been hovering
Up into the emptiness of the eternal ether.
Long ago disturbing my thought,
I heard the sound of your sweet name;
In the days of bliss in paradise
It was you alone who were lacking.

Tamara
Fear is in my soul, and hell in my heart!
Unhappy me, unhappy me!
O heavens, help me!
I am exhausted …
(*to the Demon*)
I am in your hands …
But be merciful …

Demon
Oh, instant of love, instant of renovation!
(*The Demon kisses Tamara. The Angel appears.)*

Angels
Tamara!
(*Tamara utters a cry and falls dead.*)

Epilogue and Apotheosis

The Angel proclaims that Tamara has been redeemed by her suffering, while the Demon is damned to eternal solitude. Tamara's soul is carried to Heaven accompanied by angels.

COMMENT

The tussle between good and evil, between the Devil and God, between beauty and temptation, and the final resonance of a transcendental bliss—at least for Tamara—are themes explored here. There is the Demon's declaration that his soul, since the beginning of the world, had the imprint of her image within the emptiness of the eternal time and space: even her name was known in Paradise.

Midnight is the time of the devil's influence, as it is with the Demon here, the moon again visible in the songs. The kiss, just before Tamara's assumption into heaven, alters everything (like that of Kundry in *Parsifal*). As with Kundry, Tamara is redeemed. Tamara knows that only the grave can bring her and the Prince together, a repetitive theme in the operas of Sudden Death. The Demon yearns for love, but there is no reconciliation with God, and although he can enter the convent, and overcome the Angel's resistance, his goal is defeated. The Demon, abandoned and alone, will never again experience the passion and power of love.

The Demon (like Faust) is drawn to the purity and innocence of Tamara (like Marguerite), and seeks out Tamara even in her monastic seclusion. Tamara is not able to resist her deep attraction for him. He kisses her, but when an Angel reveals the ghost of the murdered Sinodal, she frees herself from the Demon and dies a Love-Death.

The theme is Faustian, with the Demon embodying a *Weltschmerz*—explored by Goethe, Byron, Nikolaus Lenau (1802-1850) and Mikhail Lermontov—with the Demon and Tamara showing elective affinities despite their different moral locations. The plot, marked by betrayal, murder, and suffering, leads to celestial transcendence. In the last scene the Demon's pleading is eloquent and even rhapsodically through-composed, Tamara's resistance strangely muted, indicating her intuitive attraction. The distant chorus of nuns, in Russian Orthodox modality that periodically underscores the colloquy, is a reminder of her holy calling and a prelude to her angelic 'rescue'. This is an act of grace, not a freely undertaken moral stance. Her celestial elevation is directly related to that of Marguerite in the Faust story and its various musical settings, with the multiple harps and the high-lying choral parts. The ending, both thematically and musically, also has analogies with that of *Robert le Diable* and looks forward to *Parsifal*. Rubinstein also used similar musical tropes in his biblical operas *Moses* (1887) and *Christus* (1888).

LES CONTES D'HOFFMANN (The Tales of Hoffmann) (1881)

Opera in three acts by Jacques Offenbach (1819-1880). Librettists: Jules Barbier & Michel Carré, based on three short stories by E.T.A. Hoffmann (1776-1822): *Der Sandmann, Geschichte vom verlorenen Spiegelbilde, Rat Krespel*. English text by John Gutman. Frist performance: Paris, Opéra-Comique, 10 February 1881.[11]

Prologue

A crowd of merry students are assembled in Luther's tavern. In the adjacent opera house a performance of *Don Giovanni* is about to start. La Stella, Hoffman's latest idol, is to sing in it. He is dejected and sings for his friends the old song of the dwarf Kleinzach, whose legs made a noise, 'crick-crack'. He is transported by the vision of one of his past loves and asks his friends whether they would like to hear the story of his three loves. Councillor Lindorf has been watching the scene. He intends to take Stella away from Hoffmann. Hoffmann begins: "The, first I ever worshipped was called Olympi".

Lindorf
When it comes to the role of the lover,
I know that I rouse only pity,
and yet I am three times as witty
as the devil!
My eyes are throwing sparks of light,
and in my aspect physical
there's something rather quizzical,
and, therefore, at my sight
you feel something almost metaphysical!

Hoffmann
My mistress? ... Stella, yes—
Three loves I had, never requited—
in Stella I find them united:
an artist—a young girl—a lovely
courtesan ...
not one mistress, my friends—
there were three that enthralled me:

each one of them was an enchantress,
capricious—haughty—one so frail ...
Would you like me to tell you
a mad lover's tale?
The first I ever worshipped
was called Olympia ...

Act 1

Spalanzani, a famous physicist, is giving an elegant party to celebrate the debut of his daughter Olympia. Hoffmann is among his guests. He has fallen desperately in love with Olympia, who charms him with her beauty, her grace, and her singing. Dr. Coppelius (alias Lindorf) enters. He has been cheated by his friend Spalanzani and in revenge, has come to destroy Olympia who, as he well knows, is nothing but an automaton invented and built by Spalanzani with Coppelius's assistance: he supplied her eyes. Olympia sings, and whenever her voice grows weak, Cochenille touches her shoulder and the sound of a winding spring is heard. When Hoffmann rushes off to Olympia's room, he finds only the shattered remains of a lifeless puppet.

Coppelius
I am Coppelius,
All these tools of my science
will make giants
of my clients!
(*shuffling his collection of various eyeglasses*)
These glasses make you see
the world as black as night;
put those on—
everything looks as white as ermine!
Use this pair everyone looks like vermin!
Through those the world appears
a world of delight.
My eyes penetrate the secrets of the kindest and
the meanest!
Yes, my eyes can read the heart
who never had one from the start!
If you want to know how she treats
you, if she loves you, or if she cheats you,
or if you want to fool your sight,
that white is black and black is white,
buy your eyes from me,
and then wait and see!

(*Hoffmann, with a pair of Coppelius's glasses glances, gazing through the curtains*)

Hoffmann (*Alone with Olympia*)
We are ... alone!
And I can say: I love you dearly.
Olympia of my heart:
let me look into your eyes,
and let them give me hope
that you will answer to my sighs ...
Our love bears us to heaven,
heavenward
where but joy and eternal love remain.
Love like the stars above you
will break through darkest night.

Act 2

Venice. A palace on the Grand Canal.

In Venice, Giulietta, a beautiful courtesan, is giving a party for her admirers, among them her present suitor, Schlemihl. Dappertutto (alias Lindorf) through the means of a magic diamond, compels Giulietta to make Hoffmann fall in love with her. As a token of his love she demands from him that he give her his reflection, just as earlier she had stolen Schlemihl's shadow. When Hoffmann asks Schlemihl to hand him the key to Giulietta's room, Schlemihl refuses and is killed in the ensuing fight. Hoffmann takes the key from Schlemihl's lifeless body but he finds Giulietta's room empty. He returns just in time to see her gliding away in a gondola, in the arms of Pitichinaccio.

Giulietta
A silvery kindly moon smiles on all our embraces.
The water's lazy tune echoes from the lagoon ...
(*to Hoffman as she takes a mirror from the table*)
As token of your love,
I want your own reflection
as seen here in this glass,
with your face and your features ...
Whether wisdom or madness—
you will do this for me!
If cruel fate should make us part,
give me, for my soul's consolation,
the dear reflection of your heart.
Still today we are earthbound,
but tomorrow, tomorrow we'll be free!

Hoffmann
My heart is filled with trepidation,
Love is damnation.
Yet my reason reels under her spell.
If her eyes shine in jubilation,
It's the triumph of darkest hell.

Chorus
Dream of love—enchanted night,
a night for sweet embraces.
Sky so dark and stars so bright
look down on our delight.
Night of endless delight ...
night of tender embraces ...
night of love—enchanted night!

Act 3

Antonia's father Crespel has brought her back to Munich so that she can forget her love for Hoffmann. He has begged his daughter to give up her singing since it was this that caused her mother's death. But Hoffmann has followed them. Frantz, the servant, announces the visit of Dr. Miracle (alias Lindorf). He allegedly has come out of anxiety for Antonia's health; but Crespel wants to prevent him from seeing his daughter since it was he, Miracle, who killed Crespel's wife. However, Miracle has a way of entering rooms without using doors. He uses his evil magic power to induce Antonia to sing again, finally even by conjuring up a portrait of Antonia's dead mother to come to life, to assist him in his devilish enterprise. When Hoffmann returns, he finds Antonia dead.

Antonia (*seated at the piano*)
Flown away, dove that I cherished ...
Oh, memory of love, of love that long has vanished …
how my heart aches for you.
Like a cloud, floating and gliding,
thus my song soars above,
if you know where he is hiding,
little cloud, give him my love.
And though the world should be
Dividing what higher fate has joined above,
ever yours is my heart,
ever yours is my love.
I thought I had heard the voice of my mother ...
My heart, while I sang, replied to her song.

Hoffmann
And in my heart of hearts
I knew you thought of me.
Now that at last I found you
I'm folding my arms around you:
When we are lovers, we are divine—
I am yours, you are mine!

Antonia and Hoffmann
And even the darkest night
can never darken this light—
you and I will be in love with love.
Song of my love I will sing you again
with words so tender: It's love's
old eternal song!

Dr. Miracle (*gesture of feeling her pulse*)
Come on—and give me your hand.
(*looking at his watch*)
Hush! Let me count her pulse.
Her pulse—irregular and fast ...
bad indication. Now sing!

Crespel
Not that, not that!
I will not let her sing.
(*Antonia's voice is heard*)
Do you not see how she blushed,
how her eyes shine with joy,
how she raises her hands
to her passionate heart?

Hoffmann
As love grows deeper and grows stronger,
There is nothing in my life that matters any longer!
(*the portrait of Antonia's mother seems to come to life*)

Antonia
Then love will be my song? I have agreed—
I'll never sing again ...

Dr. Miracle
It's her voice! Obey her word: she is your mother!
who has left you a gift
that the world must not lose.
Antonia—your mother is calling—

Sing for your mother, Antonia!
(*she falls—dying*)

Antonia
It is love's old eternal song! (*she dies*).

Epilogue

In Luther's tavern. The students have listened to Hoffmann's tales. Councillor Lindorf is pleased to see a drunken Hoffmann. The performance of *Don Giovanni* has finished and Stella arrives. Lindorf rushes off with Stella, Hoffmann is too drunk to get up from his table.

Hoffmann
And thus ends the tales of Hoffmann,
mad as a dream: I shall always cherish
those loves—strange though it may seem ...
Oh—I am mad, but madness alone is
divine!
Let us drink, let us drown all our sorrow in wine!
So let us be raving, leave reason behind us —

COMMENTS

In this opera, E.T.A. Hoffmann, the German Romantic poet, tells the sad story of his three loves, a puppet, a prostitute and a performer. All three were unhappy and all three were destroyed by an evil spirit. Transported by alcohol, there are drinking songs here, induced madness and contempt for reason. Eyes are beguiling and evil (Lindorf) and can be made of glass with an ability to change light to dark, and the beautiful to the ugly. There is magic, mirrors of transportation which lead to the loss of the soul. Olympia never existed, but was pulled apart in an act of financial revenge; Giulietta yields her love to the spell of magic and fades into the dark waters of Venice; Antonia must obey the dictates of her father, yet her voice is too exceptional to be silenced: her mother's beauty returns to her as she sings, and she dies. Dr. Miracle has already picked up her heart arrythmia, too much for Love's old eternal song. Antonia sings herself to death as, in a sense, Isolde does at the end of Wagner's opera.

Offenbach's masterly score is able to provide a symbolic fusion of the apparently disparate elements of the various short stories by E.T.A. Hoffmann. This is not so much by recurrent themes but rather in successful focus on key ideas in the three central tales, and the engagement of each with death in some form or another. If the Guiraud arrangement is followed, the stories move from a mechanical delusion (Olympia the doll), through the loss of identity or 'reflection' (Giulietta the courtesan), to actual death through illness and artistic endeavour in a rapturous if fraught love-death (Antonia, the self-sacrificing singer). In each case it is the recurrence of the evil genius (whether as Counsellor Lindorf, the inventor Dr. Coppelius, the magician Dappertutto, the destructive Dr. Miracle) who initiates the fatal temptation of the hero and the destruction of his object of (misplaced) love. The drinking song (Prologue), the waltz (Act 1) , the barcarolle (Act 2), and the *obbligato concertante* (Act 3), respectively become musical metaphors for the apprehension of some sort of loss, dying or death that comments on illusion and reality.

THAÏS (1894)

Opera in three acts by Jules Massenet. Librettist: Louis Gallet, based on the novel by Anatole France (1890). First performance: Paris Opéra, 16 March 1894. [12]

Act 1

Scene 1

The Thebaid Desert

Palemon and the Coenobite monks await the return of Athanaël from Alexandria. He arrives with the news that debauchery reigns in the city, led by the courtesan Thaïs. He recalls how as a youth he had been tempted by Thaïs but had resisted and joined the Coenobite order. Now he wishes to reclaim Thaïs from sin. He has a vision of the near-naked Thaïs on stage as Venus, from which he wakes in terror.

Athanaël
(*who has been gradually awakening, now rises full of terror and anger*)
Shame! Horror! Eternal darkness!
Lord, help me!
Merciful God, praise be to thee!
I have understood the lesson of the darkness,
and I shall give her back to thee for life eternal!

Scene 2

The terrace of Nicias's house in Alexandria

Thaïs arrives and Athanaël tells her that he has come to teach her "contempt for the flesh and love of pain". She taunts him with a parting shot: "Dare to come, you who defy Venus!"

Act 2

Scene 1

Thaïs expresses dissatisfaction with her empty life, and muses that one day old age will destroy her beauty. Athanaël enters and tells her that he loves her according to the spirit not the flesh, and that his love will last forever instead of a single night. He nearly succumbs to her physical charm, but tells her that if she converts, she will gain eternal life.

Thaïs
(*white with terror, holds her hands together, crying and moaning, and throws herself at his feet.*)
Tell me what you want of me. O no!
For pity's sake, be silent.
I chose neither my fate nor disposition
and it is not my fault if I am beautiful.
Have mercy, do not make me die!
Ah, I fear death so much!

Athanaël (*eagerly*)
No. I have told you, you shall live the life eternal;

Thaïs (*with ardour and joy*)
My burdened soul feels refreshed!
I shudder and remain spell-bound! ...
Ah! What power is this he has?

Scene 2

Thaïs has resolved to follow Athanaël into the desert. He orders her to burn down her house and possessions in order to destroy all traces of her wicked past. She agrees but asks if she can keep a statuette of Eros, the god of love. Nicias, with a group of revellers, appears, and threatens the couple. Athanaël smashes the statuette. They escape.

Thaïs
I will keep nothing from my past, nothing but this ...
(*She takes the statue in her arms and brings it back to Athanaël.*)
This ivory statue,
this child, wrought with ancient and marvellous
art, it is Eros, it is love!
Consider, O father,
that we cannot treat him harshly.
Love is a rare virtue.
He forbids that a woman should give herself
to whoever does not come in his name:
and because of this law, he should be honoured.

Thaïs (*hanging her head and trembling*)
Let all that I was return to dust and eternal
oblivion! Come!

Athanaël and Thaïs *(standing close together and calmly looking at the menacing crowd)*
Ah, let us die, if the hour is come!
Let us buy in one moment
eternal bliss and pay for it with our blood.

Act 3

Scene 1

Thaïs and Athanaël travel through the desert. Thaïs is exhausted, but the monk forces her to keep going. Athanaël begins to feel pity rather than disgust for her. They reach the convent where Thaïs is to stay. Athanaël realizes that he has accomplished his mission, and that he will never see her again.

Thaïs
The fiery sun crushes me,
'tis too heavy a burden.
Ah! I give way under the weight of the day!
Let us stop!

Athanaël
No! Walk on!
Shatter your body! Destroy your flesh!

Thaïs
Father, you speak truly!
I offer up my sufferings to the divine redeemer!

(*She is almost fainting. Athanaël supports her in his arms and helps her to sit down in the shade*)

Thaïs
My flesh is bleeding, but my heart is full of joy.
A light breeze bathes my brow.
My spirit, set free from this earth,
is already soaring towards heaven.
Everything intoxicates me!
Farewell, for evermore.

Scene 2

The Cenobite monks express anxiety over Athanaël's morose behaviour since his return from Alexandria. He confesses to Palémon that he has sexual longing for Thaïs. He has an erotic vision of Thaïs. He tries to seize her, but she rebukes him. Then, a second vision in which Thaïs is dying.

Scene 3

Athanaël rushes to the convent and finds Thaïs on her deathbed. He tells her that all he taught her was a lie, that "nothing is true but life and the love of human beings" and that he loves her. Blissfully unaware, she describes the heavens opening and the angels welcoming her into their midst. She dies, and Athanaël collapses in despair.

Albine
God calls her and tonight, the whiteness of the shroud
will have veiled her pure face!
For three months, she has kept vigils, prayed, wept;
her body is destroyed by penance
but her sins have been redeemed.
Having done what your pure spirit bade her do,
she is now about to see the eternal light!

Thaïs
Above all, do you remember your holy words
on that day when, through you, I came to know the only love? …

Athanaël (*anxiously*)
When I spoke, I lied to you ...

Thaïs
And now, here is the dawn.

And here are the roses of the eternal morn.

Athanaël
No! The heavens ... nothing exists...
nothing is true but life and the love of two beings ...
I love you!

Thaïs
The heavens are opening up!
Here are the angels, the prophets ... the saints!
They come with a smile,
their hands full of flowers.
(*standing up, shivering*)
Two white-winged seraphim
are soaring in the azure skies!
And, as you said, the sweet comforter,
touching my eyes with his fingers of light,
wipes away their tears forever!

Athanaël
Come, you belong to me!
O my Thaïs, I love you!
Come tell me I will live.

Thaïs
The sound of golden harps delights me!
I breathe in sweet fragrances!
I feel an exquisite bliss
lulling all my pains to sleep!
Ah! The heavens! I see God.
(*She dies.*)

Athanaël (*with a terrible cry, throwing himself in front of her*)
Dead! Ah, pity me!

COMMENTS

The opera contrasts the city of Alexandria, locus of sin, with the parched dry desert, the place of self-denial, the ascetic quest for heavenly transcendence. There are important dreams and visions in this opera. Early on Athanaël has a vision of the near-naked Thaïs as Venus, from which he wakes in terror. In a second dream-sequence he perceives that Thaïs is dying. She is baptised with water, yet as she reappears to Athanaël, she is Helen, Phryne, Venus Astarte, the *Ewig-Weibliche*. The exhausted Thaïs, weak and shuddering, dies with the knowledge she will meet with

Athanaël in the heavenly city, the finale resounding with sounds of heavenly music, visions of God and redemption. The link of Sudden Death to Wagnerian themes is clear.

The last act of this opera which sees the inversion of roles, aspirations and values is focused in the famous prelude to the act, the Meditation, with its long affecting solo violin line. The violin, like Lucia's flute, becomes a correlative of the issues of erotic and spiritual love, of the fallen monk, a recurring topos of literature, from folktale to modern times (as in Shakespeare's *Measure for Measure* [1604-05], Matthew Lewis's [1775-1818] Gothic novel *The Monk* [1795], the Faust stories and Mikhail Bulgakov's [1891-1940] *The Master and Margarita* [1966]). Athanaël eventually succeeds in converting Thaïs, whose depth of rebirth leads her to become a celibate nun, one consecrated to chastity, while Athanaël conversely loses his heart to her. Thaïs's vulnerability and new purity of heart are symbolized in the Meditation, which is the chief theme underscoring the final meeting between the monk and his pupil. His agonized love for Thaïs and her death become a transferred metaphor for his own loss of identity, or perhaps his own lost soul.

DIE GEZEICHNETEN (The Branded/Stigmatized Ones) (1918)

Opera in three acts by Franz Schreker (1878-1934), libretto by the composer. First performance: Frankfurt, 25 April 1918.[13]

The setting is in 16th-century Genoa

Act 1

Alviano Salvago, a hunchbacked and deformed Genoese nobleman, wants to donate to the people an island paradise called 'Elysium', which he has created. A group of dissolute young noblemen have been using an underground grotto on the island for orgies where all the dreams of secret nights—Oriental dreams born of ecstasy—are fulfilled with young women abducted from prominent Genoese families. One of them, Count Tamare, wants Carlotta, daughter of the Podestà, but she rejects him, as she is only interested in Salvago, whose soul she wants to paint.

Salvago
The Devil, why did nature give—me—
With this ugliness and this hump,
Such feelings, such desire!

Act 2

Infuriated, Tamare swears to Adorno, Duke of Genoa, that he will take Carlotta by force and reveals the secret of the grotto to Adorno. Not wanting Salvago to become more popular than he because of the gift, Adorno uses the existence of the secret grotto as an excuse to veto the transfer.

Tamare
In the rapture of the orgy
The ugly becomes beautiful
And beauty becomes ugly.
Opposites disappear in ecstasy.

Salvago is sitting for Carlotta, who complains that she cannot paint his soul if she cannot see his eyes, as he is avoiding looking at her. He responds that, ugly as he is, he still has the feelings of a man in the presence of a beautiful woman. Carlotta sees his eyes, confesses that she loves him, but faints in his arms as both are overcome with emotion. He holds her unconscious, kissing her hands, in a ‘strange timid embrace’. This leads to the uncovering of another painting of a pale corpse-like hand holding a shining object: she reveals it is the hand of a woman who never found true love: Salvago understands it is Carlotta’s hand.

Salvago (*to Carlotta***)**
In spite of your sweet face
And your voice that sounds
As an angel from heaven, you are—
a she-devil

Act 3

The citizens of Genoa go to the island for the first time and are awed by what they see—strange marble shapes, fountains and erotic statues, most from Greek mythology; fauns, naiads and Bacchantes parade around. A procession with Apollo in his sun chariot comes on. The artist is at first banned from this, but then is accepted as Venus and the Bacchantes become wild and unrestrained with Carlotta in their midst.

Carlotta
I set foot on this island
full of secret anxiety!
as if I had drunk wine,
strong and mixed with bewitching herbs,

or one of those mysterious potions brewed
from the blood of a love-sick maiden!
Like the limbs of a beautiful woman—
modestly left covered
through long years of yearning—
suddenly sprawling out in nakedness
I want to crawl deep,
Go deep into the shining darkness of a warm summer night
I love night and its shadows.

Salvago asks the Podestà for Carlotta's hand in marriage. She evades him, and in the grotto succumbs to Tamare, who is wearing a mask. She sings "Feel my heart how it beats, for you my sweet!" The Duke accuses Salvago of masterminding the abductions and the Captain of Justice accuses him of being an evil spirit in the power of Satan. Salvago, worrying for Carlotta, leads everyone to the underground grotto where she lies senseless on a rose-covered bed: Tamare prides himself on his conquest. Salvago stabs him. Carlotta awakens, Salvago rushes to her side, but with her dying breath she rejects him and calls for Tamare. Salvago, completely deranged, stumbles over Tamare's motionless body as he makes his way through the stunned crowd.

Salvago
She may not wake up
She will sleep on into the other world
In my whole life she was the only good thing
The only beauty …

Tamare
I do not know which is worth more—
a joyless life, a long wilting away—
or death in rapture and transfiguration
a happy dying in a lustful embrace!

To the sparkling dance
In her laughing eyes
Was added the wild
Ecstatic song of her lips:
"Give me death: shouted her gaze—
Give me happiness" pleaded her words.

Carlotta
(*weakly*)
Give me water
No give me wine

And my beloved will come
My handsome Tamare—before—I die—
(*she lies still and rigid*)

Salvago
Here lies—a corpse

COMMENT

Schreker composed several operas, but is best known for *Der ferne Klang*, *Der Schatzgräber* and *Die Gezeichtenen*. He is renowned for portraying sexual aberrations. His protagonists represent artists who cope poorly with social conventions, but also, like Schoenberg, Hindemith and Weill, he was anti-Wagnerian, with concepts of redemption not much in sight. In fact, there is no redemption even for the artist. The pagan world is set against the social mores of the Renaissance. Carlotta has autonomic tensions: fainting and then just dying, she comes into the category of a *femme fatale*. The Greek backdrop is nonetheless overt: Achilles preferred an early glorious death as opposed to lingering unheard of in the shades. Here Carlotta and Tamare pose the same aporia (for Romanticism), with passion ignited by potions, dark and light contrasted, Apollo and Dionysus contested.

In his earlier operas, apart for the use of *Leitmotif*, Schreker is relatively free of the influence of Wagner and Richard Strauss. Harmony is used for the purposes of colour. In the later works the composer moved to a more opulent Wagnerian style, with a distinctly ambiguous use of harmonic language. His orchestration is masterful, capable of broad full writing and delicate chamber music effects, an element of expression similar to that of Debussy. The various and tumultuous last act of *Die Gezeichneten* is a major statement of musical Expressionism in its bewildering onslaught of kaleidoscopic intellectual, symbolic and musical elements.

DAS WUNDER DER HELIANE (The Miracle of Heliane) (1927)

Opera in three acts by Erich Wolfgang Korngold (1897-1957). Librettist: Hans Muller, after a mystery play by Hans Kaltneker (1895-1919). Translation: Gery Bramalp. First performance: Hamburg, Staatsoper, 10 October 1927.[14]

Setting: In the realm of a ruler, in an unspecified mythological period.

Act 1

The Ruler, unable to win the love of his wife Heliane, considers his subjects should suffer. A Stranger has appeared who attempts to bring joy to the people but has been thrown into prison and sentenced to death; he is due to be executed the following day. The Stranger's pleas for reprieve go unheard, although he is permitted to spend his last night on earth unchained.

Voices
Blessed are they that love.
Those who have loved shall not die.
And those who died for love shall rise again.

Heliane comes into the cell, and her feelings of pity gradually turn to those of love. The stranger yearns for a final taste of beauty, which she reveals by becoming naked. She refuses to give herself to him, and goes to chapel to pray for him. The Ruler proposes that if the Stranger can teach Heliane to love him, he will offer the Stranger not only his life but also Heliane herself. Heliane returns, still naked. The Ruler orders the Stranger's death and Heliane's immediate trial. (In this act there are several stage directions to both Heliane and the Stranger trembling.)

Heliane
A young angel with trembling spread wings
Dances towards you, out of the sweet scent of wine.

The Stranger
Your hair must be golden and must pour down your brow like cool gold
Like eternal light pouring upon the altar.
Give, give me your hair, so that with its flood, I may forget tomorrow.

Act 2

The Ruler and the Messenger (a female who has lost her beauty) await the arrival of the executioner, six judges and the blind Chief Justice. The latter asks if it is midnight. The Ruler accuses Heliane of adultery, but she insists that she gave herself to the Stranger in thought only. The Ruler tells her that her paramour has been strangled by him, and Heliane turns very pale: her reaction is a certain sign of her guilt. The Ruler instructs that she should kill herself with a dagger. The Stranger asks to be left alone with Heliane.

(*They gaze at each other, trembling. His eyes sink into her body, Heliane stammering.*)

The Stranger
I came here to die for you—but on your mouth! In you! Intoxicated!
From your arms into eternity
I love you! Let me consummate in you!

He kisses her, she goes into a trance-like state; he takes the dagger and kills himself. As it is now impossible for the Ruler to discover what really happened in the prison, he puts Heliane to the test before God: If she in not guilty, as she claims, she will be able to bring the Stranger back to life. In a trance, Heliane says that she is willing to undergo this trial.

(*Her snow-white face and trembling mouth are suffused by an expression of ineffable longing of her desire.*)

Heliane
May God take me into heaven if I do not swear the truth …
In sorrow I became his.
The eternal Lord's mercy will flow into me with creative force
And from me into the dead man's breast.
(*Heliane collapses in a faint.*)

Act 3

The blind Chief Justice arrives to witness Heliane's attempt to bring the dead Stranger back to life. A tremendous booming of bells is heard. The Messenger incites the people to violence against Heliane. Heliane cannot lie: she breaks down and admits that she did indeed love Stranger. The Ruler is anxious to save Heliane, but on the condition that she will finally be his. Heliane refuses and is led away to the stake. The excited crowd is startled by a clap thunder, stars appear in the sky, the corpse of the Stranger rises up from the funeral bier, transfigured. Heliane runs into his arms and the Ruler plunges his sword into her breast. The Stranger blesses the people and dismisses the Ruler, whose power has now been broken. Heliane and the Stranger, united in unshakeable love, rise in bliss to heaven.

Heliane
(*In front of the bier, in a trance, the beating of her heart almost visible*)
I tell you in the name of God, stand … stand … *(collapses with a scream)*
(Heliane, uncontrolled, her hair cascading down)
Rise up—rise up! I love you!

Kiss me, kiss me, I love you!

(Heliane in front of the Ruler, trembles all over and sways, the Messenger and the crowd bay for her death. A clap of thunder heralds stars from what had been an overcast sky, the Stranger rises from the bier, clothed in white and gold, hair radiating light, from his darkly shining eyes issue an unearthly flame. Shafts of light stream down to earth, the air trembles with celestial music, the song of the spheres is joined with angelic voices.)

Angelic Voices
And they will rise again who died for love

Heliane
I will give myself to you, God has touched me through your eyes!
I shall live and die if death leads me to you.

(Heliane sinks upon the bier, a smile on her white lips.)

Scene 4

(*A day of unearthly beauty, a snow of blossom falls, Heliane and the Stranger together. They embrace, and the vault of a huge, magnificent dome is visible.*)

The Stranger/ Heliane (*in disembodied bliss*)
Not seeking my own self, I found you—
now you and I flow in the same stream.
We go to our death-
We go to our life—
Fear has fallen away—
Time stands still—
Guilt is buried deep and far away:
Only hearts have this power—
And love alone is eternity, eternity.

(*They enter heaven in a close embrace, the curtain falls on light and beauty.*)

COMMENT

Their first meeting is through the eyes, which have mystical symbolism, for example, those of the Stranger rising from the bier. Heliane's long golden hair, which falls all the way to her feet, is given an erotic charge, just like Mélisande's in Debussy's opera. The moments when she takes off all her clothes revealing simple beauty captivates the Stranger entirely.

Sound effects include peels of bells, claps of thunder, angelic voices like a chorus commenting on the action, and dramatic scenes of resurrection and ascension. Heliane is a Madonna figure, The Ruler is but a devil with some of the characteristics of Rubinstein's Demon and, as in that opera, here love triumphs over evil. The *Liebestod* of the finale is bound up with the religious nature of the work and resonates with some of the operas of Wagner.

The music is in the mode of Richard Strauss's *Elektra* (1911). It streams forth relentlessly, in a torrent of surging orchestration, lush harmony, with voices pushed to the limits of their *tessitura* over the flood of sound. The final section of the opera features the resurrection and transfiguration of the Stranger and Heliane. The resurrection theme, heard in various forms through the first two acts, is now transformed into a rapturous diatonic melody as the couple enter heaven, the theme scored for the whole orchestra in a grandiose bitonal statement (a harmonic technique favoured by the composer). The first three notes of the opera end the work, proclaiming a realm of light and beauty.

Notes

[1] Bellini, *I Capuleti e I Montecchi*, recommended recording, with Beverley Sills, Janet Baker and Nicolai Gedda, New Philharmonia Orchestra, cond. Giuseppe Patanè (EMI).
[2] Donizetti, *Anna Bolena*, with Elena Suliotis, Marilyn Horne, John Alexander and Nicolai Ghiaurov, Wiener Opernorchester, cond. Silvio Varviso (Decca). Other versions feature Maria Callas, Joan Sutherland, Beverley Sills.
[3] Donizetti, *Lucia di Lammermoor*. Both Maria Callas (EMI) and Joan Sutherland (Decca) made two recordings. Other versions are by Anna Moffo, Beverley Sills, Monserrat Caballe, Edita Gruberova, Natalie Dessay, Diana Damrau.
[4] Donizetti, *L'Ange di Nisida*, reconstructed and recorded by Opera Rara: *La Favorita* (in Italian), with Giulietta Simionato and Gianni Poggi (Decca); with Fiorenza Cossotto and Luciano Pavarotti (Decca); *La Favorite* (in French), with Vesselina Kasarova and Ramón Vargas (RCA Sony Classical Opera).
[5] Auber, *Manon Lescaut*, with Mady Mesplé and Peter-Christoph Runge, cond. Pierre Marty (EMI); with Elizabeth Vidal and Alain Gabriel, cond. Pierre Jourdan (Chant du Monde).
[6] Massenet, *Manon* There are many recordings; recommended: Victoria de los Angeles and Henri Legay (EMI); Beverley Sills and Nicolai Gedda (EMI).
[7] Puccini, *Manon Lescaut*. There are many recordings; recommended: Maria Callas and Giuseppe di Stefano (EMI); Monserrat Caballe and Placido Domingo (EMI).
[8] Félix Clément & Pierre Larousse. *Dictionnaire des opéras* (Paris: Librairie Larousse, 1905), 1:695.
[9] Charles Malherbe, *Auber* (Paris: Librairie Renouard, 1911), 54.

[10] Rubinstein, *The Demon*, with Anatoly Lochak and Marina Mescheriakova (Marco Polo); with Alexander Poliakov and Nina Lebedeva (Melodia).
[11] Offenbach, *Les Contes d'Hoffmann*. The classic Guiraud version has Nicolai Gedda with Gianna D'Angelo, Elisabeth Schwarzkopf, Victoria de los Angeles, George London and Ernest Blanc, cond. André Cluytens (EMI); the Oeser version is best recommended with Placido Domingo, Joan Sutherland and Gabriel Bacquier, cond. Richard Bonynge (Decca).
[12] Massenet, *Thaïs*, with Anna Moffo and Gabriel Bacquier, cond. Julius Rudel (RCA); with Beverley Sills and Sherill Milnes, cond. Lorin Maazel (EMI).
[13] Schreker, *Die Gezeichneten*, with Marilyn Schmiege and William Cochrane, Dutch Radio Philharmonic Orchestra, cond. Edo de Waart (Marco Polo).
[14] Korngold, *Das Wunder der Heliane*, with Anna Tomova-Sintow and Nicolai Gedda, RSO Berlin, cond. John Mauceri (Decca).

CHAPTER 15

OPERAS: FEMALE WITHOUT AUTONOMIC REFERENCES

Shto nasha zhizn? Igra! ("What is our life? A game!")

(Modest Tchaikovsky, *Pikovaya Dama*, Act 3)

PIKOVAYA DAMA (The Queen of Spades) (1890)

Opera in three acts by Pyotr Ilyich Tchaikovsky. Librettist: Modest Tchaikovsky. Based on a short story of the same name by Alexander Pushkin (1834), the librettist changing some of the text and adding his own lyrics to two arias (Russian: *Pikovaya Dama*, French: *La Dame de Pique*). First performance: St Petersburg, Maryinsky Theatre, 19 December 1890. [1]

The close of the 18th century: St. Petersburg, Russia.

Act 1

Scene 1

Officer Herman is obsessed with the gaming table but never bets, being frugal. He is in love with a girl above his class but does not know her name. Prince Yeletsky, another officer recently engaged, declares his happiness, Herman is envious and curses him. Lisa appears with her grandmother, the old Countess. Herman realizes that Lisa is his unknown beloved. The Countess, known as the Queen of Spades and formerly as the Muscovite Venus, due to her beauty, succeeded at gambling in her youth by trading amorous favours for the winning formula of Count St. Germain in Paris. Only two men, her husband and, later, her young lover, ever learned her secret. She was warned by an apparition to beware a "third suitor" who would try to force it from her. Knowing such a combination would solve his problems, Herman vows to learn the Countess's secret.

Herman
I do not know myself what has come over me,
I seem to have lost myself.
Indignant with my own weakness,
I am no longer my own master
I am in love! In love! …
I am poisoned, as though I were drunk,
I am sick, sick—I am in love! ...
I am alive and suffering, but when the terrible
Moment comes and I realise that she
Will never be mine, then I shall
Have only one thing left ... death.

Countess
A fatal apparition, the prey of some wild passion.
What does he want, following me?
Why is he again in my path?
I feel afraid, as though I were in the power
Of his eyes' sinister fire!
I feel afraid, afraid!

Count Tomsky
You will receive your death-blow
From the third who, impelled by burning passion,
Comes to force from you the knowledge
Of those three cards, those three dread cards!

Scene 2

Lisa is unhappy with her engagement, stirred by the romantic look of Herman. Herman appears, saying he is about to shoot himself over her betrothal to another. Lisa asks Herman to leave but they embrace.

Lisa
I am alone here, all around me lie sleeping ...
Oh! Hear me, night!
To you alone can I confide the secret
Of my heart! It is as dark,
Dark as the melancholy look of those eyes
That rob me of my happiness and peace of mind ...
Queen Night! Like you, great beauty,
Like a fallen angel he is handsome,
In his eyes is the fire of glowing passion.
He beckons to me like some wonderful dream
And all my soul is in his power!
O night ... night!

Act 2

Scene 1

Lisa gives Herman the key to her grandmother's room

Scene 2

Herman in Countess' room and sees the portrait of the Muscovite Venus and how their fates, he feels, are linked: one of them will die because of the other. She returns, sleeps then awakens in horror to find Herman in front of her. He pleads with her to tell him her secret. When she remains speechless, he grows desperate and threatens her with a pistol—at which she dies of fright. Lisa rushes in realising that the lover to whom she gave her heart was only interested in the Countess's secret.

Herman (*beneath the portrait of the Countess*)
So, there she stands—
'la Venus Moscovite'! Some secret force
Binds our destinies together.
Which of us is the victim, you or I?
All I know is, one of us
Will be the other's doom!
I may stare at you and stare in hatred,
But mv eyes can never feast their fill!
I would escape, but do not have the power ...
My searching gaze cannot tear itself away
From that terrible and fascinating face!
No, we can never part without a fateful meeting. …
You know three cards...
(*The Countess raises herself in her chair*)
For whom are you guarding this secret?
What benefit is it to you?
Perhaps it is linked with some terrible misdeed,
With the loss of eternal salvation,
With some diabolical pact?
Think well,
You are old and have not long to live,
And I am ready to take your guilt upon me!
Confide in me!
Speak!
(The Countess draws herself up and glares threateningly at Herman)
You old witch! I will force you to answer then!
(He takes out a revolver)

(*The Countess nods with her head, raises her hands to protect herself from the shot and falls dead)*.
She's dead!
It's all over and I never learned her secret.
(*He stands as though turned to stone*)
Dead! dead!

Act 3

Scene 1

In his room at the barracks, as the winter wind howls, Herman reads a letter from Lisa, who wants him to meet her at midnight by the river bank. There is a knock at the window: it is the ghost of the countess. She tells him the secret so that he can marry and save Lisa. Dazed, Herman repeats the three cards—"three, seven, ace".

Scene 2

Lisa waits for Herman: it is already near midnight. She hopes that he still loves her, but she sees her youth and happiness swallowed in darkness. Herman appears but talks wildly about the Countess and her secret. No longer even recognizing Lisa, he rushes away. Realizing that all is lost, she commits suicide by throwing herself into the river.

Lisa
Oh! I am weary and worn!
Life promised me nothing but happiness.
I am worn and weary with suffering!
Anguish gnaws and consumes me ...
And what if the midnight chimes answer
That he is a murderer, a seducer?
Oh, I am afraid, afraid!
(The clock strikes from the fortress tower)…
Then it was true, then, that I have linked
my fate to an evil-doer!
My very soul belongs for ever
To a murderer, a monster!
My life and my honour
Lie in his wicked hands,
By heaven's own decree
I share the curse upon the murderer!

Herman
Oh yes! I learned from you the secret

Of three, seven, ace!
(He bursts out laughing and pushes Lisa away from him)
Go away!
Who are you? I do not know you!
Away! Away!

Lisa
Oh! He is lost! Lost! And I with him!
(She runs to the bank and throws herself into the river).

Scene 3

At a gambling house. Herman arrives, wild and distracted. He starts betting extravagantly. He bets the three and wins: he bets the seven and wins again. At this he takes a wine glass and declares that "life is but a game". Herman bets everything he has on the ace but when he shows his card he is told he is holding the queen of spades. Seeing the Countess's ghost laughing at her vengeance, Herman takes his own life and asks Yeletsky's and Lisa's forgiveness. The others pray for his tormented soul.

Herman
What is our life? A game!
Good and evil—no more than dreams! ...
(*The Countess's ghost appears. All stand back from Herman, terrified*)
Old witch! You! You here!
What are you laughing for?
You drove me out of my mind.
A curse upon you!
What's that? What do you want?
A life?
My, life? Take it, then, take it!
(*He stabs himself. The ghost vanishes. A number of the guests rush to the prostrate Herman*).

Herman
Prince! Prince, forgive me!
I am in agony, in agony! Dying!
Who is that there? Lisa? Oh! God!
Why? Why? And you forgive! Yes?
And do not curse me?
Ah! How I love you, my angel!
My beauty! My, goddess!
Ah!
(*He dies*)

Chorus
Lord! pardon him!
And give rest to his
Turbulent troubled spirit!

COMMENT

In this opera there are three deaths, but only one *Unexpected Death*. The Countess dies, suddenly, after Herman has drawn his pistol. She is an elderly lady, with her own guilty past, and a sudden heart attack would be a reasonable explanation. In contrast Lisa and Herman commit suicide.

This opera contains several Gothic features: light and dark, storms and thunder, ghosts, magic numbers, chimes of midnight, madness, curses, the Queen of the Night, and the final transcendence, when Herman appeals to Lisa, beauty and angel, for forgiveness as the chorus ask for redemption. The dark supernatural undertones of St Petersburg society are reflected in Pushkin's tale, much changed as Tchaikovsky makes Herman such a real character. The composer himself told how he cried writing the music for Herman's death.

Prince Yeletsky's famous aria "Ya vas lublyu" ("Yes, I love you") occurs in Act 2. "Life is but a game", the quote from Act 3 of the opera, became a proverb in Russian.

Tchaikovsky's score responds to the surreal and claustrophobic nature of Pushkin's short story. The prelude announces the theme of doomed love, and Herman's brooding nature is presented on his first entry (his D minor arioso with ruminating bassoons). The atmosphere of superstitious dread is infused into the narrative by Tomsky's Ballad of the Three Cards (the recurring motif a simple triadic e4-g4-e4 in E minor) recounting the nodal story of the old Countess and her young life in Versailles and the fatal consequences of her involvement with cards. The theme recurs throughout the score.

Herman's complex psychology is explored in his emotional engagement with Lisa and her grandmother. Is he genuinely drawn to Lisa? Or does he use her to gain access to the old Countess and her secret of the cards? Each scene develops this ambiguous situation, with Herman's secret nocturnal confrontation with the Countess, sustained with gripping atmospheric suggestiveness, leading to her sudden death. So too in the later scene between the two when Herman, alone in the barracks during a winter storm, is visited by the Countess's ghost. The combination of the stormy natural tone poem, the atmosphere of night and dread, the evocation of the Countess's funeral with church choirs, the waking nightmare of her appearance and imparting of the secret of the cards (three-seven-ace, a dull A minor monotone on f3), create a mental and Gothic landscape of vivid intensity. The atmosphere of brittle gaiety introduced in the final scene in the gambling den only intensifies the inexorability of Herman's movement into total delusion and suicide, with his famous and tragically ironic and fatal question. His death is heralded by the re-appearance of the Countess's ghost to the eerie A minor music of the Barracks Scene, which shifts to the clear and assuring C major with high tremolos as he dies professing love for Lisa. An obituary follows, a harmonically rich four-part male chorus singing in hushed *pianissimo* (*Andante sostenuto* in D-flat major, the favoured key of intense emotional declaration), with the postlude restating the theme of Herman and Lisa's doomed love now in the major, in high arching octaves leading into a descending first inversion sequence over tremolos, all generating great affectivity. [**Figs. 15.1-3**]

LA NONNE SANGLANTE (The Bleeding Nun) (1854)

Opera in five acts by Charles Gounod, libretto by Eugène Scribe and Germain Delavigne. It was written between 1852 and 1854. The story is partially based on a ghost story in *The Monk* (1795), a Gothic novel by Matthew Lewis. First performance: Paris Opéra, 18 October 1854.[2]

Act 1

Agnès is in love with Rodolphe, although the two families are at war with each other. Moldaw's castle is said to be haunted by the bloody figure of a nun murdered by Luddolf, Rodolphe's father.

Act 2

To bring the families together, Pierre, a hermit, proposes that Agnès should marry Rodolphe's brother Théobald. The lovers agree that to escape, Agnès should disguise herself as the ghost of the 'Nonne Sanglante' and elope with Rodolphe at midnight. Rodolphe goes off with the apparent ghost of the nun, believing it to be Agnès. Rodolphe's ancestors magically appear in the restored family castle to witness the marriage.

Act 3

Théobald has been killed in battle, Rodolphe is in theory free to marry Agnès. Rodolphe is nightly haunted by the Nun who reminds him of his vows. She eventually reveals that Rodolphe can only be released by killing the man who murdered her.

Act 4

At the marriage, the Nun's ghost appears at midnight and reveals Luddolf was her murderer. Rodolphe abandons the ceremony in horror, re-igniting the ancient feud between the families.

Act 5

To save his son, Luddolf presents himself to Moldaw's men as his son, and is murdered, dying in his son's arms at the tomb of the murdered Nun. This act of expiation is acceptable to the ghost, who ascends to heaven praying for Luddolf.

Rodolphe
O cruel disgrace
To die far from her!

Chorus
Death to Rodolphe!

Agnes
Death to Rodolphe, they say.

Rodolphe
What does it matter?
They ask for my life.
I will give it to them.
Ah, whoever be the author,
I will punish this crime.

Chorus
Rodolphe, heavens!
So who then fell under our blows?

Luddolf
I did!

Chorus
His father!

Luddolf
A willing victim of their daggers!
I implore, all-powerful God:
To him be the recompense,
On me be the punishment!

The Nun
In your breast
ineffable clemency
Unite us two.
The only virtue of the guilty one
Lies in his repentance.

Chorus
O ineffable mercy
Deign to receive them
The virtue of the guilty one
Is in repentance!

COMMENT

This story is Gothic, and involves murder, crime and punishment, foiled love, and forgiveness as key to celestial transcendence. Ghosts, visions and events at midnight are within the libretto but there are no references to any autonomic events.

The murder of the remorseful and self-sacrificing Luddolf takes place during a fairly perfunctory G minor chorus. The nodal moment follows, where the restless spirit of the Bleeding Nun finds the moment of transformation and celestial translation. The key enharmonically changes to C major and trance-like utterance on repeated G's and D's over high octave tremolos centred on E. This, over serene rising and falling chords in fifths and octaves, adds a moment of 'otherworldliness' before the confirmatory chorus over high reiterated C major triplets brings the work to an end.

THE WRECKERS (1906)

Opera in three acts by Ethel Smyth (1858-1944). Librettist: H. B. Laforestier (Henry Brewster).[3]

Act 1

A Cornish fishing village. Sunday evening.

The villagers make a living by wrecking ships. Lawrence, the lighthouse keeper, has seen beacons burning on the cliffs and is sure someone is warning ships of the danger. Thirza, wife of the preacher Pascoe, is in love with Mark, who is loved by Avis, the daughter of the lighthouse keeper. Thirza cannot continue life in the village of the wreckers. She denounces Pascoe as the traitor who has been warning the ships of danger.

Act 2

Thirza's house

Mark is the one responsible for the warning beacons. Thirza warns him that if he lights the fire the villagers will see the flames and trap him. The lovers embrace. Mark begs her to leave Pascoe and run away with him. Pascoe arrives just in time to see the lovers escaping. Pascoe sees his wife's face in the moonlight and collapses. The villagers find Pascoe near the beacon and believe him to be the traitor.

Act 3

The interior of a large cave

An impromptu court has been convened. Avis declares that he is the victim of witchcraft, under the spell of Thirza. Mark confesses, Thirza also accepts the blame. Avis tries to save Mark by claiming he spent the night with her, but the lovers are determined to meet their fate together. They are to be left chained as the incoming tide gradually fills the cave. The villagers leave as the waters begin to rise and ecstatically the lovers face death in each other's arms.

Thirza
When the rivers upwards shall be flowing,
When the dead wood green with sap hath proved,
Ye may see my heart repentance showing
For the sin of having lived and loved.
O'er black gulfs my spirit circles, soaring
Past all Time, above Death's dark decline.
Then shall I for mercy kneel imploring,
I who knew through him the joy divine?

Lawrence
Lift up your voices; let the psalm we sing
When souls are passing in their doomed ears ring.

Mark
Turn those eyes towards me, fall of sky and sea
Where love unquenchable, undimmed doth shine!

Thirza
Vain, vain are all things save thy love and mine,
And now we die together.
(*the gate clangs to*)
Victory!

Both
Love, mouth to mouth we soon shall calm be sleeping
As we have longed to sleep so oft before . . .
Hear the fierce nuptial song the waves sing, leaping
Herald of that blest kiss that ends no more.
Hark the winds shrilling notes! they wail and quiver,
Love 'tis the viols whose clear tones rejoice;
Below the ocean's booming mighty voice
Peals like the organ, thundering on for ever.

(*A shaft of sunlight passes through a cleft in the rocks.*)

Here comes the golden sun to light our dying,
Our endless joy, vast as the azure sky;
So, on the sea's great heart together lying
Let us be swept to our death ecstasy.

(*A larger wave dashes in; the curtain falls.*)

COMMENT

The sudden deaths here are accompanied by a *Liebestod*, with all the echoes of the sea, divinity, and sacrifice (for the evils of the village). The associations looking forward to Peter Grimes in Benjamin Britten's opera are undeniable. Thirza sings that the sound of the waves is her bridal song and Mark adds that the cry of the wind is like the dancing and singing of spirits. Clutching one another, they wait for the final embrace of the sea to compete their ecstasy.

The final scene begins with a description of the tidal cave where the guilty lovers are to be incarcerated. The music constantly conjures up the sea in harp runs and glissandi. A great chorus of judgement initiates the fate of Mark and Thirza, in which the sea figures constantly, featuring in the orchestra and the imprecations of the chorus. Great booming bell-effects toll in the bass, while the chorus now intones music of the Requiem, using the modalities of Anglican church music. The doomed lovers, tied to their rock, begin their final prayer-like *Liebestod*, to chamber effects on the harp and horn. They sing in unison with enriched orchestral accompaniment, with the entry of high strings, building to a great instrumental climax with the chorus. The sea imagery intensifies, leading to final fanfares and crashing timpani.

ELEKTRA (1909)

Opera in seven scenes by Richard Strauss (1864-1949). Librettist: Hugo von Hofmannsthal (1874-1929), after Sophocles's tragedy (411/410 BC). English version by Alfred Kalisch. First performance: Dresden, Hofoper, 25 January 1909.[4]

Synopsis

The curse of the House of Atreus, King of Mycenae, had descended to his sons Menelaus and Agamemnon. The latter had four children by Clytemnestra: a son Orestes, and three daughters: Iphigenia, Chrysothemis and Elektra. Agamemnon has been murdered by Klytemnestra and her lover Aegisthus.

Scene: Mykene.

The inner courtyard bounded by the back of the palace.

Elektra is distraught over the murder of her father, and seeks revenge. She has visions of triumphal dances and obsequies for him.

> **Elektra**
> Therefore must their blood
> Descend to do thee homage meet; and we,
> Thy son Orestes and thy daughters twain,
> We three, when all these things are done, and steam
> Of blood has veiled the murky air with palls
> Of crimson, which the sun sucks upwards,
> Then dance we, all thy blood, around thy tomb
> (*In ecstatic pathos*)
> And o'er the corpses piled, high will I lift,
> High with each step, my limbs; and all the folk

Who see me dance. Yea all who from afar
My shadow see, will say: "For a great King
All of his flesh and blood high festival
And solemn revel hold; and blessed he
That children hath who round his holy tomb
Will dance such royal dance of Victory!"
Agamemnon! Agamemnon!

Klytemnestra is plagued by bad dreams, which Elektra attributes to the fear that Orestes will come to avenge his father's death. Elektra threatens Klytemnestra, but Chrysothemis brings news that Orestes is dead: Klytemnestra is overjoyed.

Elektra and her sister must now avenge the death of Agamemnon. Orestes returns, at first in disguise. He recognises Elektra, as she does him.

Elektra
A corpse am I, that erstwhile was thy sister,
At my touch, that boasted once a kingly father!
Methinks I once was fair: when from my mirror
Turning, the lamp I darkened, knowledge came
With thrills of wonder. I felt then
How the slender rays of moonlight,
Seeking my body's whiteness out, did rest on it
And linger, loath to leave it. And my hair,
Such hair it was as maketh men to tremble,
This hair, now so unkempt, besmirch'd and matted.
Dost hear me, brother? All that I had and all
I was, the gods took from me. Maiden shame
E'en flung I far from me, the shame, that treasure
That passeth all, which like the silvery film,
Of moonlight, unto every woman clinging,
Doth from her body drive, and from her soul,
All horror, all uncleanness. Hear'st thou, brother?
All these thrills of sweetness did my father
As expiation claim. Think'st thou not
When in my beauty I rejoiced, his moans
Resounded oft, his sighs resounded
In my chamber?

Orestes enters the palace. Klytemnestra's death-wails are heard. Elektra runs around like a wild animal. Aegisthus arrives and she dances around him. His cries are heard from the palace. Elektra dances in triumph, and collapses dead.

(Elektra descends from the threshold. She has flung back her head like a Maenad. She flings her knees and arms about.)

Elektra
(*stays motionless, gazing at Chrysothemis*)
Say naught and dance on. All must come
To my side! Here take your place! The burden of joy
I carry, and I lead the sacred dance.
Who happy is as we, can do but this:
Say naught and dance on!
(*Elektra makes a few more steps of uncontrolled triumph and falls lifeless.*)

COMMENTS

Richard Strauss had already triumphed with *Salome*. In this opera the tranquil concepts of Winckelmann's Greece and Goethe's humanism are overwhelmed by vengeance and tragedy in this and the later opera. The staging takes place in a dismal courtyard, not a palace, and the 'daemonic' is not below the surface. Elektra is much more isolated than in her representation in Aesculus, Sophocles, or Euripides: the metaphysics of the Greeks is missing. There are Klytemnestra's sinister dreams, and the question of their meaning. Klytemnestra asks Elektra how to stop them. What did they mean? Can one have love without hate, two central *Leitmotifs* of human emotions and behavioural actions: the crime of matricide and the love of her father—the Electra Complex a name which came from Carl Jung, to counterpoise with an Oedipus Complex.

The opera centres around three women. Klytemnestra and Elektra die. Elektra becomes hysterical, her music and movements come from only her. Dedicated to Dionysus, she dances herself to death, an ending far removed from the classical plays, yet triumphal to Chrysothemis:

> A blackened corpse once was I among the living and this glad hour the flame of life hath made me, and my fierce flame consumes the gloom of all the world. And my face must glow far whiter than the moonlight when it glows most white. Do you see that light that from me shines?

Agamemnon dominates the opera. His D minor motif forms the basis of this tragedy. It initiates the opera—harsh, passionate, accusatory; and it returns at the end after Elektra's death, inflexible and menacing. It infuses all Elektra's utterances, sometimes moving splendidly into the major mode. This harsh heroic theme is contrasted with a fervent expansive melody in E-flat symbolizing Elektra's deep attachment to her father, and the emotion that binds his children to his memory. Chrysothemis's music, by contrast, is characterized by beautiful melodies and rich harmonies, never sentimental, but resolute in sharing the heroic nature of the dead father.

The climax of the opera, both musical and psychological, comes in the scene with Klytemnestra. Her entrance is heralded by savage, driving music, and a similar excitement pervades her dialogue with Elektra. The repressed evil of past deeds, and the corruption of suppressed emotion grow during their quarrel into mutual hatred, with overwhelming power in the orchestral accompaniment. A Wagnerian sense of *Leitmotif* and use of an augmented orchestra are carried forward in this grandiose and savage score. Motivic abruptness is contrasted with a flowering of cantilena passages when tender feelings break through. The harmonic language was of unprecedented daring, with dissonances, with interweaving and clashing, with variation and integration that cannot be taken in without repetition. The orchestration requires 115 players, the largest possible number for an opera orchestra.

LA VIDA BREVE (The Short Life/Life is Short) (1913)

Opera in two acts by Manuel de Falla (1876-1946). Librettist: Carlos Fernández-Shaw. Translated by Tess Knighton. The opera was written between August 1904 and March 1905, but not produced until 1 April 1913 in Nice.[5]

Time: The 20th century, Granada.

Act 1

The young Gypsy, Salud, is passionately in love with a young well-to-do man named Paco. She does not know that he is already engaged to a woman of his social class. Her uncle, Sarvaor [Salvador], and her grandmother (*La abuela*) have discovered this, and they try to prevent Salud from interrupting Paco's wedding after she learns the truth.

Act 2

Wedding festivities are in progress.

Salud and Sarvaor gate-crash the festivities. Paco is caught off guard: he utters Salud's name before denying he knows her and ordering her ejection. Salud falls dead at his feet. The grandmother and Sarvaor shout out "Judas".

Salud
To me you!
I'm drowning! ... I'm dying! ...
Paco! ...
(*falls dead*)

Uncle Sarvaor
Dead!

The Grandmother
(*popping up and screaming*)
Health! Baby! My glory!
My soul!
How awful!
(*To Paco*)
Ah, infamous! False! Judas!
(*falls dead*)

COMMENT

This surely is an *Unexpected Death*, but without the autonomic references, although Salud is described as 'delicate'. It surely reflects deception, rejection and loss of love and showing the social limitations between the classes. Salud dies of a broken heart.

Musically the splendid party of the second act contrasts with the delicate atmosphere of the first. A passionate bolero initiates the dancing which becomes more frenzied until it is a mad confusion that ends in a single cry. The scenes with Salud, exploring her happy but tragic love, are detached from the crowd scenes, and present strong, glowing vocal and dramatic climaxes. Both the drama and the depiction of popular life are energized by the vigour of Spanish folk music, vested in its compulsive dance rhythms and its recurrent melodic melismatism.

THE MAKROPULOS CASE (1926)

Opera in 3 acts, with music and libretto by Leoš Janáček (1854-1928), based on the play *Vec Makropulos* by Karel Čapek. It was composed between 1923 and 1925. First performance: Brno, 18 December 1926.[6]

Act 1

Kolenatý's law office, Prague, 1922.

Vitek, Kolenatý's clerk, notes that the probate case of Gregor v. Prus has been going on for almost a century. Kolenatý represents the middle-class Gregors against the wealthy and aristocratic Prus family. Albert Gregor comes in to ask about the case. Vitek's daughter, Kristina, enters. She is a young singer and praises Emilia Marty, a famous singer she has seen rehearsing and admits that she will never be the artist Emilia Marty is.

Baron Joseph Ferdinand Prus died in 1827, leaving no will or legitimate children. His cousin claimed the estate, but so did Albert's ancestor, Ferdinand Gregor, who asserted that the Baron had promised the estate to him. Neither party could proffer an actual will, which Kolenatý answers is essential. Emilia says that there is a will. Albert tells Emilia that if he does not get the estate, he will be penniless and shoot himself. The will is found. Albert has to prove that Ferdinand Gregor was the Baron's son out-of-wedlock. Emilia says that she can also prove that.

> **Gregor** (*to Emilia*)
> I've so much to say to you. You treated me so
> harshly, that sends a man off his balance. I feel
> a hot wind searing, scorching. What is it? Men
> sense it and turn at bay, just like wild beasts. You
> arouse something, something terrible. Were you
> never told that? (*goes nearer to her*) Emilia, you
> surely know that you are lovely!

Act 2

The empty stage of the opera house.

Emilia enters, the old Count Hauk-Šendorf enters, and thinks he recognizes Emilia as Eugenia Montez, a Romanian woman with whom he had an affair half a century before. Jaroslav Prus reveals that the mother of the Baron's child was recorded as Elina Makropulos, who might be the same as Ellian MacGregor, whose love letters he has read. He continues, saying that only a descendant of Ferdinand Makropulos can claim the estate. Emilia offers to buy a mysterious document found with the will.

> **Stage Hand**
> One thing I would like to know. Does a woman
> of her sort have many lovers?

Cleaner
Oh yes, oh yes, oh I should say so indeed …

Prus (*who has read the letters*)
They contain allusions to strange practices, intimate things. I am no novice Miss Marty, but I have to admit that the most accomplished rake hasn't got as much experience in certain matters as this lady.

Marty
What you mean is, as this harlot?

Prus
There it's only E.M. No more.

Marty
That obviously stands for Ellian MacGregor.

Prus
It could stand just as well for Emilia Marty, Eugenia Montez, or thousands of others.

Marty
But it is Ellian MacGregor.

Prus
Or rather, Elina Makropulos, nationality Greek ….

Gregor
Emilia I warn you! You treat me harshly, you horrify me, even so that can make me happy. I'd like to strangle you when you humiliate me. I'd like … Emilia, don't make me kill you! In you there's something that repels me. You're base, evil, terrible. A beast without feeling.
Nothing seems to move you. You are cold as steel. As if you'd risen from the grave. It's perversity loving you. And yet I love you so, I'd tear the flesh from my body.

Marty
How do you like the name Makropulos?
Stupid fool. See that, that scar on my throat?
He was another who said he'd kill me, and if I stripped myself naked in front of you then you

could see all my other souvenirs! Why is it men
feel that they must kill me?

Gregor
I'm in love with you!

Marty
Then kill yourself! If only you knew that I am
past all caring. (*sadly*) If only you knew.
(*wrings her hands*)
Unhappy, unhappy Elina!

Act 3

Emila's hotel room the next morning. Emilia and Jaroslav have spent the night together. Though disappointed by Emilia's coldness, Jaroslav nonetheless gives her the envelope containing the document. News arrives that Janek, son of Jaroslav, has committed suicide because of his infatuation with Emilia. Jaroslav grieves, but Emilia is indifferent. Count Hauk-Šendorf enters saying he has left his wife and wants to elope with Emilia to Spain. Kolenatý has noticed that Emilia's handwriting matches that of Ellian MacGregor and suspects her of forgery. She leaves. Jaroslav says that the handwriting of Elina Makropulos on Ferdinand's birth certificate also matches that of Emilia.

Emilia decides to tell the truth. She is Elina Makropulos, born in 1585, daughter of Hieronymus Makropulos, an alchemist in the court of Emperor Rudolf II, who ordered him to prepare a potion that would extend his life. When the potion was ready, the Emperor ordered his alchemist to test it on his daughter first. She fell into a coma, and Hieronymus was sent to prison. After a week, Elina woke up and fled with the formula, which proved successful. She has since lived an itinerant life for three centuries, becoming one of the best singers of all time. She has assumed many identities, including 'Eugenia Montez', 'Ekaterina Myshkin', and 'Ellian McGregor'. She confided her secret to Baron Joseph and gave him the formula, which he attached to his will for his son. However, the document was lost among the Baron's papers after his death.

The potion is finally wearing off. Elina wanted the formula to gain another 300 years of life. As the first signs of old age appear on her face, the others, initially disbelieving her story, come to believe her, and to feel pity for her. Elina has realized that perpetual youth has led her to exhausted apathy and resolves to allow death to come naturally to her, understanding that a sense of transcendence and purpose come from a naturally short span of life. Aging rapidly before the eyes of the astonished onlookers, she offers Kristina the formula so she now can become a great artist herself. However, Kristina burns the parchment in a candle flame. Elina collapses as she recites the first words of the Lord's Prayer in Greek.

Kolenatý
How old are you now?

Marty
I'm growing older. Three hundred and twenty-seven.
Because I'm at the end of things. Feel my hands Bertie, how cold they are. Feel them, they're colder than ice! …

Kolenatý
Yes, now we've got her. What's your real name?

Marty (*falling*)
Elina Makropulos.

Kolenatý (*lets her fall to the ground*)
Damn it, she's not lying!
Fetch the doctor

Marty
Dying or living it's all one, it's the same thing.
It's a great mistake to live so long! Oh, if you
could only know how easy life is for you! You
are so close to life! You see in life some meaning!
Life has for you some value! Fools, how happy you all are.
And it's due to the paltry chance that you will all die soon.
You believe in mankind, love, virtue, progress.
There's nothing more that you can want.
But in me life has come to a standstill, O Jesù
Christe, I cannot go on. How dreadful this
Loneliness is! In the end it's the same, Kristina,
singing and silence. There's no joy in goodness,
there is no joy in evil. Joyless the earth, joyless
the sky! When you know that then your soul dies
within you.
(*Kristina takes the document and holds it over the flame.*)

Marty
Pater hemon (*breaks down*).
(*The document is completely burnt*).

COMMENT

This strange opera, based on a lawsuit that has extended for many generations, brings to question identity (just who is Emilia?), transformations over time and generations, potions, reflections of the Eternal Feminine (mother and lover of many), with encapsulating beauty and danger and mystery. As the effect of the potion wears off and Emilia ages rapidly, life for her has at last come to a standstill. Cold and lonely, joyless without feelings, the terrible loss of the transcendent view on life and love is not possible without death. She dies uttering the Greek words *Pater hemon*—"Our Father". The opera can be seen as female betrayal and restoration of the natural order through sacrifice.[7]

The end of the opera presents Marty's extraordinary soliloquy in dying. She appears preceded by a motif first heard in the prelude, and from which of the later material of the opera derives. Its variation, played on the solo violin, suggests a new sense of resignation in Marty, an acceptance of death, and therefore a type of salvation. She now resembles a ghost, and her old harshness of manner has softened, as though she has now come to understand how desperately she had always wanted to die (her words "It is miraculous how Death has touched me!" echoed by an unseen chorus). To the recurrence of the second fanfare theme of the prelude (representing the period of Emperor Rudolph II into which she was born, and hence the formula of her deathlessness) she passes on the Makropulos secret to Kristina as the youngest. As the document burns, Marty sinks gradually into death, gasping that her enduring persona, Elina Makrupolos, can and indeed must die. The translucent texture of the music transcends the details of the scene, and underscores the import of the drama, that fulfilment in life is preferable to length of life, and that fear of death should be replaced by a sense of its necessity in the shaping of the human story. The analogy of Čapek's drama to Rider Haggard's (1856-1925) *She* (1887) is striking. [**Figs. 15.1-6**]

Notes

[1] Tchaikovsky, *The Queen of Spades*, with Zurab Anjaparidze, Tamara Milashkina and Valentina Levko, Bolshoi Theatre Orchestra, cond. Boris Khaikin (EMI/Melodya); Wieslaw Ochman, Penka Dilova and Stefka Exstatieva, Sofia Festival Orchestra, cond. Emil Tchakarov (Sony Classical).

[2] Gounod, *La Nonne sanglante*, with Yoonki Baek and Natalia Atamanchuk, Theater Osnabrück, cond. Peter Bäumer (CPO).

[3] Smyth, *The Wreckers*, with Anne-Marie Owens and Justin Lavender, BBC Philharmonic, cond. Odaline de la Martinez (Conifer Classics).
[4] Strauss, *Elektra*, with Birgit Nilsson, Leonie Rysnek, Marie Collier and Tom Krause, Vienna Philharmonic, cond. George Solti (Decca).
[5] De Falla, *La vida breve*, with Victoria de los Angeles, cond. Raphael Frübeck de Burgos (EMI).
[6] Janáček, *The Makropulos Affair*, with Elisabeth Söderström, Wiener Philharmoniker, cond. Charles Mackerras (Decca).
[7] See Caryl Emerson, "Čapek, Janáček, that Makropulos Thing, and a Word about Sacrificed Women in 20th-Century Slavic Opera" in Craig Cravens, Masako U. Fidler, & Susan C. Kresin (eds), *Between Texts, Languages and Culture. A Festschrift for Michael Henry Heim* (Bloomington, Indiana, 2008). "An innocent-maiden, Elina Makropulos, is betrayed by two ambitious men (one of them her own father) for the sake of power and knowledge. Here is the story of Faust, but in a far more interesting and complicated plot than that of carnal love: the very stuff of tragedy. But despite this betrayal, in the final act, through the heroine's courage, the natural order is restored. And this restoration, much more than humor or even love, is everywhere the basic mission and obligation of comedy."

CHAPTER 16

OPERAS: MALE WITH AND WITHOUT AUTONOMIC REFERENCES

I'm suffocating here—
I feel how all my blood
is rushing to my head.

(Modest Mussorgsky, *Boris Godounov*, Act 2)

Lord! Lord! Lord illumine me
with your Presence! Deliver me, enrapture me!

(Olivier Messiaen, *Saint François d'Assise*, Act 3)

1. With Autonomic References

BORIS GODUNOV (1874)

Opera in four acts with a prologue by Modest Petrovich Mussorgsky, libretto by the composer, adapted from the tragedy by Alexander Pushkin, *The Comedy of the Distress of the Muscovite State, of Tsar Boris, and of Grishka Oprepiev* (1826) and Karamzin's *History of the Russian Empire* (1829). First given in St Petersburg, charity performance, 17 Febraury 1873.[1]

Time: The years 1598 to 1605

Prologue

Boris has been appointed regent after the death of Ivan the Terrible (1584). The latter's son Dmitry died aged eight in mysterious circumstances. Boris, suspected of involvement in the murder, is the best candidate to be the next Tsar. Initially Boris refuses the throne, but then accepts and is crowned in the Kremlin.

Act 1

Boris proves to be a good, wise ruler, but after a few years famine blights the country, and some blame Boris and divine punishment linked with the death of Dmitry. The novice Grigory plans to stir a rebellion.

Boris's son Fyodor hears from his father all that the Tsar has achieved, but Boris is made aware of a plan, emanating from Lithuania, of a Pretender bearing the name Tsarevich Dmitry.

Act 2

Boris seeks information that Dmitry really did die, although is told that even his dead body may have miraculous powers. Boris suffers remorse and guilt, and hallucinates he can see the dead Dmitry. He denies the crime, but collapses praying God will have mercy on his guilty soul.

Boris
I'm suffocating here —
I feel how all my blood
is rushing to my head,
it's raging in my temples.
A guilty conscience
is a cruel punishment.
(*It is getting darker; a carillon begins to play.*)
If you did
but once in life
an evil deed,
and though it was your fate
that made you do it—
your soul is doomed,
your heart is drowned in poison—
The furies
haunt and mock you—
like hammer blows
falls on your ears
the thunder
of damnation.
My head is reeling—
reeling—
and all my strength has left me—
I see ... the child ...

I see it lying there ...
(*The clock strikes eight. A ray of moonlight falls on the moving figures.*)

Act 3

The Pretender plots with the Polish Princess Marina, encouraged by the Jesuit Rangoni.

Act 4

In the Kromy Forest the people rise in revolt against Boris. The Pretender passes en route for Moscow, leaving a Simpleton grieving for Russia.

At the Kremlin it is agreed that Grigory and his followers should be executed. Still hallucinated, Boris is told that Tsarevich Dmitry has become a saint from beyond the grave. Boris begs his son never to ask how he, Boris, became Tsar, seeks forgiveness from his son, and dies.

Boris
I'm choking—choking! Help me!
I want to see my son … I cannot breathe ...
(*Fyodor enters and throws his arms around his father)*
Leave us alone—my son and me!
The plaint of death ...
give me ... the cloister's vestments—
the Tsar withdraws to God!
Heaven! Heaven! I am lost.
Oh Lord—forgive me for my sins!
Oh fearful Death, how cruel
is your torture—It is not time yet —
I still am Tsar — I still am Tsar ...
Heaven — death — forgive me, all—
He ... he now is ... Tsar!
Forgive me—forgive me. (*Boris dies*)

COMMENT

In this opera, of which there are several versions, Boris dies suffering pangs of guilt and seeking redemption not only from his son, but also from God. The truth of the murder story has long been the subject of discussion, Boris being portrayed as a tragic hero, or as one responsible for the death of Dmitry. The latter was Pushkin's view: his tale focussed on the psychological torments of the guilty Boris. Did he die of a guilty conscience, because his failure as Tsar to maintain his peoples' love and trust, or was he pushed to despair by the political plots which were gathering around him? In the libretto there are references to heart problems. Boris speaks of his “aching heart” at the very moment of assuming power. His guilt assumes an increasingly psychosomatic dimension with intensifying autonomic features, as when the issues of life are stirred and excited by the striking of a clock. The chiming clock emphasizes the past and present for Boris, and his death knell is heralded by bells (echoing the earlier Coronation Scene) followed by the hallucinations, a familiar trope, as with the guilt-ridden hallucinations of Verdi’s *Macbeth* (1847) or Herman’s visions of the dead Countess in Tchaikovsky’s *Queen of Spades*. The story contains other Byronic characters, from Schumann’s *Manfred* to Liszt’s *Oberman:* Boris finally pleads for the protection of his guilty soul.

The Tsar’s coronation already sees the manifestation of his psychological uncertainty, his self-doubt, founded on his remorse. The scene is held together by the continuous repetition of a main musical idea and by a majestic theme on the ostinato bells that slowly rises to an overwhelming clamour in its gradual rhythmic variations on the sequence c5-a5-e5-g5 over huge triadic chords (c2-c3, c3-c4). Bells and chimes are used throughout the opera to reflect the Tsar’s changing psychology.

Boris’s great monologue is the main feature of Act 2, which is dominated by an expressive theme in the extreme key of G-flat (“How heavy is the hand of God in his wrath”) which depicts the grandeur of the monarch despite his inner turmoil. The latter is reflected in the harmonic dissolution of the powerful opening in a falling chromatic figure. The conclusion of the act resumes this mood in which the aria ends and sees the return of the bells.

The chimes of a great clock run through the music, almost like the beat of an anguished heart, over a chromatic figure in the strings, strident clashes in the woodwinds, and a muted tremolo in the basses, adding a threatening sense of unease. Similar figures and orchestral colours return in the scene of Boris’s death, but unfold more slowly, without such relentless hardness. The ecclesial chorus and tolling bells sustain something of the conciliation of a Requiem. Rarely does any score capture the interplay of musical gesture as correlative of autonomic dysfunction and collapse.

DEATH IN VENICE (1973)

Opera in two acts by Benjamin Britten, his last. Librettist: Myfanwy Piper, based on the novella *Der Tod in Venedig* (1911) by Thomas Mann. It was first performed at Snape Maltings, near Aldeburgh, England, on 16 June 1973.[2]

Place: Venice and Munich, 1911

Act 1

Scene 1. Munich

Aschenbach, a famous German novelist, is weary and bemoaning the fading of his artistic inspiration. He catches sight of a traveller "from beyond the Alps by his looks" and is moved to travel South in the hope of refreshing his artistic imagination.

Aschenbach
My mind beats, and no words come.
Why am I now at a loss?
I reject the words. Called forth by passion …
Should I go too beyond the mountains?
Should I let impulse be my guide?
Should I give up the fruitless struggle with the word?

Scene 3. The Journey to the Lido

Aschenbach
Ah, Serenissima!
What lies in wait for me here.
Ambiguous Venice, where water is married to stone
And passion confuses the senses?
Mysterious gondola,
A different world surrounds you,
A timeless, legendary world
Of dark, lawless errands
In the watery night.
How black a gondola is—
Black coffin black,
A vision of death itself
And the last silent voyage.

Scene 4. The First Evening at the Hotel

Aschenbach is enamoured of a young Polish boy, Tadzio, in whom he sees a special beauty.

Aschenbach
Poles, I should think, Governess, with her children—
A beautiful young creature, the boy.
Surely the soul of Greece
Lies in that bright perfection, a golden look,
A timeless air.
Mortal child with more than mortal grace.

Scene 5: On the Beach

Aschenbach observes Tadzio playing on the sands. As a Pole, the boy hates the Russian guests: Aschenbach concludes that there must be a dark side to his perfection.

Scene 6. The Foiled Departure

Walking the streets of Venice, seeing rubbish on the streets and smelling the foul water of the canals, Aschenbach feels unwell, and decides that he must leave Venice. On arriving at the station, he finds that his luggage has been sent on the wrong train. He is angry. On seeing Tadzio again he realises that the boy was the cause of his regret at leaving.

Scene 7. The Games of Apollo: The Feast of the Sun

Aschenbach on the Lido beach, watches Tadzio and his friends play. Aschenbach's thoughts (voiced by the chorus) are of the gods Phaedrus, Apollo and Hyacinthus, their actions mirroring those of Tadzio. Aschenbach is inspired artistically by the boy's beauty. Aschenbach wants to speak to the boy but when the opportunity arises, he cannot bring himself to speak. Almost choking on the words, Aschenbach realises the truth:

The Voice of Apollo
He who loves beauty, worships me.
Love that beauty causes is frenzy, god-inspired
Nearer to the gods than sanity.

Aschenbach
The boy, Tadzio, shall inspire me.
His pure lines shall form my style.
The power of beauty sets me free.
I will write what the world waits for
rejoicing in this presence.
When thought becomes feeling, feeling thought ...
When the mind bows low before beauty ...

When nature perceives the ecstatic moment ...
When genius leads contemplation for one moment of reality ...
Then Eros is in the world.
I – love you.

Act 2

Sitting with a book but distracted by his own thoughts, Aschenbach decides to accept his feeling for the boy as it is, while seeming ridiculous but not dishonourable. He deliberates on beauty and creativity.

Aschenbach
Overcome by beauty I tried to use the emotion released for my own creation. My beating heart and trembling limbs refused to obey my will. So I had to mock myself as the crestfallen lover. Who really understands the workings of the creative mind? Nonetheless 'so be it'. This 'l love you' must be accepted; ridiculous and sacred too and no, not dishonourable, even in these circumstances.

Scene 8. The Hotel Barber's Shop

Aschenbach
The city's secret, growing darker, every day, like the secret in my own heart.

Scene 9. The Pursuit

Aschenbach
Tadzio, Eros, charmer, see I am past all fear,
blind to danger, drunken, powerless,
sunk in the bliss of madness.

Scene 11. The Travel Bureau

Aschenbach asks a clerk about the plague and is told that the city is in the grip of Asiatic cholera. He advises Aschenbach to leave immediately before a blockade is imposed.

Scene 12. The Lady of the Pearls

Aschenbach decides to warn Tadzio's mother of the danger posed by the plague, but fails, and asks himself

What if all were dead,
and only we two left alive?

Scene 13. The Dream

Aschenbach dreams of the gods Apollo and Dionysus, who argue their respective viewpoints of reason and beauty versus chaos and ecstasy. Apollo is overwhelmed and leaves Dionysus to a wild dance. Aschenbach wakes and realises how little of his former intellectual rigour and detachment remains. He is resigned to the change.

Dionysus
Receive the stranger god.

Apollo
No! Reject the abyss.

Dionysus
Do not turn away from life.

Apollo
No! Adjure the knowledge that forgives.

Dionysus
Do not refuse the mysteries.

Apollo
No! Love beauty, reason, form.

Dionysus
He who denies the god, denies his nature.

Apollo
Be ruled by me and by my laws.

Dionysus
Come! Beat on the drums.
Taste it, taste the sacrifice.
Join the worshippers,
Embrace, laugh, cry,
To honour the god.
I am he!

Aschenbach
Let the gods do what they will with me.

Scene 15. The Hotel Barber's Shop (ii)

The barber works at beautifying Aschenbach with make-up and hair dye.

Barber
A masterpiece, a masterpiece!
Now the Signore can fall in love with a good grace.

Scene 16. The Last Visit to Venice

Aschenbach sits down, tired and ill, and bitterly mocks himself as an old fop. He recites some of Plato's dialogue between the old philosopher Socrates and the boy Phaedrus. The subject of the dialogue is the paradoxical, dangerous relationship between the artist and his subject.

Aschenbach
Does beauty lead to wisdom, Phaedrus?
Yes, but through the sense.
Can poets take this way then
For senses lead to passion, Phaedrus.
Passion leads to knowledge
Knowledge to forgiveness
To compassion with the abyss.
Should we the reject it, Phaedrus.
The wisdom poets crave.
Seeking only form and pure detachment
Simplicity and discipline?

Scene 17. The Departure

Aschenbach sits in his usual chair on the deserted beach where Tadzio and another boy, Jaschiu, are playing. The game becomes rougher, Jaschiu dominates, pushing Tadzio's face into the sand. In an attempt to assist, Aschenbach tries to get up but is too weak. Jaschiu runs away, leaving Tadzio on the beach alone with Aschenbach. Tadzio beckons the author, but he slumps in his chair. Tadzio continues walking far out to sea.

Aschenbach
Ah, no!

Chorus (*off*)
Adziú, Adziú, Adziú!

Aschenbach
Tadziú!

COMMENT

At the time of composing *Death in Venice*, Britten had cardiac failure, which required surgery, and he expressed concern that he might die before finishing the work. *Death in Venice* was his final opera.

The score is marked by some haunting soundscapes of 'ambiguous Venice'. Sea and Serenissima (Venice), water and fluidity of boundaries, Apollo and Dionysus all provide the settings for the death of Aschenbach, as Tadzio beckons him towards the sea.

The setting of this opera and story, in a decaying and cholera infected Venice, starts with a Faustian ennui, a scholarly writer suffering writer's block, who visits Venice envisioning being liberated and uplifted. Overcome with the beauty of Tadzio, he cannot bring himself to speak, he fears insanity, and his thoughts are overcome with feelings, Eros. Dionysus has claimed him, soul and body, since his "beating heart and trembling limbs refused to obey my will".

We sense all along that the ageing Aschenbach (like the Elderly Fop he sees on the journey to Venice) is physically frail, and the hint that his heart may not stand the emotional experiences aroused by Tadzio is implied in the libretto.

The story closely follows Thomas Mann's *Novelle*. Tadzio is quite a pampered specimen (*ein verzärteltes Vorzugskind*), not the dancer of grace and perfection. Knowledge derived solely through the senses leads (cf. Phaedrus) into the abyss, with echoes of Euripides's *The Bacchae.* It is a path of dangerous charm, and as such can be the downfall of creative writers.

The boy embodies the Dionysian side of Aschenbach's character, which to this point has been suppressed beneath his cool detached Apollonian view of life. (The situation is analogous with that of Athanaël in *Thaïs*.) Scene 7 is called *The Games of Apollo*, and features the voice of the god ("Beneath dazzling sky the sea rolls silken-white… he who loves beauty worships me"). Scene 13 *The Dream* features the Voice of Dionysius ("Receive the stranger god"), opposed by Apollo ("No reject the abyss"). Tadzio's beauty is the salve to Aschenbach's writer's block, yet as Mann implies, it is perhaps best not to know the sources of artists' creativity.

An orchestral postlude recalls the ideal of beauty celebrated in the Hymn to Apollo, and counterpointed by bright tinkling music associated with

Tadzio. Like Britten's *The Turn of the Screw* (1954), the opera is divided into scenes connected by interludes. The score makes use of dance (with wine, the quintessentially Dionysian medium, while dance is the means of Tadzio's self-communication). As with *The Tales of Hoffman*, there is a recurring, morphing role (The Traveller, The Elderly Fop, The Old Gondolier, The Hotel Manager, The Hotel Barber, The Leader of the Strolling Players, the Voice of Dionysius) that contributes to Aschenbach's downfall. Like Dr Miracle, these roles are all taken by the same baritone. [**Figs. 16.1-5**]

2. Male Without Autonomic References

BÁNK BÁN (1861)

Opera in 3 acts by Ferenc Erkel (1810-1893). Librettist: Béni Egressy, based on a stage play of the same name by József Katona. First performance: Budapest, 8 March 1861.[3]

Act 1

King Endre II is the monarch of a poverty-stricken country, Gertrude is his queen and Bán Bánk the king's deputy. Otto, the Queen's younger brother, is trying to seduce Bánk's beautiful wife, Melinda.

Act 2

Bánk, distraught, prays over his nation and his good name. Tiborc, an old peasant, tells Bánk about the desperate state of the country. It is revealed that Tiborc, a vassal of the Bán, saved his life at a battle long ago; Bánk promises his aid. Otto's seduction is unsuccessful, and he drugs and rapes Melinda. She goes to her husband half-insane with shame. Bánk calls the Queen to account for plunging the country into poverty, and for the honour of his betrayed wife. Gertrúd scorns him and draws a dagger. Bánk wrests the dagger from her but in the scuffle she is fatally stabbed.

(*The chapel in royal palace*)

Bánk Bán
Torn by the storm without, and
Self-reproach within
One great feeling has kept me alive
Through all that care and grief.
My reputation is lost! ...

My homeland, my homeland, my everything!
I know I owe my life to you.
Golden fields, silver rivers running
With heroes' blood flooding with tears …
My Magyar homeland, I bless you!
'Tis sweet to live and to die for you, …
God made you in high spirits
Your face outshines the chorus of angels.
You were my heaven,
Everything of my soul. …

Bánk Bán
Cursed is the air that you breathe!
Cursed the land that bore you,
Cursed is your family,
Cursed! Vile Meranian!
(*Bánk stabs the queen, Gertrud dies.*)

Act 3

Insane, Melinda throws herself and her son into the waves of the Tisza River, within view of the helpless Tiborc. King Endre returns. Standing by his queen's funeral bier, Bánk admits that, convinced of her guilt, which was known to all, he killed the Queen. Tiborc arrives with the corpses of Melinda and the child. Bánk collapses dead over the bodies of his wife and son. The nobles and retainers pray for the repose of all the dead.

COMMENT

Bánk kills Gertrud, and he simply dies. Note the curses, and the things that Bánk has lost, also the contrast between the death of the Queen and that of Bánk. Confrontation between Bán Bánk and Gertrude is in the form of an impassioned duet in melodramatic mode, with surging lyricism suddenly cut short by the stabbing, the flow halted with strong orchestral commentary, especially in the striking writing for the deep brass, and notably the trombones and tuba.

Beauty is damned, and Bánk reveals the end through "the vengeance that trembles to honour itself".

The Act-2 finale begins with the brilliant colour provided by Melinda's highly decorated line. But this turns into a death march as the nobles confront the sequence of violent actions. In Act 3 the demise of Gertrude is underlined by the almost mystical entry of Tiborc with the news of

Melinda's mortal tragedy. Bánk Bán's death is depicted in his broken recitative with choral interjections. His lament ensues, with its prominent harp accompaniment, and, like Edgardo in *Lucia di Lammermoor*, he hopes for peace in heaven for Melinda and himself. His death is marked by a brief choral obituary.

DIE KÖNIGIN VON SABA (The Queen of Sheba) (1875)

Opera in four acts by Karl Goldmark (1830-1915). Librettist: Hermann Salomon Mosenthal. First performance: Vienna, Court Opera, 10 March 1875.[4]

Act 1

A hall in Solomon's palace

Sulamith, the daughter of the High Priest, and Assad, a diplomat, are due to marry the next day. They are planning the arrival of the Queen of Sheba to the court of King Solomon. Assad meets with Solomon and reveals he has fallen in love with a mysterious woman among the cedar forests of Lebanon, and does not love Sulamith. The Queen of Sheba arrives. Celebrations in her honour include a ballet, with a performance of *Bienentanz der Almeen*, and a bacchanal. As she greets the king, she pulls back her veil, revealing to Assad that she is the mysterious woman he had met on his journey. She pretends not to know Assad, to his confusion. Solomon tells Assad to not pursue his infatuation but to continue with his marriage to Sulamith.

Act 2

The garden of the palace at night

The Queen of Sheba has left the social gathering held in her honour. Astaroth, her slave, informs her that Assad is nearby and proceeds to lure Assad to her mistress with seductive oriental music ("Magische Töne"). Assad and the Queen embrace.

Assad and Sulamith are about to be married in front of the Ark of the Covenant when the Queen appears to give a wedding present. She continues to treat Assad like a stranger, which throws him into distress, and he blasphemes by referring to the Queen as his god, causing an uproar. Assad is led off to await punishment, most likely his execution.

Act 3

The court of King Solomon

Worried for Assad's fate, the Queen and Sulamith, at different times, plead for Solomon to give Assad mercy, but he refuses. The Queen leaves plotting revenge, and Sulamith departs for the desert to bewail her future.

Act 4

The vicinity of Sulamith's desert retreat

Solomon has banished Assad to the desert. The Queen of Sheba seeks him to convince him to come back with her to her kingdom. She finds him alone, not too far from Sulamith's retreat, and tries to seduce him. He rejects her advances and reveals his regret and desire for a death that might redeem his offense against God. Assad then prays for Sulamith, during which time he is engulfed in a violent sandstorm. He is later found barely alive by Sulamith and her companions. He begs for her forgiveness, which she bestows just before he dies in her arms.

COMMENT

This opera contains many important features of the Romantic era. There are pre-echoes of *Der ferne Klang* (magic tones), with a mysterious woman enchanting the protagonist; we have a forest (*Pelléas*), and an ending with redemption. The last act of presents a Love-Death situation similar to other operas. Here the hero Assad seeks out his first love, Sulamith, who, believing herself deserted, has sought refuge in the desert. Assad begs her forgiveness but has been exhausted by elemental forces and dies locked in her embrace. The plotline resembles *La Favorite*, where the repentant Leonore seeks out Fernand in his monastery and, begging forgiveness, dies in his arms. This is a pattern also seen in reverse in *Thaïs* where the monk seeks out the heroine, now a nun in a desert convent, who dies as he professes his undying love. The movement is also similar to that of *Tristan und Isolde,* where Isolde comes to the wounded Tristan in his Breton seclusion only for him to succumb to his injuries while she professes the transforming mystery of their love.

As in all these scenarios, Goldmark's score provides a *Liebestod* built around an extended duet. A solemn prelude leads into a fervent address from Sulamith, and this becomes a colloquy with Assad. A solo violin initiates a passionate declaration by Assad and the dramatic exchange between the two. An elegiac chorus opens Assad's final words, with focus on the notion of *Erlösing* (release). His death is marked by a choral postlude characterized by serene orchestral harmonies. This is one of several operas where *Unexpected Death* features in the finale in the context of some form of cataclysmic event.

LE VILLI (1884)

Opera-ballet in two acts by Giacomo Puccini. Librettist: Ferdinando Fontana. It is based on the short story *Les Willis* by Jean-Baptiste Alphonse Karr. Karr's story was in turn based on the Central European legend of the Vila, also used in the ballet *Giselle*. First performance: Milan, Teatro dal Verme, 31 May 1884.[5]

Act 1

Spring

Family and guests dance at a celebration of the engagement in marriage of Roberto and Anna. Roberto must leave before the ceremony to collect an inheritance, and Anna worries that she will never see him again. Anna confides to Roberto her dreams of him dying. Roberto tells Anna that she should not worry about his love failing: that she may doubt her God but not his love for her.

The scene is set for a celebration.

Mountaineers
The old woman of Mainz
Made Roberto her heir!
The treasures she hoarded
Are many indeed!
So Roberto tonight
Will leave
A poor man, but will come back
Rich to marry his lover!
Long live, long live,
Long live the newly betrothed!
Turn, now turn, spin and turn!
Jump, now jump, leap and turn!
The music throbs in delirium,
The dance propels us on!
Oh, how fast the hours fly by
As we dance on nimble feet!
Dancing is the rival of love

And speeds the beating of the heart!
Turn, now turn, spin and turn!
Jump, now jump, leap and turn!

Intermezzo

Roberto is enchanted by a Siren and forgets Anna. Anna waits through the summer and the autumn, and in the winter dies in his absence. The legend of the fairies (*le villi*) is that when a woman dies of a broken heart, the fairies force the heart-breaker to dance until death.

Narrator
In those days in the town of Mainz
A siren bewitched old and young alike.
She lured Roberto to the indecent orgy
And there he forgot his love for Anna.
Meanwhile, striken with ineffable anguish,
The betrayed maiden waited for him.
But she waited in vain: when winter fell
She closed her eyes in eternal sleep.

Act 2

Winter

Anna's father, Guglielmo, holds Roberto responsible for Anna's death and calls upon the Villi to take vengeance on Roberto. The Villi call upon Anna's ghost to lure a penniless Roberto into the forest. Roberto has been abandoned by the Siren but prays for forgiveness. This cannot succeed because of the curse put upon him by the Villi. As Roberto bewails his fate, Anna appears to him and tells him of the suffering that she had to endure. Roberto begs for forgiveness and he too feels the pain of Anna burning in his heart. Roberto is not forgiven and Anna calls upon the Villi, who with Anna will dance with Roberto until he dies of exhaustion at Anna's feet.

Roberto
I forgot her, I betrayed her,
And because of me she died.
But what a frightful sorrow
I shall have to suffer!
With remorse in my heart
I feel I am dying!

Roberto approaches Anna urged by an unknown force. Anna opens her arms and draws him to her. The Villi gather around the couple, dance furiously, then disappear.

Spirits *(offstage)*
Here we await you, traitor!
From us, expect no mercy!

Spirits and Villi
A man who in life was deaf to love
Shall find no forgiveness in death!
Traitor, we await you!
Turn! Jump! Turn! Jump!

Roberto sees the Villi pursuing him, his path barred by Anna, who grasps him and leads him into a wild dance among onrushing Villi.

Roberto
falling exhausted at Anna's feet
Anna, mercy!

Anna (*as she disappears*)
You are mine!

Spirits and Villi
Hosannah! Hosannah! Hosannah!

COMMENT

This is an instance of sudden death predicted by the Narrator. The Witches' Sabbath also portends the death of Roberto. Anna worries that she will never see him again ("Se come voi piccina"), and dies suddenly, possibly worn out by grief. Exhaustion and heat feature in the libretto and culminate in Roberto's final collapse. It is one of three operas in which dancing is fatal, the others being *Elektra* and *La vida breve*.

True to its generic designation, this opera has a strong orchestral-symphonic style. The theme of the story, deriving from the old German-Slavic legend of the Wilis (discussed in Chapter 9), further emphasizes its conscious appropriation of the motifs of this tale by imitation of its most famous expression in Adolphe Adam's ballet *Giselle* (1841). The themes of deception, betrayal and mental fragility are endemic to the story, with Anna's anxieties over Roberto's departure, and premonitions of a disastrous separation underpinning the pervasive sense of tragic irony that shapes the plotline.

The vocal dimension is dominated by Anna's opening aria of concern, Roberto's Act 2 outpouring of remorse, and the two duets for the lovers, the first developing Anna's relentless doubts; the second, part of the finale, a lyrical outpouring of the sadness of their thwarted love—with Anna now a supernatural being, and the whole an allegoric externalization of Roberto's mortal remorse. Both the arias and duets have the overtones of the mad scene about them, the duo-finale in particular having the quality and implication of a type of Love-Death, with the elemental being drawing the human victim into a delirious form of dying (the Undine-Rusalka syndrome).

The frenzied dance (*tragenda*) by which the Villi propel Roberto to death thematically dominates the second act, from the prelude to the finale, and adds to the literature of the *opéra-ballet* (*La Muette de Portici*, *Le Dieu et la Bayadère*) and the *ballet-blanc* where maleficent supernatural female beings exercise a fatal nocturnal-lunar hold over some unfortunate male who falls into their enchanted circle (*Robert le Diable*, *La Sylphide*, *Giselle*, *La Bayadère*, *Swan Lake*). Dance becomes the trajectory of physical frenzy, psychic rapture and sacrificial death (as in *Elektra*).

OSUD (1904)

Opera in three acts by Leoš Janáček. Librettist: Fedora Bartošová. The opera was written in 1904, broadcast in 1934, but first produced only at the Brno State Opera, 28 October1958.[6]

Act 1

Míla and the composer Živný were once lovers, but Míla's mother forbad the marriage, although Míla was pregnant. She and Živný meet again and rekindle their love, but her mother tracks them down in the crowd and predicts disaster.

> **Živný**
> Isn't a composer's function *ironically* to write postmortems on youthful passion?
> In anger, bitterness, in the face of an abject betrayal?
> (*uneasily*)
> I am here to seek an echo of the cry that exposed my spirit's emptiness.
> See how our souls are reaching out to us, calling to us from the darkness,

From that insane abyss where the hopes that once we had are lying, drowning in bitterness
Oh how glorious the harmony of human passion, when two spirits seek and find each other. We lived in a world of our own creation made of dreams.

Act 2

Four years later

Živný and Míla are married, but her mother lives with them. She becomes mad. The couple read through the composition of the unfinished opera Živný began during their separation. It portrays Mila as faithless. Míla's mother snatches the music from them before trying to throw herself off a balcony. Attempting to stop her, Míla too is pulled over, and both are killed.

Mila's mother
Oh, my heart is full of uncertainty, yours is always demanding, my heart is never free from anguish, yours is calling me. 'Come, oh come to me, Destiny'! ... But the songbird fell from the earth. You will see what I mean, after I am dead and gone.
(*She runs through the door to the flight of steps. Mila hurries after her and in the course of a struggle is pulled over the balcony as her mother throws herself off.*)

Zivny
Silence, Silence! Silence, silence, and a blast from a clear sky. With nothing to warn us of lightning. To warn us of it. Why was there no thunder? Why was there no thunder?
(*He sees the others carrying Mila's mother upstairs, and falls to the floor at Mila's side*).

Živný
Yet this is the music I wrote when first we fell in love.
I wrote the whole of this opera, and only the last act remains incomplete.
Oh, cruelty, such cruelty!
(*glancing at the score*)
This is how I took revenge. Wanting to tear out the heart from your body, and reveal your wounds in public, drowning your death knell, howling with laughter. I wanted to bathe in the tears you shed, and strike at your breast with a knife, hacking till I had killed you. Bring you to judgment; show you for a lying harlot! Lay bare your pretences, and tear off the mask that concealed your debauchery.
(*laying the body on the sofa suddenly falls to the floor at Míla's side*)

Act 3

Eleven years later

Živný's unfinished opera is finally to be performed. He rehearses a chorus from the opera with his students, among them his son, now a young man. Another student, Verva, guesses that the hero of the opera is the composer himself. Through the music, Živný again relives his love for Míla and his cruelty to her. Tormented by regret, he asks his son to bring him a glass of water and then collapses. The end of the opera must remain in God's hands.

Verva (*playing the piano*)
Well then, that's how the opera ends. He'd brought the whole of the opera except for the incomplete finale, which was "still in God's hands and there it would stay".

(*Živný enters unobserved, pale*)

How happy he was then! He felt that God was smiling on him. Confidence blossomed within him; no longer lonely, his passion embraced the all of human kind.

Živný
Where once their spirit shared a life of perfect unison, now their existence was always threatened by strident rhythms and by shrillness. Crashing and beating down on them, and in the end, strings that had bound them snapped in two.

(*Streaks of lightning Živný stands by trees in storms to watch them shake in terror. In the lightning's flash he saw the gates of infinity*)

It seems to be so clear, where the silver lightning shatters into fragments there is a white face, so pale and so sad! In the ecstasy of madness, now I see you once more!

Your heavenly face, your golden hair that falls around your brow, and your shining eyes smile at me…

(*Živný falls in a faint and lies unconscious: raising himself slightly*)

Can't you hear it too? That terrible sound! intones wordlessly. Listen now! intoning again expressively. That is her weeping passionately. Can't you hear it too?

Verva (*colleague, music lecturer*)

Could that be the music for the final scene?

Živný *(hearing him, and raising himself briskly)*

Music for the final scene? That is still in God's hands, and there it will stay! (*He clings to Doubek, who supports him with difficulty.*)

(The students crouch behind the organ. Doubek turns from Živný and looks at the lightning outside in alarm.)

COMMENT

There are no clear references of autonomic activity, yet this is a tale of *Unexpected Death*, based on love, revenge and a search for redemption on behalf of the composer, who cannot finish his own opera. The music of his previous harmonies reflected Mila's beauty, which assumed perfection of form. When their love came to an end, his libretto conveyed laughter and joy, but above all his pain. Sunshine is contrasted with tempests, and there is much on the beauties of nature, yet disturbed like the trees by storms. Živný has to silence the music forever. There are three deaths, but that of Živný has transcendental meanings. In respect of his hallucinations, his music for the final scene has a yearning for *Der ferne Klang* and the *Ewig-Weibliche*.

The last scene of the second act, with its themes of tacit recrimination, madness, suicide and accidental, involuntary death, begins in lyrical serenity, affection generating echoing effects in the musical texture. There is a sudden change on the entrance of the mad mother-in-law, with agitated strings and brass. The string writing becomes even more disturbed, and heavy treading chords assuming a type of ground bass. This builds into a passionate outburst, becoming louder and louder and fuller, reinforced by the deep brass. At her fall from the balcony, the high strings play breathless iterations (familiar from the Shower Scene in Alfred Hitchcock's film *Psycho*, 1960) with rapid figures in the brass instruments. All suddenly quietens, before the anguished outburst from Živný, followed by a huge crescendo, marked by a powerful and repeated motif for the deep brass.

The last act, set in the Conservatory during rehearsals for the new opera, perhaps uses an idiom out of proportion to the events depicted. It is better suited to a more epic scenario. The composer arrives among the excited students and begins to explain his music. He relates to his own experiences, and gradually the autobiographical element asserts itself, and begins to take him over. Soon his description becomes an ever-impassioned monologue in which the artistic creation and the summoning up of the past become intertwined, until the past is actual in the present, with external correlative in the storm. The emotional strain exerts a psychosomatic effect, and he is gradually overtaken to the point of personal paroxysm and succumbs to a physical collapse. The music has all of Janáček's sharp and lucid orchestral colouring. The commentary grows incrementally until the moment of seizure. The lower strings provide a sinister iterative commentary, like life ebbing away, broken by terrific blasts of the higher bright brass, before the death, which is punctuated by the immense enunciation of the fate motif on the lower brass. The analogies with Marty's final monologue in *The Makropulos Affair* are striking.

DER FERNE KLANG (The Distant Sound) (1912)

Opera in three acts by Franz Schreker (1878-1934). Words by the composer. First performance: Frankfurt, 18 August 1912.[7]

Act 1

Fritz, a composer, wants to marry Grete, but before that happens he has to write a piece of music and discover the mysterious far distant sound (*der ferne Klang*) which he hears within him. Fritz leaves her and goes in search the distant sound (*Leb wohl*).

> When the wind with ghostly hand
> strikes over harps from a far …
> and I seek the craftsman who touches the harps
> And seek the harps which give birth to the sound
> and I stop the sound, I will be rich and free,
> a master artist of God's grace … then I can come back.

As Grete is returning to her house, she meets a strange old woman, who asks the surprised girl about Fritz and promises to help Grete if she needs it. Grete continues her way home.

Grete's mother, Frau Graumann, tells Grete about the family debts. Grete decides she should get a job and complains that her father drinks too much. Frau Graumann is disturbed her daughter should work. Grete's father, Graumann, arrives with an actor and Dr. Vigelius. Graumann has just gambled his daughter away to his landlord in a game of skittles and they have come to collect the debt. When Grete refuses, her father becomes furious.

Grete pretends to be happy to marry the landlord. But when her mother leaves her alone in the room, she jumps out of the window and goes off to find Fritz. Alone in a wood she cannot find him. She thinks of drowning herself, but the moon comes out and she is aware of the beauty of nature at night. She falls asleep, dreaming of their love. The old woman, in reality a prostitute, appears and promises to bring her help.

Act 2

Ten years later: an island in the Gulf of Venice. Grete still thinks of Fritz, but whoever can touch her heart the most deeply with a song will win her. The Count sings "In einem Lande ein bleicher König", a sad but beautiful song about a King with a burning crown which torments him, which he throws into the sea from which a pale woman emerges to drag him into the waves. A stranger appears in the midst (*Der König* from the ballad). It is Fritz, who recognizes Grete immediately and tells of his vain search for the Distant Sound. He wants to make her his wife but suddenly realises that she is now a prostitute and leaves. She, in despair, falls into the arms of the Count.

Act 3

Five years later, Fritz has completed his opera, *Die Harfe*. During the premiere, the first act goes well but the second act is a failure.

> The disappointment! a pitiful end! …
> the harp would not sound.

Grete, meanwhile, now a common streetwalker, hears that the composer is ill, and faints.

Fritz is at home, old and depressed, and laments his lost love. It is spring. His friend Rudolf tries to cheer him up and reminds him that there is still time to rewrite the opera. Fritz tells him that he is near the end of his life and only wants to see Grete. She is found by Dr Vigelius, who explains the wager and brings Grete to Fritz. They embrace, and at last Fritz hears the harp music he has searched for so long. He starts to compose a new ending for his opera, but lies dying in the arms of his beloved.

> **Fritz**
> How strange it is! I have become old—
> tired in fruitless bonds, gray from worries
> sick from bitter torments and ripe for the grave.
> *(alone with Grete)*
> Your cheeks, your arms are so pale—

Grete
It is joy, my dearest, that takes the blood from my cheeks
And your eyes shine as in a fever, and your heart beats so strongly.

Fritz
Do you hear the sound? The harp rings to me, as the spheres sound—
powerful and rushing—and there on the mountains fire is flaming upwards
.

Grete
The smoldering spark that never goes out,
the intense embers that death only cools,
the desire for love …
Fritz be at peace, soon will you be better,
then will we at last finally be happy [*glücklich sein*].
Sleep my dearest—do you hear—
you are so aroused—on my heart you will find peace [*Friede*].

Fritz
It is finished, the last unsuccessful act, now I have found you.

COMMENT

The themes in the opera bring in nature, the night, the moon, an interlude of the King's crown (cf. Goethe's *Ballad of the King of Thule*—as in the Faust works by Berlioz and Gounod), the final compositional revelation like that of Živný in Janáček's *Osud*. It also contains dreams. It is a tale of the search for love, lost and refounded through music, of rejection and redemption, and the transcendental death of Fritz, old and lonely, as the music dies and the flames arise before him in the arms of Grete—in all, a *Liebestod* to be sure.

This is Schreker's best-known work, and an early example of musical Expressionism. The final moments begin with Fritz singing over arpeggiated figures on the celesta (an instrument that became very fashionable in the late 19th century for parts formerly reserved for the harp). Excitement grows in the duet with Grete, sometimes *a due*, before subsiding. Fritz's part becomes increasingly restrained, detached, almost mystical, with the harp and the cor anglais prominent. The celesta and harp become consistent, the harp playing wide arpeggiated upward chords. Grete's part assumes the nature of a lullaby, softer horns heard as from the distance. Fritz now speaks as he hears the distant sound, the celesta and triangle approximating the magical tones as he slips away. Grete cries out as she realizes he is dead. The theme and the nature of the music relate to the myth of the Ptolemaic spheres. There is also a link to Gustave Holst's suite *The Planets* (1920), to the ethereal music for Neptune, sending the soul out into infinite space.

LULU (1937)

Opera in three acts by Alban Berg (1885-1935). The libretto was adapted by the composer from Frank Wedekind's (1864-1918) two *Lulu* plays, *Erdgeist* (*Earth Spirit*, 1895) and *Die Büchse der Pandora* (*Pandora's Box*, 1904). First performance: Zürich, 2 June 1937.[8]

In the Prologue a circus animal tamer welcomes the audience. Lulu comes on dressed as a Pierrot, and is referred to as evil, fated to murder.

Act 1

Scene 1

The artist (Walter Schwarz) is painting Lulu's portrait. Dr. Schön is watching, and is joined by his son, Alwa. They leave. The artist pursues Lulu, refers to her as Eve, she gives way to his advances. Her husband Dr Göll disturbs them and drops down dead. Realising her husband has died, Lulu states that she is now rich. She and Schwarz sing a duet in which he questions her beliefs, but the answer is always the same.

> "A question: Can you tell the truth?"—"I don't know."

Scene 2

Schwarz says he has sold several paintings of Lulu, now his wife. She is visited by Schigolch a beggar (who, in an unspecified way, has somehow featured in her past). He asks for money, which she gives him, and when he calls her "Lulu", she states she has not been called that in a long time. Lulu is disturbed to read that Dr Schön is engaged. Schön arrives and recognizes the beggar, referring to him as Lulu's father, which she does not deny. Dr. Schön asks Lulu to no longer see him, to avoid a scandal and to maintain his respectable marriage, but she sings

> If I belong to any man in this world, I belong to you

Schön implies that he has had a longstanding affair with Lulu, since she was twelve, when she was a street girl. The artist is distressed, realizing that he knows little about Lulu, not even her name, which appears to be different for every lover. Schön confirms that Schigolch, after the death of his wife, arranged to marry her off to Dr Göll. The artist locks himself in his studio. Lulu brings a hatchet, the door is forced open, but the artist is dead, having cut his throat. Alwa arrives, and reports that a revolution has broken out in Paris. Lulu is unmoved by the tragedy. Schön calls her a monster, but she hints that she and Dr. Schön will be married after all.

Scene 3: Lulu's dressing room in the theatre

Schön has put Lulu (now as Mignon) on the stage. She creates a scene over his fiancée forcing him to write a letter breaking off the engagement. She faints.

Act 2

Scene 1. House of Dr Schön

Now married to Mignon, Schön finds Lulu declaring love to his stepmother, with Countess Geschwitz drooling over her. He hands her a gun and tells her to shoot herself, but she shoots him and is arrested.

Scene 2

She is imprisoned, but escapes after switching places with Countess Geschwitz. Alwa, Schön's son, and Lulu flee to Paris, then to London.

Act 3

Scene 2 London

Alwa has syphilitic insanity, Lulu again becomes a street walker. She is murdered by Jack the Ripper.

COMMENT

In this opera there are several deaths, and one of them at least rates as *Unexpected Death*. Each of the three men with whose deaths she is involved becomes one of Lulu's lovers. Lulu's first husband has a fatal stroke on catching his wife *in fragrante delicto*. Her second one cuts his throat unexpectedly (a suicide), and she shoots Dr Schön. The part of Jack the Ripper is sung by the baritone who earlier was Dr Schön.

How does one understand this melange of tragedy and death, and the role of Lulu with her other pseudonyms? A love goddess with destructive powers? a nymphomaniac? a victim of childhood sexual abuse? or a product of nature (part of the circus menagerie)? Lulu is also a Pierrot,

Petrouchka: Alwa refers to her as "a soul that will rub the sleep from its eyes in the next world". Is she related to the biblical Lilith (Book of Isaiah 34:14), linked to chaos, but credited as the first wife of Adam, the dark side of Eve, and identified as the serpent in the Garden of Eden. Lilith features in Goethe's *Faust* (Part 1) on *Walpurgissnacht.* The *femme fatale*, Salome, Thaïs and Turandot are embedded within her character. Lulu thus seems to combine the polarities of the rapacious Lilith (as night monster and vampire of Babylonian origin, haunting the wilderness) and the tragically fated Patrushka (a bitter-sweet jester from Russian folklore).[9]

Lulu, the character being the theme of the opera itself, is the personification of the evil generated and perpetuated in the world by the sexual drive, a power seen as a wild and infernal fire. Wedekind's play presents situations of pain and hopelessness, but Berg's adaptation provides positive glimpses of life and optimism.[10]

This dual concept infuses the music: it is generally bitter and disillusioned, but there are also deeply lyrical moments, especially for Lulu herself, as though providing glimpses of some redemption for her tragic and devilish character. The orchestra also generates rich subtleties and various moods, sustained by instrumental patterns. The twelve-tone system (dodecaphony) is used with care and consistency, very different from the rawer idiom of Schoenberg, and is consistently linked with the lyrical elements.

Berg was preoccupied with symmetry and he uses the technique of the palindrome (a theme the same forwards as backwards) in various ways to underscores the symbolism operative in the work. The whole opera is perceived as a mirror reflecting on Lulu herself: her affluent popularity in Act 1 is mirrored darkly by the squalor and degradation of Act 3, with her husband in the first act played by the same singers who portray her sexual clients in the third, and with the motifs of each character repeated. This sense of repetition and echoing, of eternal recurrence, is further emphasized in the Film Interlude in Act 2 at the centre of the work. Here the events depicted in the film mirror the structure of the work as a whole, with Lulu entering prison then leaving (as if caught up in an inescapable cycle). The accompaniment to the film is an exact musical palindrome, marked at its midpoint by a piano arpeggio, first rising then falling.

THE TURN OF THE SCREW (1954)

Opera by Benjamin Britten. Librettist: Myfanwy Piper, after a short story by Henry James (1898). The opera is shaped by 15 orchestral variations on a developing theme, each variation a 'turn' of the ever-tightening 'screw'. First performance: Venice, Teatro La Fenice, 14 September 1954.[11]

(*The action takes place in and around Bly, a country-house in the East of England, in the middle of the 19th century.*)

Prologue

(The Prologue is sung in front of a drop curtain, and tells the backdrop.)

A singer tells about a young governess (who remains unnamed) who had charge of two children at Bly House. She was hired by their uncle and guardian in London who was too busy to care for them. He imposed three stipulations on the Governess: never to write to him concerning the children; never to inquire about the history of Bly House; and never to abandon the children.

Act 1

The Governess is anxious about her new position. Mrs. Grose is the housekeeper. When the Governess meets Miles, the boy, they look into each other's eyes and the Governess feels a strange connection with him. A letter from Miles's school arrives, informing her that the boy has been expelled but providing no reason. The Governess is certain that Miles (like his sister Flora) is too innocent to have done anything bad. She resolves to disregard the letter.

The Governess is very much troubled by footsteps outside her door, and also by cries she has heard in the night. She observes a very pale man positioned on a tower of the house. Later she observes the same pale man looking in through a window. Mrs. Grose tells her it may be Peter Quint, the former valet at Bly House. Mrs. Grose infers that Quint may have preyed on Miles, and that he had a sexual relationship with Miss Jessel, the former beautiful young governess. Mrs. Grose also hints that Miss Jessel had an inappropriate relationship with the children. Miss Jessel left her post and soon died. Shortly afterwards, Quint also died under mysterious circumstances on an icy road near Bly House.

The next morning, while the Governess teaches Miles Latin, he enters into a trance-like state and sings a song “Malo”.

Later that day at the lakeside, Flora compares the Dead Sea with Bly House. The Governess observes a strange woman across the lake, who is seemingly watching Flora. The Governess is horrified, concerned that the woman is the ghost of Miss Jessel, who has returned to claim Flora. That night, Miles and Flora meet Miss Jessel and Peter Quint in the woods and fantasise about a world where dreams come true. The spirits vanish. Miles sings a song about how he has been a bad boy.

Variation IV

Governess
I have been frightened.
His hair was red, close-curling,
a long, pale face, small eyes.
His look was sharp, fixed and strange.
He was tall, clean-shaven, yes,
even handsome.
But a horror!

Mrs. Grose
Quint, Peter Quint,
the master’s valet.
I had only to see to the house.
But I saw things elsewhere I did not like.
When Quint was free with everyone—
With little Master Miles—
No, Mr. Quint,
I did not like your ways!
And then she went.
She couldn’t stay, not then.
She went away to die.

Governess
To die?
And Quint?

Mrs. Grose
He died too.
Fell on the icy road—
struck his head, lay
there till morning, dead!

Governess
I know nothing of these things.
Is this sheltered place
the wicked world
where things unspoken of can be?

Variation V

Scene 6. The Lesson (The Governess is teaching Miles Latin)

Miles
Malo: I would rather be
Malo: in an apple-tree
Malo: than a naughty boy
Malo: in adversity.

Governess
Why, Miles, what a funny song!
Did I teach you that?

Miles
No, I found it.
I like it.
Do you?
Malo, Malo, Malo ...
(*The scene fades.*)

Variation VII

Scene 8. At Night

Quint
(unseen)
Miles!
Miles!
Miles!

(The lights fade in on the front of the house and the tower. Quint is on the tower. Miles in the garden)

Miles
I'm here ... O I'm here!

Quint
I am all things strange and bold,
The riderless horse snorting,

stamping on the hard sea sand,
The hero-highwayman
plundering the land.
I am King Midas with gold in his hand.

Miles
Gold, O yes, gold!

Quint
I am the smooth world's double face,
Mercury's heels feathered with mischief
and a God's deceit.
The brittle blandishment of counterfeit.
In me secrets,
and half-formed desires meet.

Miles
Secrets, O secrets!

Quint
I am the hidden life that stirs
When the candle is out;
Upstairs and down, the footsteps
barely heard.
The unknown gesture, and the soft,
persistent word,
The long sighing light of the
night-winged bird.

(*Quint and Miss Jessel disappear. The Governess runs to Miles.)*

Mrs Grose
Why, whatever's going on?
Miss Flora out of bed!

Governess
Miles!
What are you doing here?
(*Mrs. Grose takes Flora away.*)

Miles
You see, I am bad, I am bad,
aren't I?
(*Miles goes into house followed by the Governess as the lights fade.*)

Act 2

The ghosts of Peter Quint and Miss Jessel reappear. They discuss who harmed whom first when they were alive, and accuse one another of not acting quickly enough to take possession of the children.

The next morning, the family attends church. Mrs. Grose declares that nothing can be amiss if the children sing and behave as sweetly as they do. But when Miles mentions the ghosts of Quint and Jessel, the Governess realizes things are much more serious than they seem.

The Governess goes into the children's schoolroom where she sees the ghost of Miss Jessel seated at the teacher's desk. The spectre laments her fate, and speaks of her sufferings in the afterlife. That night the voice of Quint calls out to Miles, terrifying him. The lights go out, and the ghost hovers over the terrified child. Flora begins to rave about unspeakable horrors. The Governess confronts Miles alone. As she questions him, the ghost of Quint appears and pressures Miles not to betray him. Miles blurts out Quint's name and Quint's ghost vanishes. Miles falls on the floor, the Governess cradles the dead child in her arms, singing of her grief and wondering if she did the right thing after all.

Variation VIII

Scene 1. Colloquy and Soliloquy

(*The lights fade in on Quint and Miss Jessel—nowhere.*)

Quint
I seek a friend
Obedient to follow where I lead,
he shall feed
my mounting power.
Then to his bright subservience
I'll expound
the desperate passions
of a haunted heart,
and in that hour
"The ceremony
of innocence is drowned".

Miss Jessel
I too must have a soul to share my woe.

"The ceremony
of innocence is drowned."
(*The Ghosts come together.*)

Variation XV

Governess
Miles, dear little Miles,
who is it you see?
Who do you wait for, watch for?

Quint
Do not betray our secrets.
Beware, beware of her!
Miles, you're mine!
You must be free.
On the banks, by the walls,
remember Quint.
At the window.
On the tower,
when the candle is out,
remember, Quint.
He leads, he watches,
he waits, he waits.

Miles
Peter Quint, you devil.
(*He runs into the Governess's arms.*)

Governess
Ah, Miles, you are saved,
now all will be well.
Together we have destroyed him.

Quint
Ah Miles, we have failed.
Now I must go.
Farewell.
Farewell, Miles, farewell.
(*Quint slowly disappears*)

The governess jumps up, reduced to the mere blind movement of getting hold of Miles. Miles continues to be uneasy and muddled and seems to think of something far off. Miles has pain and shortness of breath and, as he poses a question, he is quite detached and helpless, he struggles for air. The governess's violence contributes to his sudden death.

Governess
Ah! What is it?
Miles, speak to me, speak to me.
Why don't you answer?
(*She realizes that the boy is dead.*)
Miles, Miles, Miles!
Don't leave me now!
(*She lays him down on the ground.*)
Ah! Miles!
Malo, Malo!
Malo than a naughty boy.
Malo, Malo in adversity.
What have we done between us?
Malo, Malo, Malo, Malo, Malo...

COMMENT

In contrast to *Death in Venice*, the plot of *The Turn of the Screw* is much more complex. The Governess has several 'turns', and the 'pure children' are anything but innocent. Her attacks comprise hallucinatory experiences, visual apparitions of a sinister nature accompanied by a strange sensation of dread and death, associated with feelings of unfamiliarity. These dream-like states describe not only *déjà vu* reminiscences, but a strange loss of temporal awareness. In one of her attacks she recounts how she fell to the ground only to regain awareness some time later in the day. At the climax of the story, the Governess is alone with the boy Miles, she has another 'turn', but in order to protect the boy from the spectre "in a blind moment", with a tremor of the hands and falling, she gets hold of him and he dies in her arms.

Henry James remained secretive about the nature of the Governess's visions, not wanting to spoil the mystery he had created. His father had hallucinatory experiences, and his sister Alice was a chronic invalid, having episodes which at the time were called 'hysteria'. His brother William was the prominent psychologist and was well acquainted with strange mental states; we have noted his transcendental credentials earlier.

The neurologist James Purdon Martin (1893-1994) was in little doubt that Henry James would have been acquainted with the writings of his brother William, and the latter's knowledge of the ideas of the famous London neurologist John Hughlings Jackson (1835-1911) on epilepsy. He speculated that Henry James visited the latter and learned that the subtle but sometimes dangerous psychological and motor manifestations of temporal lobe epilepsy could have been related to the death of Miles, namely by strangulation from the grip of the Governess in a seizure. The link to a 'turn' seems obvious, and the implication was made clear in Jonathan Miller's production of the opera in which on stage she suffocated him while in a seizure (ENO, 1979).

Whereas in *Death in Venice* the elder man is the one seduced, and not the seducer, in the *Turn of the Screw* the plot has no such clarity. The evil nature of Quint needs to be set alongside the perhaps paranoid experiences of the Governess. Yet red-headed Quint, as James wrote in the novella "was much too free. Too free with everyone". He has the world's double face, secrets with half-formed desires operating when the candle is out. "The ceremony of innocence is drowned" sing Quint and Jessel together, the line taken from a poem by W. B. Yeats (1865-1939) ("How Far to Bethlehem"). Incest between the children, the sinister nursery rhymes, the guilt of Miles' expulsion from school, his reply to Quint "I'm here... O I'm here!", and Miles's eerie refrain ("Malo"), reinforce the Gothic nature of this opera, and the unusual nature of the *Unexpected Death*.

The two acts and eight scenes are connected by interludes in the form of variations on a short theme which opens the opera after the sung prologue. The musical structure of the work is given strength and unity by this scheme. As with *The Queen of Spades*, the music generates and grippingly portrays an eerie, oppressive atmosphere of the supernatural. The variations unfolded in the interludes serve to intensify, like a tightening screw, the course of the drama and the inexorable closing-in of the tragic (and evil) outcome. Scene 8 is titled 'Miles' and begins with Variation XV. The final interlude and scene are shaped in the form of a passacaglia, with a segment of the principal theme in the bass, and uses tumultuous timpani. After the struggle with the ghost of Quint, and despite Quint's desperate imploring (with themes from the Act 1 finale recurring, and the principal theme now unfolded in the bass in its entirety), Miles names his tormentor, and the Governess pronounces, "Now you are saved!" Quint disappears, fades from sound. But Miles is dead. The Governess lapses into a singular, restrained lullaby over the boy's corpse, using his lament "Malo", which becomes his requiem. It shares in the generic qualities of a mad scene and is reminiscent of the Simpleton's epilogue in *Boris Godunov*. [**Figs. 16.6-10**]

SAINT FRANÇOIS D'ASSISE (St Francis of Assisi) (1983)

Opera by Olivier Messiaen (1908-1992). Libretto by the composer, with the poetry of St Francis of Assisi. First performance: Paris Opéra, 28 November 1983.[12]

Place: Italy, Time: 13th century.

Act 1

Francis is travelling with one of his monks, Brother Leo, and teaching him the meaning of 'perfect joy', which he says is to be found in the acceptance of suffering while thinking of the Crucified Christ.

Then Francis and his community are discovered at prayer, and the saint sings verses from his *Lodi delle creature*.

Finally he meets a Leper in great physical and spiritual distress. Francis tries to teach him acceptance, but Francis is rejected until an Angel appears and sings that God is love. Francis, realizing that he has not loved the Leper sufficiently, embraces him and brings about a miraculous cure. The act ends with a choral epilogue.

Act 2

The Angel returns to the monastery door and questions the monks about predestination. One responds angrily; another gives a different answer, and the Angel leaves.

The Angel appears to Francis, who is at prayer. On his viol he plays a celestial melody at which Francis faints. Three brothers come to find him, and Francis awakens, saying that if the music had continued longer his soul would have parted from his body.

Francis is with Brother Lasseo, identifying the birds, to which he preaches, praising their gifts of flight, of freedom, and of being able to sing wordless music, like the angels. He blesses them, and after a moment of silence there begins a great bird concert, with different instruments and ensembles playing different songs out of synchrony. The birds then fly off, making a great cross in the sky.

Act 3

It is at night, and the orchestra becomes sombre, falling to owl calls and severe 12-note patterns. Francis prays that before his death he may feel in his body and heart the anguish of the Crucified Christ, and the chorus responds with Christ's words of acceptance of his sacrifice.

Francis
Lord! Lord!
Lord! Lord! Music and Poetry
Have led me to you: by image, by symbol,
And by lack of Truth.
Lord! Lord!
Lord! Lord! Lord illumine me
with your Presence! Deliver me, enrapture me,
dazzle me always by the overflowing of your
Truth…
(*He dies.*)

COMMENT

The composer reflects the psychological and emotional state of the characters through the use of *Leitmotif* and birdsong. Several motifs exist in the orchestral score, most of which connect to one or more characters: Death (or "j'ai peur"); Perfect Joy ("la joie parfait"); Solemnity; Grace. The large orchestra requires more than 110 musicians. Pitched percussion instruments are also used: a xylophone, a xylorimba, a marimba, a glockenspiel and a vibraphone, as well as three *Ondes Martenot*. The opera requires a 10-part, 150-voice choir, which serves a twofold role: Greek chorus and divine presence.

Throughout the piece the chorus comments on Francis' spiritual journey and his autonomic disturbance. Messiaen's synaesthesia caused a perception of colours associated with particular harmonies or musical scale degrees. In Act 3 scene 8 (*Death and the New Life*) an introduction with Francis' theme is played in the interior of the little Portinuncula Chapel, Assisi, where he lies dying among his brethren and birds, and bids farewell to all he has loved. The Angel and the Leper appear, now richly clad. All go out: it becomes dark. Only one intense light illuminates the spot where Francis died. The chorus sings the last hymn on the text of the Resurrection. The light grows ever brighter until it becomes blinding and unbearable. The final moments (*Francis' Death and Ascent into Heaven*) are underscored on a C major chord structure, providing the musical realization of the burst of white light. [**Fig. 16.11-12**, St Francis]

Notes

[1] Mussorgsky, *Boris Godounov*, with Boris Christoff and Nicolai Gedda, Orchestre du Conservatoire Paris, cond. André Cluytens (EMI).
[2] Britten, *Death in Venice*, with Peter Pears and John Shirley-Quirk, English Chamber Orchestra, cond. Stuart Bedford (Decca).
[3] Erkel, *Bánk-Bán*, with Atilla Kiss and Eva Marton. Hungarian Millennium Orchestra, cond. Tamás Pál (Hungaraton).
[4] Goldmark, *Die Königin von Saba*, with Klára Takács and Siegfried Jerusalem, cond. Adam Fischer (Hungaroton).
[5] Puccini, *Le Villi*, with Placido Domingo and Renata Scotto, National Philharmonic, cond. Lorin Maazel (CBS Masterworks).
[6] Janáček, *Osud,* with Philip Langridge, Helen Field and Kathryn Harries, Welsh National Opera, cond. Charles Mackerras (Chandos).
[7] Schreker, *Der ferne Klang*, with Thomas Moser and Gabriele Schnaut, Radio-Symphonie Orchester Berlin, cond. Gerd Albrecht (Capriccio).
[8] Berg, *Lulu,* with Teresa Stratas, Orchestre de l'Opéra de Paris, cond. Pierre Boulez (DG).
[9] The word *lilit* (or *lilith*) only appears once in the Hebrew Bible, in a prophecy regarding the fate of Edom. Her name is often translated "the Night Hag", as in the RSV. Quoting from Isaiah 34 (NAB):

> (12) Her nobles shall be no more, nor shall kings be proclaimed there; all her princes are gone. (13) Her castles shall be overgrown with thorns, her fortresses with thistles and briers. She shall become an abode for jackals and a haunt for ostriches. (14) Wildcats shall meet with desert beasts, satyrs shall call to one another; There shall the Lilith repose, and find for herself a place to rest. (15) There the hoot owl shall nest and lay eggs, hatch them out and gather them in her shadow; There shall the kites assemble, none shall be missing its mate. (16) Look in the book of the LORD and read: No one of these shall be lacking, For the mouth of the LORD has ordered it, and His spirit shall gather them there. (17) It is He who casts the lot for them, and with His hands He marks off their shares of her; They shall possess her forever, and dwell there from generation to generation.

In Hebrew legend she was the first wife of Adam and became a night phantom and the enemy of childbirth and the new-born. She is symbolic of the Terrible Mother, and linked to the Greek figure of Hecate, with her demands for human sacrifice. See J. E. Cirlot, *A Dictionary of Symbol*s (New York: Philosophical Library, 1962), 180.
[10] "'Marriage as Prostitution', the fifth chapter, comes from dos Santos's previous writings on Lulu and her fraught relationships. A number of cultural and intellectual currents drove Berg as he composed; the ideologies of Lucka, Wagner,

Karl Kraus, and Otto Weininger contributed to his portrayal of husbands and joins them as one and the same. Political realities also helped to shape Berg's sense of identity and agency, and, in turn, his music. Without the influence of his activist sister Smaragda, for instance, Berg might not have given the moving final words of *Lulu*—a Wagnerian *Liebestod*, no less—to the Countess Geschwitz, whose sexuality would have warranted imprisonment in Austria at the time. The final chapter of dos Santos's monograph focuses on Lulu's most faithful follower, whose very twelve-tone *leitmotiv* speaks to her characteristic androgyny. In the end, it is her unrequited love for Lulu that transcends death; not various suitors' attempts to own Lulu, nor Alwa's metaphysical longings." Nicholas Stevens in reviewing Silvio J. dos Santos, *Narratives of Identity in Alban Berg's 'Lulu'* (Rochester, NY: University of Rochester Press, 2014).

[11] Britten, *The Turn of the Screw*, with Jennifer Vyvyan and Peter Pears, English Opera Chamber Group, cond. Benjamin Britten (Decca).

[12] Messiaen, *Saint François d'Assise*, with Dietrich Fischer-Diskau, Radio-Symphonie Orchester, Radio Wien, cond. Lathor Zagrosek (Orfeo).

1.1 Gustave Moreau (1826-1898), *Salome the Head Hunter*, even while dancing

1.2 Aubrey Vincent Beardsley (1872-1898), *Salome*

1.3 Orpheus, Roman mosaic

1.4 Orpheus, Roman mosaic

1.5 François Perrier (1590-1650), *Orpheus before Hades and Persphone*

1.6 Federigo Cervelli (1625-1700), *Orfeo ed Euridice*

1.7 Jacopo Vignale (1592-1654), *Orfeo*

1.8 Émile Lévy (1826-1890), *The Death of Orpheus surrounded by Bacchantes*

1.9 Jean Delville (1867-1953), *The Death of Orpheus*

1.10 Gustave Moreau (1826-1898) Orpheus on the tomb of Eurydice

1.11 Friedrich Nietzsche (1844-1900)

1.12 Arthur Schopenhauer (1788-1860)

1.13 Sappho, Archaeological Museum, Pompei

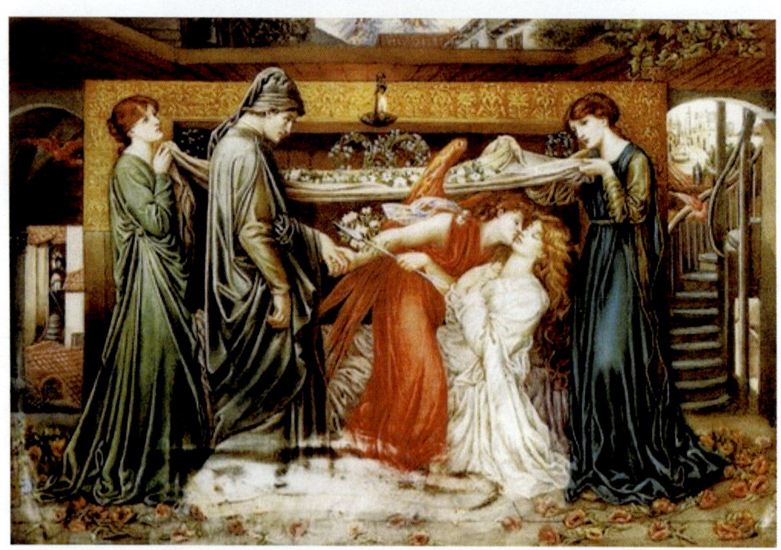

1.14 Dante Gabriel Rossetti (1828-1882), *Dante's Dream at the Time of the Death of Beatrice* (1871-81). Love leads him to her bier, leaning over to kiss Beatrice

1.15 Joseph Noel Paton (1828-1901), *Dante Meditating on the Episode of Francesca and Paolo Malatesta*, 1871

1.16 Dante Gabriel Rossetti, Beata *Beatrice*. Dante is behind her, as light forms a halo around her, her transcendent state suggested by her eyes, in a rapture of impending death

1.17 Dante Gabriel Rossetti, *The Story of Paolo and Francesca*

1.18 William Blake (1757-37), *The Lovers' Whirlwind, Paolo and Beatrice*

1.19 Raphael (1483-1520), *The School of Athens*

1.20 Gianlorenzo Bernini (1598-1680), *The Ecstasy of St Teresa*

2.1 Johann Heinrich Tischbein (1751-1829), *Goethe in the Roman Campania*

2.2a Novalis (Friedrich Leopold von Hardenberg) (1772-1801)

2.2b Novalis, Sophie and her gravestone

2.3 Casper David Friedrich (1774-1840), The Cross in the Mountains

2.4 Wolfgang Amadeus Mozart (1756-1791)

3.1 Théâtrephone

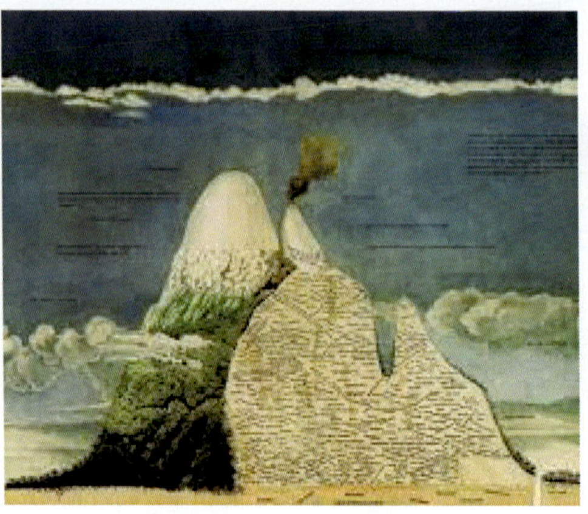

4.1 Alexander von Humbolt (1769-1859), *Naturgemälde*

4.2 William Blake, *Newton*, a rationalist preoccupied with calculations

4.3 Goethe's *Urpflanze*, two-leaved, joined

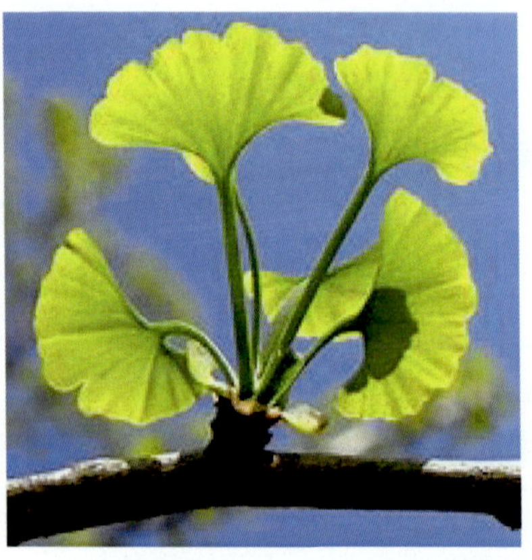

4.4 Goethe identified the *gingo biloba* or *ginko biloba*, associated with the *Urpflanze*

4.5 André Brouillet (1857-1914), *Les fascines de la Charité, service du Dr Luys*, 1880

5.1 Jules-Bastin Lepage (1848-1889), *Joan of Arc* (1884). Her out of body image is in the background

6.1 Francisco Goya (1746-1828), *Love and Death* in the collection *Los Caprichos*, 1799

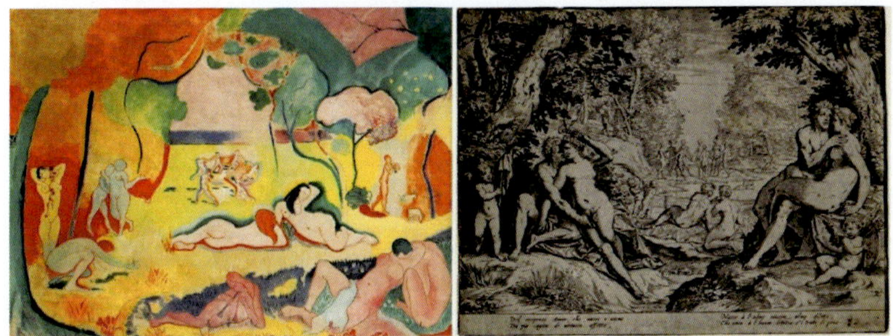

6.2 Henri Matisse (1859-1954), *Joie de vivre*; Renaissance print, *Erotic Love*

6.3 Sir Edward Coley Burne-Jones (1833-1898), *Laus Veneris* (In Praise of Venus) (1873-75), reflecting the hopeless Venus oblique

6.4 Evard Munch (1863-1944), *Madonna*,1895

7.1 John William Waterhouse (1849-1917), *Undine*, 1872

7.2 Thomas Cooper Gotch (1854-1931), *The Bride Death raising her hand in a gesture of welcome*

9.1 German Romantics Friedrich de la Motte-Fouqué in Husarenuniform

9.2 Alexander Pushkin (1799-1837)

9.3 Heinrich Heine (1797-1856)

9.4 Paul Gaugin (1848-1903), *In the Waves*, 1889

9.5 Thomas Mille Dow (1848-1919), *The Kelpie* (1895), a watersprite with vipers dripping from her hair

9.6 Edward Burne-Jones, *The Depths of the Sea*

9.7 William Blake, *Oh Flames and Furious Desires*, from *Urizen* plate3, 1794-96

9.8 William Reynolds-Stephens (1862-1943), *Summer—Women and Waves*, 1891

10.1 Auber, *La Muette de Portici*, Adolphe Nourrit as Masaniello

10.2 Rossini, *Guillaume Tell*, trio (Act3)

10.3 Bizet, *Les Pecheurs des Perles*, duet (Act 1)

11.1 Edward Burne-Jones, *The Golden Stars*

11.2 Beethoven, *Fidelio*, Wilhelmine Schröder-Devrient as Leonore

11.3 Mozart, *Die Zauberflöte*, stage design by Kark Friedrich Schinkel (1781-1841)

11.4 Weber, *Der Freischütz*, stage design for the Wolf's Glen

11.5 Halévy, *La Juive*, stage design for Act 1 (1835)

11.6 Donizetti, *Lucia di Lammermoor*, Fanny Persiani as Lucy

11.7 Meyerbeer, *Les Huguenots*, Enrico Caruso as Raoul

11.8 Verdi, *La Traviata*, Maria Callas as Violetta

12.1 Bellini, *Norma*, Giuditta Pasta in the title role

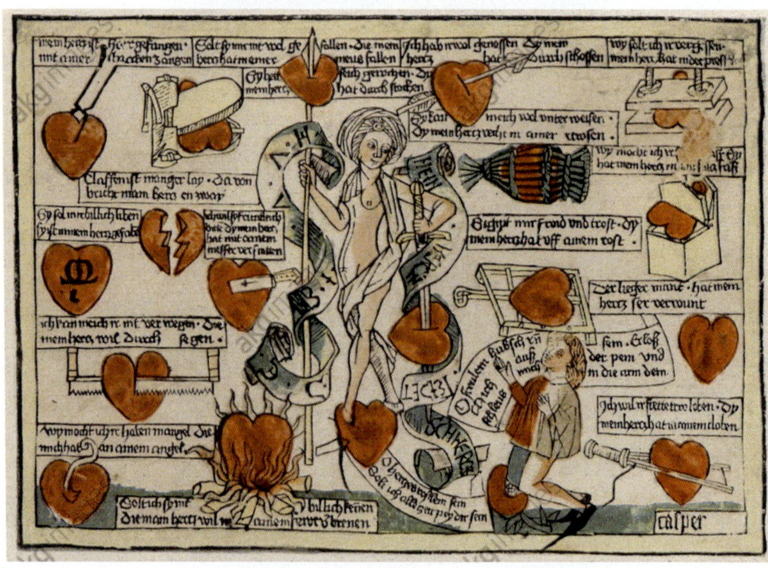

13.1 Frau Minna

13.2 Louis-Léopold Boilly (1761-1845), melodrama seen in an opera house

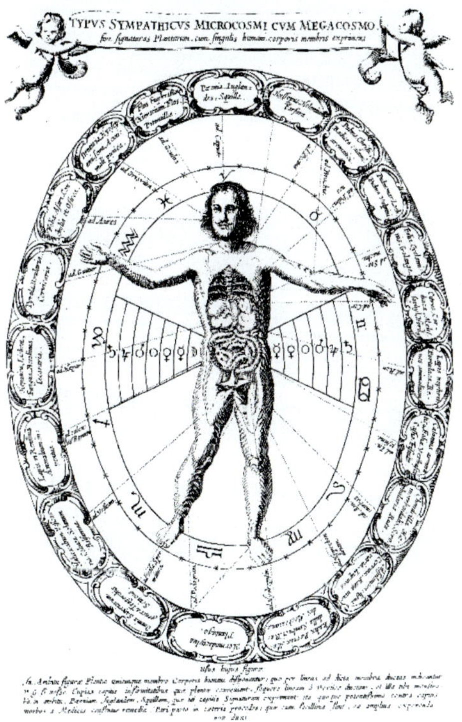

13.3 *The Solar Plexus*. The sympathies between the microcosm and megacosm, from Athanasius Kircher (1602-1680), *Mundus subterraneus* (1650, vol. s)

13.4 Stendhal (Henri Beyle) (1788-1842)

14.1 Vincenzo Bellini (1801-1835)

14.2 Gaetano Donizetti (1797-1848)

14.3 John Everett Millais, *The Bride of Lammermoor*

14.4 Daniel Auber (1782-1871) by Hortense Haudecourt-Lescot (1784-1845)

14.5 Jules Massenet (1842-1912)

14.6 Giacomo Puccini (1858-1924)

14.7 Massenet, *Manon*, playbill, 1884

14.8 Sibyl Sanderson as Manon

14.9 Manon at the Inn

14.10 Offenbach, *Orphée aux enfers*, playbill, 1874

14.11 *Les Contes d'Hoffmann* a. Antonia

THE TALES OF HOFFMANN, ACT II.
The scene at the conclusion of the second of the Tales. The infatuated Hoffmann returns from Giulietta's room in time to see Pitichinaccio handing Giulietta into a gondola. Dapertutto picks up the sword which Hoffmann dropped after the slaying of Schlemil.

14.12 Offenbach, *Les Contes d'Hoffmann* b. Giulietta

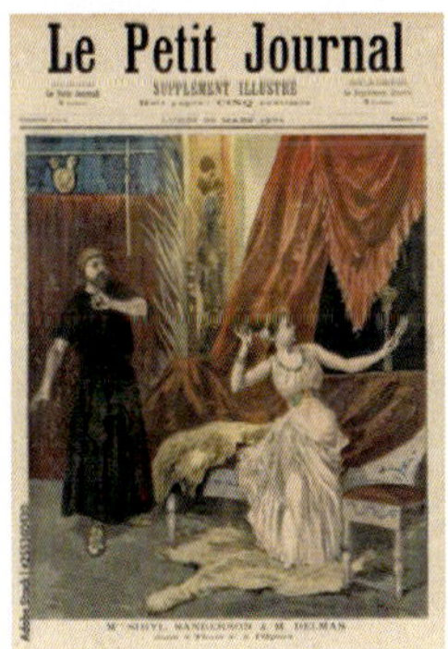

14.13 Thaïs and her Mirror, *Le Petit Journal* (26 March 1894)

14.14 Erich Wolfgang Korngold (1987-1957), photo

15.1 Tchaikovsky with Medeya Mei and Nikolai Figner, the creators of Lisa and Herman

15.2 *The Queen of Spades*, Herman and the Countess

15.3 *The Queen of Spades*, The Gambling Scene

15.4 Leos Janáček (1854-1928), relief sculpture

15.5 *The Makropulos Case*, BBC Proms

15.6 *The Makropulos Case*, Vienna

16.1 Modeste Mussorgsky (1839-1881) by Ilya Repin (1844-1930)

16.2 *Boris Godunov*, Boris's death

16.3 Benjamin Britten (1913-1976), photo

16.4 *Death in Venice*, Aschenbach on the train

16.5 *Death in Venice*, Aschenbach and Tadzio

16.6 Franz Schreker (1878-1934)

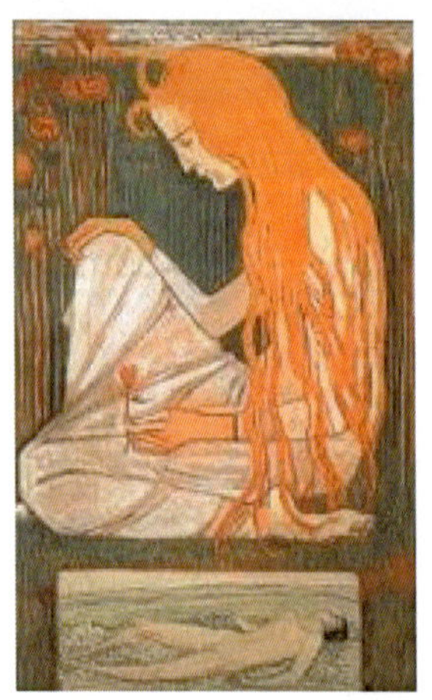

16.7 *Der ferne Klang*, poster

16.8 Alban Berg (1885-1935), photo

16.9 *Lulu*, poster

16.10 *Lulu*, playbill Châtelet

16.11 Carlo Saraceni (1579-1620), *The Ecstasy of St Francis*

16.12 Giovanni Bellini (1429-1516), *St Francis in the Desert. St Francis received the stigmata.* The sun comes out on the left while the saint looks to the divine light on the right.

17.1 *La Muette de Portici*, Masaniello and Fenella

17.2 Arthur Hughes (1832-1915) Ophelia 1852. Her impending doom referenced by the bat approaching in the gloom. When down the weedy trophies and herself fell in the weeping brook.

17.3 *Hamlet* John Everett Millais (1829-1896), *Ophelia, with poppies, violets and forget-me-nots*, 1852

17.4 Pietro del Pollaiolo (1443-1496) attrib., *Apollo and Daphne*

18.1 Arthur Rackham (1867-1939), *Undine*, 1909

18.2 E.T.A. Hoffman (1776-1822) by Johann Passini, after Wilhelm Hensel, 1822

18.3 *Undine*, stage design by Schinkel: The Castle

18.4 *Undine*, stage design by Schinkel: The Liebestod

18.5 Albert Lortzing (1801-1851)

18.6 Alexander Dargomyzhsky (1813-1869)

18.7 Antonin Dvořák (1841-1904), photo

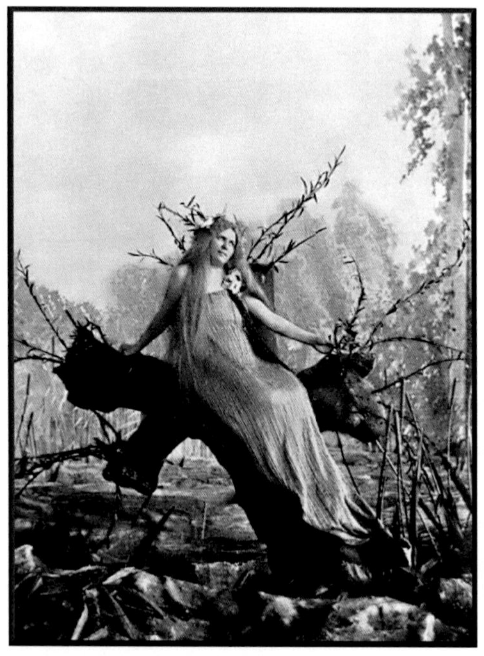

18.8 *Rusalka*, Ruzena Maturova, premiere (Prague, 1901)

18.9 *Undine*, German postcard

18.10 Hector Berlioz (1803-1869)

18.11 Charles Gounod (1818-1893)

18.12 Arrigo Boito (1842-1918), photo

18.13 *Mephistopheles* by Ferdinand Victor Eugène Delacroix (1798-1863), *Illustrations from Faust*

18.14 *Faust* Vision of Marguerite, Scene from the Prologue in Gounod's opera

18.15 *Faust meets Marguerite* from Delacroix, *Illustrations*

18.16 *Marguerite at the Spinning Wheel*, from Delacroix, *Illustrations*

18.17 *Faust and Marguerite in the Garden*, Kupferstich by Moritz von Schwind (1804-1871)

18.18 Ary Scheffer (1795-1858), *Marguerite and Mephistopheles in Church*

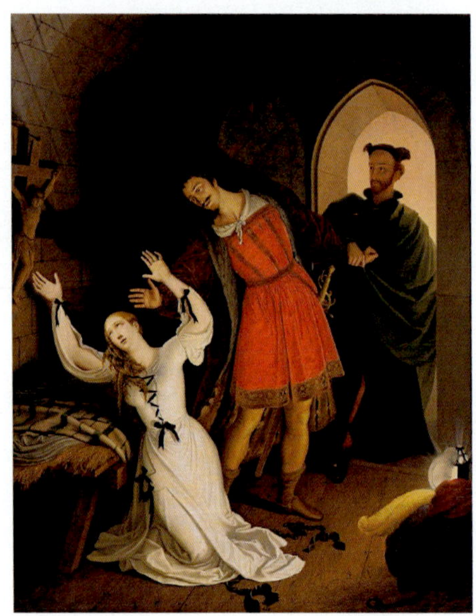

18.19 Ludwig Ferdinand Schnorr von Carolsfeld (1794-1872), *Faust and Marguerite in Prison*

MEPHISTOPHELES: Away! Else both of you are lost! Away!
Such useless chatter! Talk, talk and delay!
My horses are shivering and shaking!
Dawn is breaking!

18.20 *The Prison Scen*e, from Delacroix, *Illustrations*

18.21 Ferruccio Busoni (1866-1924)

18.22 Pelléas and Mélisande at the Well Edmund Blair Leighton (1852-1922)

18.23 Pelléas and Mélisande at the Tower

18.24 The Death of Mélisande

19.1 Wagner by Franz von Lenbach, 1871

19.2 *Rienzi*, Victorious entry into Rome, Dresden 1842

19.3 *Der fliegende Hollander*, Senta's sacrifice, Dresden 1843

19.4 Tannhäuser in the Venusberg, Dresden 1845

19.5 Laurence Koe (active 1888-1904), *Venus and Tannhäuser*, 1896

19.6 *Tannhäuser*, St Elisabeth

19.7 Robinet Testard (1470-1523), *The Allegory of Music*

19.8 *Lohengrin*, Elsa von Brabant

19.9 *Lohengrin*, Elsa and Lohengrin parting

19.10 Lohengrin and the Swan, Neuschwanstein

19.11 The Swan Grotto at Linderhof

19.12 Miniature from the MS of *Tristan et Iseult*

19.13 Stained glass window of Tristan and Isolde

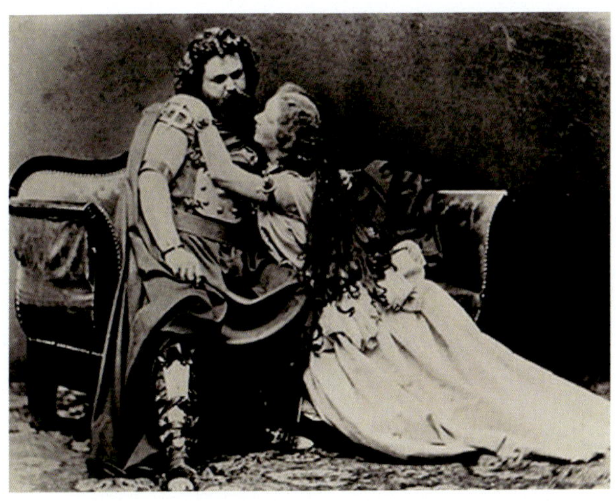

19.14 *Tristan und Isolde*, the creators Ludwig and Malwine Schnorr von Carolsfelt

19.15 John William Waterhouse, *Tristan and Isolde*

19.16 Herbert Draper (1863-1920), *Tristan und Isolde*

19.17 Ignatius Tascher (1871-1913), *Parzival*, 1901

19.18 William Morris (1834-1896), *Vision of the Holy Grail*

19.19 The Last Supper, Holy Chalice, Valencia Cathedral, carved from deep red agate

19.20 *Parsifal*, The Good Friday Spell, Bayreuth 1882

19.21 Parsifal clutches his heart at seeing the agony of Amfortas. Fidus from *Jugend*, 1900-09

19.22 Ferdinand Leeke (1839-1923), *Kundry dies with Parsifal blessing the Knights with the Grail*

19.23 Henri Fantin-Latour (1836-1904), *Das Rheingold*

19.24 Arthur Rackham, *Die Walküre*, Brünnhilde's Pleading, 1910

19.25 Franz von Metzner (1870-1919), *Siegfried*

19.26 Ludwig Habig (1872-1949), *Siegfried with the Forest Bird*, 1905

19.27 Hendrich Hermann (1854-1931), *Götterdämmerung*, Siegfried's Death, 1906

19.28 Arthur Rackham, *Götterdämmerung*, Brünnhilde's Immolation,1911

20.1 *Der Freischütz*, the Wolfs Glen, design Simon Quaglio (1795-1878), Munich 1822

20.2 The Snake-Goddess, Palace of Knossis (1600 BC)

20.3 *Rusalka*, Song to the Moon, the Metropolitan Opera

20.4 Edward Burne-Jones. *Mirror of Venus* from *The Flower Book*, 1905

20.5 Hecate. William Blake, *The Triple Hecate*, 1795

20.6 *Die Zauberflöte*, stage design with Sphinx and moon by Schinkel

20.7 Ernest Rodin (1840-1917), *The Kiss*

20.8 Giotto (1276-1337), *Two Kisses* from the Scrovengi Chapel, 1305
a. *The Legend of Joachim at the Golden Gate*

20.8 Giotto (1276-1337), *Two Kisses* from the Scrovengi Chapel, 1305
b. *The Kiss of Judas*

20.9 Anthony Frederick Augustus Sandys (1829-1904), *Helen of Troy*, 1867

20.10 Mélisande hangs her hair from the Tower

20.11 Sandro Botticelli (1445-1510), *The Birth of Venus*
a. The pagan image of Venus

20.11 Sandro Botticelli (1445-1510), *The Birth of Venus*
b. Venus reflected in the portrait of the Virgin

20.12 Donatello (1386-1466), *Saint Mary Magdalene*

21.1 *Tannhäuser*, The Hall of Song

21.2 *Tannhäuser*, The Song Contest

21.3 Geertgen tot Sint Jans, *The Glorification of the Virgin* (c. 1490–1495)

21.4 John Collier (1850-1934), *Tannhäuser in the Venusberg*

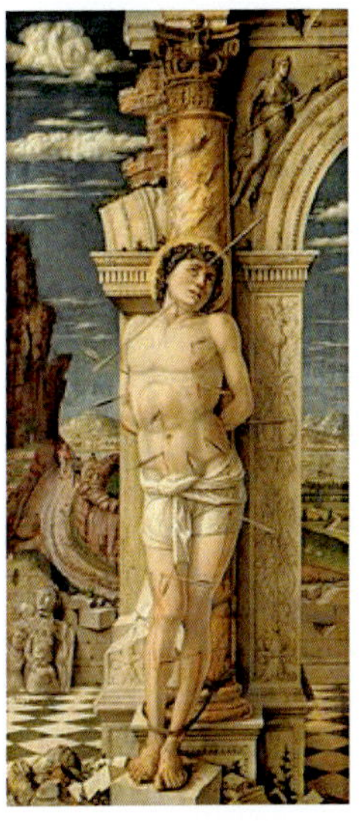

21.5 Andrea Mantegna (1456–1459), *St. Sebastian*

23.1 Jean-Jacques Rousseau (1712-1778)

23.2 *Laocoön* (1st c. BC)

23.3 The Fall of Man, *Adam and Eve* by Giuseppe Cades, after Raphael (1483-1520)

23.4 John Collier, *Lilith*

23.5 Masaccio (1401-1428), *The Expulsion of Adam and Eve from Eden*. They hand in hand with wandering steps and slow, Through Eden took their solitary way…

23.6 Giovanni Bellini, *The Blood of the Redeemer*

25.1. Villa Wahnfried, Wagner's House in Bayreuth: *Wahn* (delusion, madness) and *Friede* (peace, freedom)

25.2 Palazzo Vendramin Calergi, Venice

CHAPTER 17

OTHER OPERAS

Ah yes! Now I understand;
This is the garden of paradise.
Listen! The angels are singing.

(Frederick Delius, *A Village Romeo and Juliet*, Act 3)

LA MUETTE DI PORTICI (1828)

Grand opera in five acts by Daniel-François-Esprit Auber. Librettists: Eugène Scribe & Germain Delavigne (1790-1868). (The opera is also known by the hero's name as *Masaniello*.) First performance: Paris Opéra, 29 February 1828.[1]

Synopsis

1647-48: Neapolitan revolt against Spanish Habsburg rule, which was led by a fisherman named Masaniello from the nearby port town of Portici.

Act 1

Alfonso, son of the Viceroy of Naples, marries the Spanish Princess Elvira. Alfonso has seduced and abandoned Fenella, the mute sister of the Neapolitan fisherman Masaniello. Alfonso fears that she has committed suicide. Fenella, arrested by the Viceroy, escapes and narrates the story of her seduction by gestures, showing a scarf which her lover gave her, and intimating that love has been the cause of her misfortunes. Elvira promises to protect her. In chapel Fenella recognizes her seducer as the bridegroom of the Princess. When the newly married couple come out of the church, Elvira discovers from the mute girl's gestures, as she points to Alfonso, that he was her faithless lover. Fenella flees, leaving Alfonso and Elvira in sorrow and despair.

Act 2

The fishermen prepare their nets. Fenella appears on the hill and looks as if she is going to cast herself into the sea. She indicates to Masaniello that she was about to kill herself, but wanted to see him first to confess her wrongs and receive his pardon. Pietro, Masaniello's friend, informs them that a crowd of soldiers is approaching. Masaniello swears revenge, but Fenella, who still loves Alfonso, has not mentioned his name. Then Masaniello calls the fishermen to arms and they swear perdition to the enemy of their country.

Act 3

In the Naples marketplace Selva, the officer of the Viceroy's bodyguard, from whom Fenella has escaped, discovers her and attempts to re-arrest her. Mansaniello arrives and recognises his sister. There is a general revolt.

Act 4

Fenella comes to her brother's dwelling and mimes the horrors taking place in the town. Masaniello is distraught.

Mansaniello
Descend Oh balmy sleep
Thou friend of the unhappy!
On her from heaven descend;
And with thy smiling dreams
The cruel thoughts disperse
Of her disconsolate fate.
Descend, Oh! sleep descend
And peace and calm restore
To this celestial being.
(*Fenella falls asleep*)
Disperse the cruel thoughts
Of her afflicted heart.

Pietro enters with comrades and tries to excite Masaniello to further deeds, but he wants only liberty and shrinks from murder and cruelties.

They tell him that Alfonso has escaped and they seek to kill him. Fenella, who hears all, decides to save her lover. Alfonso and Elvira arrive at her door asking for a hiding-place. Fenella wants to avenge herself on Elvira, but pardons her for Alfonso's sake. She swears to save them both or die with them. Masaniello, re-entering, assures the strangers of his protection and even when Pietro denounces Alfonso as the Viceroy's son, he holds his promise sacred, obeying the laws of faith and hospitality. Pietro with his fellow-conspirators, angered, leaves him.

Masaniello is proclaimed King of Naples by the jubilant crowd.

Act 5

Before the Viceroy's palace, Pietro confides that he has administered poison to Masaniello, in order to punish him for his treason, and that the King will soon die. A fresh troop of soldiers is seen marching against the people, with Alfonso at their head. Masaniello alone can save them, but he is mortally ill and becoming insane. Fighting takes place and an eruption of Mount Vesuvius occurs. Masaniello recovers his senses but is slain by the crowd in the act of saving Elvira's life. On hearing this, Fenella falls senseless, then recovers, sees Alfonso, rushes out, looks to heaven and leaps from the parapet into burning lava of Vesuvius. The fate of Alfonso and Elvira is not revealed.

COMMENT

Auber's story introduces us to Fenella, the opera's eponymous heroine, taken from Sir Walter Scott's *Peveril of the Peak* (1822), a deaf and dumb dwarf of the same name in the novel. As with many of the operas we discuss, there are political and family contentions, the failure of passionate love, seduction and infidelity, love triangles and madness. Yet here the *Unexpected Death* is suicide. More relevant to our themes are those which reverberate through many of the sudden deaths. Fenella uses gesture and dance to express her joy and sorrow, not able to articulate her feelings in words. The drama of her death in flames, with the destructive yet awesome power of nature, is an evident contrast with the water-related scenes of the fishermen and suggestions of village life.

The role of Fenella is of great significance for the history of dramatic music. She is a mute character, and her role (created by the ballerina Lise Noblet, 1801-1852) is entirely realized through dance and gesture. She has no fully-fledged pieces (like the three formal dances in Acts 1 and 3), but rather a seamless sequence of orchestral music that represents her feelings, her story and her emotional life through mime.

The action of the opera is contingent on her destiny: her sexual betrayal by Prince Alfonso; her love and devotion for her brother, the protective hero Masaniello; and eventually her grief at her brother's betrayal and poisoning, leading to her suicide by throwing herself down into the lava flow of the erupting Vesuvius. The combination of drama, pathos, human political folly, natural cataclysm and Fenella's suicide (a type of Love-Death with and for her brother) provides a unique exemplar of various topoi of Romantic music drama. The nature of her emotionally changed dancing, leading into the musical depiction of the natural phenomenon of the terrestrial irruption into human affairs, is couched in powerful musical language, forming a wordless and extended epilogue to the human action. In the finale, the running strings, the huge treble chords over a chromatically surging bass provide a Romantic tone poem of dynamic influence on later ideas and developments—both dramatic and musical. [**Fig. 17.1**]

A VILLAGE ROMEO AND JULIET (1907)

Opera by Frederick Delius, the fourth of his six operas. The libretto by the composer is based on a story by Gottfried Keller, *Romeo und Julia auf dem Dorfe* (1856). Fiest performance: Berlin, Komische Oper, 21 February 1907.[2]

The Prologue *Scene 1 (Time: September)*

A strip of luxuriously overgrown land on a hill. Sali, son of the farmer Manz, and Vrenchen (Vreli), daughter of the farmer Marti, are children playing together. The Dark Fiddler is the rightful owner of this disputed land but is illegitimate and without legal rights. He warns them that the land must not be tilled. Manz and Marti dispute ownership of the land, and put a stop to the relationship between their respective children.

Act 1 *Scene 2 (Six years later).*

At Marti's now run-down house, Sali and Vrenchen plan a meeting. Since their childhood, a lawsuit about the land has ruined both Manz and Marti. The Dark Fiddler re-appears and tells them that, regardless, they will meet again. Marti sees the two lovers and takes Vrenchen away. In trying to stop Marti, Sali injures him severely. As a result, Marti loses his reason and must be confined in an asylum. Sali returns and sees Vrenchen at her house, which is to be sold. The two declare their love and decide to leave together.

Act 2 *Scene 3*

The wildland, overgrown with red poppies in full bloom, surrounded by cornfields. Sali and Vrenchen meet and clasp each other's hands laughingly.

Act 2 *Scene 4* (*The Dream of Sali and Vrenchen).*

They dream that they are being married in the old church of Seldwyla. The church bells are ringing in the distance.

Scene 5 (The Fair)

At a local fair, Sali and Vrenchen buy rings. Sali mentions an inn, the Paradise Garden, where they can dance all night, and they repair to it. The Dark Fiddler and some vagrants are drinking there. He greets the lovers and suggests they join him to share a vagabond life in the mountains

Act 3 *Scene* 6 *(The Paradise Garden)*

A beautiful garden run wild: everything shows traces of bygone beauty. The Black Fiddler appears with Vagabonds surrounding him.

> (*Vrenchen kisses Sali, a beautiful change comes over the Paradise-garden. The rising yellow moon floods the distant valley with a soft and mellow light. It almost seems as if something mysteriously beautiful had touched the garden as if by enchantment.)*
>
> **Vrenchen** (*to the sounds of the wind*)
> Ah yes! Now I understand;
> This is the garden of paradise.
> Listen! The angels are singing.
> Oh, Sali, how I love you!
> I've had that thought this many a day,
> But never dared to ask you.
> (*with conviction*)
> We can never be united,
> And without you I could not live!
> Oh, let me then die with you!
>
> **Sali**
> Aye, let us die together!
> To be happy on short moment,
> And then to die,
> Were not that eternal joy?

Sali and Vrenchen
See the moonbeams kiss the woods,
The fields, and the flowers;
And the river, softly singing,
Glides along and seem to beckon.
Listen! Far-off sounds of music
Startle trembling echoes,
Throbbing, swelling, faintly dying
In the sunset's fading glow.
Where the echoes dare to wander,
Shall we two not dare to go?

Sali
(*Painting a boat filled with hay*)
See our wedding bed awaits us!
Come, Vreli, sweet one!

(**The Black Fiddler** *appears on the veranda of the inn, playing wildly on his fiddle: Sali lifts Vrenchen into the boat*)

Vrenchen
Look, my nosegay goes before me!

(*She plucks the nosegay from her bosom and casts it into the river. Vrenchen then stoops down and pulls the plug from the bottom of the boat. The boat moves out into the stream; Sali casts the plug of the boat into the river, and then sinks down upon the hay in Vrenchen's arms. The boat drifts slowly down the river and sinks.*)

Bargeman (*In the distance*)
Travelers we a-passing by!

COMMENTS

This is a Romantic opera ending with a *Liebestod.* in the songs of Vrenchen and Sali. The *Tristan*-like deaths are described in the original story thus:

> … as they sat there aloft the boat gradually drifted out into the middle of the river, and then floated, slowly turning, downstream. The river flowed now through, high, dark forests, which over-shadowed it, and again through open fields; now past quiet villages, now past isolated huts. Here it widened out into a placid stretch, as calm as a quiet lake, where the boat almost stopped; there it rushed around rocks and rapidly left the sleeping shores behind. And as the morning glow appeared, a city with its towers emerged from the silver-gray stream. The setting moon, as red as gold,

> made a shining path up the stream, and crosswise on this the boat drifted slowly along. As it drew near the city, two pallid forms, locked in close embrace, glided down in the frost of the autumn morning from the dark mass into the cold waters.
>
> A short time afterward the boat floated, uninjured, against a bridge and stayed there. Later, the bodies were found below the city, and when it was ascertained whence they came, the papers reported that two young people, the children of two poverty-stricken, ruined families, that lived in irreconcilable enmity, had sought death in the water, after dancing and making merry together all the afternoon at a kermess. Probably—so the papers said—this occurrence had some connection with a hay-boat from that region, which had landed in the city without a crew; and the assumption was that the young people had stolen the boat to consummate their desperate and God-forsaken marriage—a fresh evidence of the increasing demoralization and savagery of the passions. Along with the dark fiddler, madness and dreams, need one say more.[3]

The whole story is a movement from opposition and futility to death freely embraced by the doomed lovers, who accept to "slowly die together". From visiting the fair at the tavern of the Paradise Garden, they make their way to the river and die with each other in the hay-barge as it slowly floats downstream. Their path of life is haunted by the mysterious figure of the Dark Fiddler, who is an embodiment, a type of death personified. His voice is heard presiding as they sail away to be submerged by the encompassing waters represented in the dreamy music. The Walk to the Paradise Garden presents the dominant theme of the score, and one that weaves its way through the last pages. The Walk becomes the path leading to a regained Eden that can be found only in the spiritual realm, attainable only in death.

HAMLET (1868)

Grand opera in five acts by Ambroise Thomas. Librettists: Michel Carré & Jules Barbier, after Shakespeare's play (1600-01). Translation: Avril Bardoni. First performed: Paris Opéra, 9 March 1868.[4]

Act 1

Scene 1. The Coronation Hall

The celebration of the marriage of Queen Gertrude and Claudius. Ophelia is ignored by Hamlet. She recognises his deep despair, but he avows his love for her, telling her "never to doubt my love".

Hamlet
The bond between us is eternal,
My soul is yours

Scene 2. The Ramparts

Horatio and Marcellus meet Hamlet. The Ghost appears, Hamlet swears to avenge his father.

Act 2

Scene 1. The Gardens

Ophelia, reading a book, is concerned at Hamlet's sudden change and indifference. The Queen worries that Hamlet is mad, and thinks Ophelia may be able to cure him.

Scene 2. The Play

The play begins, and Hamlet narrates. The play recounts a story similar to the murder of Hamlet's father. Hamlet, feigning insanity, accuses Claudius of killing his father, and snatches Claudius' crown from his head.

Ophelia
Ah! Brutal insult! Ah! Mad Frenzy!
Ah! My heart trembles with terror. I shall die, alas!
(*Hamlet collapses*)

Act 3

Closet Scene

The Queen tries to persuade Hamlet to marry Ophelia, but Hamlet, realising he cannot marry the daughter of the guilty Polonius, refuses. Ophelia returns her ring to Hamlet.

Hamlet
Go to a nunnery, go Ophelia!
And may you forget forever this fleeting dream!
She is mad who could believe that Hamlet loved her.

Ophelia
Farwell joy and happiness! Farewell to dreams of love!

Act 4

The Mad Scene

Ophelia enters, dressed in a long white gown with flowers and creepers threaded through her flowing hair. After Hamlet's rejection, Ophelia has lost her reason, and, convinced that Hamlet is her husband, drowns herself in the lake.

Ophelia
The water-nymph swims and drags you
Beneath the waters of the sleeping lake
Farwell sky farewell sweet friend!
Happy the bride in her husband's arms!
Joy supreme!
Ah! Cruel one, I love you! I die for you!
There he is! I think I hear him!
To punish him for having made me wait,
You Wilis, white water-nymphs,
Hide me among your reeds!
(*Ophelia is seen floating on the water as the current carries her away)*

Act 5

Gravediggers' Scene

Hamlet sings of remorse for his ill treatment of Ophelia. The ghost tells Hamlet to kill Claudius. Hamlet, still in despair, is proclaimed King to cries of "Long live Hamlet! Long live the King!"

Hamlet
Oh dear Ophelia! Poor child whose love, like a deadly poison,
Has withered your youth and disturbed your reason
Like a delicate flower that has bloomed upon a grave,
Your heart, by sorrow crushed, trembles and succumbs.
You have been forced to share my fatal destiny.
My spirit lies in the grave, alas and I am King!

There is a "Covent Garden" ending with no ghost. After Hamlet kills Claudius, he embraces Ophelia's body and he dies.

Hamlet
My task is accomplished, and Ophelia, I die with you.
(*He falls dead by the body of Ophelia*)

COMMENT

The story told in this version of *Hamlet* has considerable variations from Shakespeare's play, not the least the given ending, that Hamlet lives as king. But from our perspectives it has many tropes of importance, including: a ghost, a mad scene (or two, Hamlet's pretended), murder (Hamlet kills Claudius), and the Wilis, pulling Ophelia to death by drowning. This could be a suicide, and is expected (if you know Shakespeare's play). But there are questions about her death. Surely, there are some heart-related references, and it may be her death was from a broken heart, but she and Hamlet are both still bonded even unto death.

There are two endings to the opera. The first sees the reappearance of the Ghost of Hamlet's father as a type of *deus ex machina*, ensuring that the Prince is crowned King of Denmark. The second shorter version (written for Covent Garden, 1870) dispenses with the supernatural and results in the death of Hamlet. Both are endings rather sombre and mark the restoration of moral order rather any kind of personal triumph.

In the first ending, Hamlet laments that Ophelia has been forced to share his fatal destiny. He asks for forgiveness. She does not die like many women we are discussing, at the end of the opera. In the second version of the libretto (the so called "Covent Garden ending") we have a conclusion reminiscent of *Tannhäuser,* and an *Unexpected Death*.

In *Hamlet* autonomic events and states are pervasive. The themes of murder, deception, guilt, conscience, remorse, madness, suicide and revenge infuse the plot and cause extreme reactions in the various protagonists. The theme of insanity underpins the crime and its punishment. Hamlet feigns madness to pursue his ends while Ophelia, caught up in the tangled emotions of Hamlet's sense of pusillanimity and the guilt of Gertrude and Claudius, eventually succumbs to mental instability and death by drowning, whether deliberately or accidentally. The composer distinguishes between these types of behaviour with imagination. Hamlet's assumption of madness is depicted in disjointed recitative, with an arioso passage purporting to be of an antic visionary quality to convince the Court of his loss of reason. His ploy is to oversee the revenge play by the visiting troupe. The underlying Dionysian modality is revealed in his *Chant bachique*, where he urges all to drown their sorrow in the inebriating liquors of wine. The seriousness of this mode is made clear in the return of the theme of the drinking song in minor variant as it is later integrated into the very serious unmasking of the guilt of the regicides, and where the theme becomes one of triumph.

Ophelia's highly-strung personality is suggested from her first appearance and duet with Hamlet. She has her own specific theme on the clarinets and cor anglais, a wide tessitura veering to high notes and decorated passages. Her Act 2 aria uses a theme of her love for Hamlet, and develops this sense of strain. But the depiction of her mental breakdown is her mad scene in Act 4, a sustained scena in several movements like the prototypical situation in *Lucia di Lammermoor*. Each episode sustains tension and instability in the various movements. The use of high woodwind and extreme coloratura, with the song of 'the pale blond Wilis' of the water, provide an evocation and inversion of the Undine motif, preparing us for her drowning. It is, like most mad scenes, an extended variant of the *Liebestod*. This is powerfully evoked in the final movement with its dreamlike ambience, the quiet soprano line over the wordless vocalise of the chorus, the harp obbligato conjuring up the watery medium, with the theme of Ophelia's and Hamlet's doomed love emerging towards the end. It combines with the melody of the water sprite to create a remote, eerie and yet a soothing tone poem conjuring up the death by drowning.

There is a newer version of *Hamlet*, composed by Brett Dean and the librettist Matthew Jocelyn (2017). Performed at Glyndebourne, this adheres very closely to Shakespeare's verses if not his chronology. [**Figs. 17.2-3** Hamlet]

SALAMMBÔ (1890)

Grand opera in five acts by Ernest Reyer (1823-1909). Librettist: Camille du Locle (1832-1903). The story is based on the novel by Gustave Flaubert (1862). Translation: Richard Arsenty. First performance: Brussels, La Monnaie, 10 February 1890.[5]

Act 1

In ancient Carthage, the general Mathô is successful against the Romans. He and Queen Salammbô have fallen in love. Mathô refuses to lead the rebelling mercenaries who feel betrayed by the city.

Act 2

Salammbô is deeply disturbed and seeks the advice of the high priest. She wishes to see the sacred veil of Tanit (the *Zaïmph*) who safeguards the city. Mathô is brought to the temple to steal the talisman. He overhears Salammbô and appears before her draped in the veil. She is horrified at this blasphemy, but he escapes with the precious garment.

Act 3

Salammô's father, the general Hamilcar, is regarded as the hope of the city and blames Salammbô for the theft of the veil. Human sacrifices are offered to Moloch. Salammbô dresses in finery to penetrate the camp of Mathô to try and recover the veil.

Act 4

Salammbô enters Mathô's tent. The latter is disappointed to find that Salammbô has come for the *Zaïmph*, not out of love for him. The Carthaginians attack the mercenary camp and Salammbô flees with the veil. The mercenaries are defeated and captured. Salammbô appears wearing the veil. It is decided that Mathô will be offered to Tanit as a special sacrifice.

Act 5

Great celebrations are held in the forum of Carthage. Tanit is again covered in her veil. Salammbô and Hamilcar appear, and Salammbô takes her place on the throne. Mathô is brought in to be sacrificed. The people demand that Salammbô perform the ritual. She agrees, but on the point of sacrifice, Mathô lifts his face to her in love. The Queen dies, declaring that "Whoever touches you, blessed and holy veil, must die". Mathô stabs himself to join her in death.

> (*She takes the knife and walks to where Mathô is kneeling before the altar. A moment of silence between the two individuals. Mathô lifts his face towards Salammbô in a gesture of love. Salammbô makes an effort to raise the knife but her arms fall back to her side.*)
>
> **The Priests and the Chorus**
> What? ... The blade slips from her hand!
> Avenge Tanit, Salammbô! Strike!

Mathô
Strike …

Salammbô
Accept then, Tanit,
The blood about to flow! ... May your vengeance
Be appeased with this scarlet dew!
Whoever touches you, blessed and holy veil, Must die!
(*She stabs herself.*)

Everyone
O gods! ... O day of sorrow!...

Mathô
(*taking Salammbô in his arms and seizing the knife*)
Do not approach me! She is mine …
Salammbô, I adore you and am coming to join you!
(*He stabs himself and falls, holding the dying Salammbô in his arms.*)

The Chorus
Anaïtis, Derceto, Mylitta!
Whoever touches your venerated veil
Shall die, O implacable goddess!

COMMENT

The musical language of *Salammbô* attempts to fuse the traditions of *grand opéra* with the innovations of Wagnerian music drama, introducing a continuous texture, a type of *Sprechgesang*, with more self-contained choral and orchestral numbers (dance and march). The orchestral commentary is rather thin. The vocal line is recitative with moments of more lyrical *cantabile*. Sustained passages of lyricism are rare, and any autonomic moments associated with the heroine Salammbô are dealt with rather perfunctorily, whether this is fainting during the love scene with Mathô, or in the expressions of rapture, ecstasy, and weakness. The boldest passage in which the warmth of inspiration contends with novelty of form is the Tent Duet where one finds cries of fury, passionate outbursts, mystical raptures and amorous ecstasies. The Council of Elders is characterized by coloured and vibrant orchestration. Salammbô's suicide is marked in a few bars of rich woodwind writing and some high attractive harmonies, but the situation of her being joined in death by her lover is hardly developed when compared to a similar situation in Bellini's *Norma*.

DAPHNE (1938)

Opera by Richard Strauss, subtitled "Bucolic Tragedy in One Act". Librettist: Joseph Gregor. Translation: Maria Pelikan. First performance: Dresden, 15 October 1938.[6]

The chaste girl Daphne sings a hymn of praise to nature. She loves the sunlight as trees and flowers do, but she has no interest in human romance. She cannot return the love of her childhood friend Leukippos, and she refuses to put on the ceremonial clothes for the coming festival of Dionysus, leaving Leukippos with the dress she has rejected. He breaks the flute on which he was playing music to Daphne.

Daphne
Sun, with your wonderful light, let me dwell
Among my brothers the trees.

Leukippos
I'm your companion: My heart is alive

Daphne
And so is the tree. I feel its branches caressing me
(She kisses the branches).

Daphne's father Peneios advises preparing a feast to welcome Apollo. Just then a mysterious herdsman appears. Peneios sends for Daphne to care for the visitor.

(A full moon casts a light on the scene)

The herdsman tells Daphne that he has watched her from his chariot, and repeats to her phrases from the hymn to nature she sang earlier. He promises her that she need never be parted from the sun, and she accepts his embrace. But when he begins to speak of love, she becomes fearful and runs out. He refers to himself as her brother.

Apollo
For ever shall you enjoy the light of day!
Follow me Daphne! There shall be no dark of night.

Daphne *(in ecstasy)*
Oh let me be thine—Oh let me embrace you!
Of luminous gladness! Brother … Brother!
(the moon is hidden behind a cloud)

At the festival of Dionysus, Leukippos is among the women wearing Daphne's dress, and he invites her to dance. Believing him to be a woman she agrees, but the strange herdsman swings his bow and thunder is heard. There is general fear which stops the dance. Daphne realizes that both Leukippos and the stranger are in disguise. She has been deceived. The stranger reveals himself as the sun-god of light, Apollo. Daphne refuses both her suitors. Leukippos curses Apollo, more thunder, and Apollo kills him with an arrow. Daphne is transfixed, and slowly comes back to life.

Leukippos
Daphne, beloved, I have dared to love you—
And an immortal came down to slay me. *(He dies)*

Daphne mourns with the dying Leukippos. Apollo is filled with regret. He appeals to Dionysus to take Leukippos to high Olympus, and asks Zeus to give Daphne new life in the form of one of the trees she loves. Daphne is transformed, and she rejoices in her union with nature.

Daphne (*Liebestod*)
Oh my Leukippos! Beloved companion!
Grieving I hear what your sorrowful flute sang.
It no longer recalls the breeze's playing.
Now the flute reveals to me your heart! …
Why did I seek to embrace an immortal,
When I should have pleaded that he might leave us in all our frailty
And grant us grace while he travels the heavens for ever and ever.
All that I ever held dearest I want to bring you:
(She sings of a fountain, the colours of butterflies, and all the blossoms)
But Daphne, pitiful Daphne will stay in silence
Here at your graveside in deepest mourning.
(She sinks down and covers her face)

Apollo *(amazed at what he hears her singing, lost in contemplation of her)*
Am I a god still—or diminished and humbled by human emotion?
You gods of Olympus! See my guilt and my misery
My arrow struck down the fairest maiden, the innocent Daphne!
Let her be mine: The tree Daphne—the glorious laurel
Forever betrothed to Phoebus Apollo.
(*It is dark. Apollo disappears, Daphne rushes upstage, and is still as if held by magic)*

Daphne
I'm coming, I'm coming, sweetly rises the earth's sap within me!
Let me greet you with leaves and branches,
Life-giving light.

(*Daphne disappears. A tree stands in her place*)

COMMENT

The opera is based loosely on the mythological figure Daphne from Ovid's *Metamorphoses* (c. 6 AD) and includes elements taken from *The Bacchae* by Euripides (c. 406 BC). Daphne's affinity with nature is embodied, and her life is fulfilled as the laurel-tree.[7] It is also a rescue opera (by Apollo), and the finale is a *Liebestod.* There are resonances here of Isolde. Although he does not appear, Dionysus is lurking in the music. This plays its part in the jealousy of Apollo against Leukippos: his dancing with Daphne at the Dionysian festival is the cause of Leukippos's death.

The music embodies the transformation taking place: Daphne's voice changes as she becomes a tree (in F-sharp major). In the libretto it is observed: "This transformation scene, the metamorphosis, is opulently silvery in the string section. The flute of Leukippos that falls silent after Daphne rejected him can resume playing again in the orchestral textures".

The finale to *Daphne* begins with a loud roll of timpani. A low passage for the bassoons introduces the magic of the metamorphosis (a *Liebestod* induced by Apollo), deep like the earth, continuing in low ruminations in the orchestra, with the entry of trombones/tuba. Daphne's voice rises steadily, getting higher all the time, like the sprouting branches of the laurel, with ever more ethereal effects in the orchestra, but always still rooted in the deep bass. There is then a sudden change to a high radiant passage in the horns, and the Moon Music begins: waves of sound, trilled flutes and harp and high chords, with vocalisation from Daphne (as she becomes wordless) and woodwind echoes, fading into the high treble, as though the new leaves now rustle in the moonlit air. [**Fig. 17.4** Daphne]

Notes

[1] Auber, *La Muette de Portici*, with Alfredo Krause and June Andersen, Orchestre-Philharmonique de Monte Carlo, cond. Thomas Fulton (EMI).

[2] Delius, *A Village Romeo and Juliet,* with Elizabeth Harwood and Robert Tear, Royal Philharmonic Orchestra, cond. Meredith Davies (EMI).

[3] Gottfried Keller, *A village Romeo and Juliet* (1856). Trans. Paul Bernard Thomas. From volume XIV *of German Classics of the Nineteenth and Twentieth Centuries* (New York: The German Publication Society, 1856), 56.

[4] Thomas, *Hamlet*, with Joan Sutherland and Sherill Milnes, Welsh National Opera cond. Richard Bonynge (Decca) (the Covent Garden ending); with June Anderson

and Thomas Hampson, London Philharmonic Orchestra, cond. Antonio da Almeida.

[5] Reyer, *Salammbô*, with Kate Aldrich and Gilles Roagon, Orchestre de l'Opéra de Marseilles, cond. Lawrence Foster (MRS).

[6] Strauss, *Daphne*, with Hilde Güden and Fritz Wunderlich, Wiener Philharmoniker, cond. Karl Böhm (DG).

[7] "The transformation of one being or of one species into another generally relates to the broad symbolism of Inversion, but also to the essential notion of the difference between primigenial, undifferentiated Oneness and the world of manifestation. Everything may be transformed into anything else, since nothing is really any-thing." See J. E. Cirlot, *A Dictionary of Symbols* (New York: Philosophical Library, 1962), 199-200.

CHAPTER 18

SPECIAL CASES

> Oh silence! The merciful will of heaven
> has chosen him for a pure love-death.
>
> (E.T.A. Hoffmann, *Undine*, Act 3)

Introduction

In the last four chapters we have considered those operas with an *Unexpected Death* which have a bearing on our overall themes. Some are closely related to the transcendental aspects of the deaths, *Liebestode*, and the special aspects of the music, itself revelatory and transcendental, as when the loving couple die together.

Our study requires consideration of several operas which we have grouped under 'special', in the sense that they comprise either a group of operas around the same theme, namely the Rusalka and the Faust series, or have a special place in the opera caucus at the centre of our concerns. These include *Pelléas et Mélisande* and the operas of Richard Wagner. The latter are the theme of Chapter 19.

1. The Undine/Rusalka Operas

UNDINE (1816)

Romantische Zauberoper (Romantic magic opera) in three acts by E.T.A. Hoffmann. Libretto by the composer and adapted by Friedrich de la Motte Fouqué himself. First performance: Berlin, Hofoper, 3 August 1816.[1]

This opera is fundamental to our theme, introduced in earlier chapters, especially Chapter 9, *The Elemental Being*.

Act 1

Undine, a mysterious girl, born in a palace of crystal on the bed of the ocean, is brought up by a fisherman and his wife to replace their own daughter, lost as a baby at sea. When she grows up she falls in love with the handsome knight Huldbrand, who marries her.

Act 2

In her husband's castle lives the wicked Bertalda, the real daughter of the fisherman and his wife, who had been abducted by one of Undine's uncles, a marine demi-god who appears in the opera as Heilmann. Bertalda captivates Huldbrand's heart. The despairing Undine is dragged down to the bottom of the sea by the water spirits who watch over her. They decide that her faithless husband must die.

Act 3

The wedding of Huldbrand and Bertalda is celebrated in an atmosphere of sadness at the castle of Ringstetten. Against her own will, Undine is obliged to intervene by the laws of the spirits. She rises up from the depths of the water and follows a long mysterious itinerary that brings her to the well in the courtyard of the castle of Ringstetten. She enters Huldbrand's bedchamber and kisses him passionately, knowing that her kiss, following the destiny written in the world of magic, will bring him death. A *Liebestod* that looks towards *Tristan und Isolde*, is followed by a concluding hidden chorus of spirits.

> (*The wedding celebration. The fountain is opened and springs to life, painting rainbows in the sun*)
>
> **Undine**
> Have a good night.
>
> **Others**
> Woe he is, we are lost.
>
> **Huldbrand**
> Oh how lovely she smiles. Not one kiss?
>
> **Undine**
> Yes, darling, because I must, but my kiss brings you death.
>
> **Huldbrand**

When such a departure beckons
It surely means gaining salvation.[2]

Undine
Now good night.
Others
Woe he is lost.
Woe to us, down to dark gates
he is drawn by the horrible might

Heilmann
Oh silence! The merciful will of heaven
has chosen him for a pure love-death [*Liebestod*]

(Huldebrand is carried on a bier, a grey cloud of mist rises over the lake)

Final Chorus
Pure love, pure longing,
lives in sweet reflection,
earnest singing, sweet believing
looks in with devotion,
desires to stay with Undine!
Good night to all trouble and world's splendour.
Purest love
Radiant longing
Lives in sweet refulgence.
Earnest longing
Sweet supposition
Looks to there with devotion,
Longing to be with Undine.
Good night
All earthly cares and pomp.

UNDINE (1845)

Romantische Zauberoper (Romantic Magic Opera) in four acts by Albert Lortzing (1801-1851). The libretto is by the composer, who follows the Fouqué fable in all its essential features but adds new perspectives. First performance: Magdeburg, 21 April 1845.[3]

Here the hero becomes Hugo, and the all-seeing all-wise Heilmann becomes the guardian of the watery mysteries, Kühleborn.

Hugo
Oh I am fine! Useless fear,

It's nothing, nothing. Reassure yourself!
Fill the cups—cheerfulness—
No, no, not!
Start the dance!

(The company divide in pairs to dance. When Hugo tries to take Bertalda's hand, the tower clock begins to chime "twelve" in measured pauses and in dull tones.)

Hugo
(trembles at the first chime)
Ha! it strikes—midnight—

Bertalda
(whispering to him)
Get ready—have courage!

Hugo
I'm frightened!—My blood—freezes like ice!
Midnight!

(At the last chime of twelve, a terrible thunderbolt, all lights extinguish, a storm sounds from the outside, general cry of terror from those present. A bluish light shines from the side, as Undine, veiled, moves towards the middle of the hall.)

Hugo
(Staring sideways)
Ha, now I realize, my dream comes true—
You come to judge—take me!
But let me again, before I die,
Look upon your lovely face.

Undine
(lifting back the veil)

Hugo
O radiant image that delighted me so much,
Once again, give me the bliss,
Of embracing you lovingly.

(Undine opens her arms)

Hugo
You beckon to me. I'm coming!
(He rushes into her arms)

So let me die.

(He sinks down at her feet dazed. —After a second thunderbolt Hugo and Undine sink.—Through the windows, burst open by the storm, one can see flashes and hear the roar of the waves. The chorus escape on all sides, then the hall collapses under terrible noise. The water floods in, carrying flotsam and debris before them, to the proscenium, rising ever higher. One gradually perceives the flicker of shiny objects, until finally, after the waves have reach the ceiling, the Crystal Palace of the Water Prince, with sparkling stones, shells and decorations, etc. appears.—Water spirits are grouped everywhere. In the middle, on an elevation, Kühleborn, at his feet Hugo and Undine. All shimmer in shiny robes.)

Kühleborn
(*to Hugo*)

You outraged this pure innocence,
Your life was forfeited;
But the poor one suffered innocently,
So grace prevailed!
For her sake shall you be forgiven.
You will stay with us from now on!
That is your punishment.
Beware, you ensouled ones, you mock us—
But this is how the soulless take revenge!

Choir of Water Spirits
Swansong, Swan melody
Once again we hear you!
Swansong,
Where perjury never dwells,
Where only eternal peace [*ew'ger Friede*] is enthroned!

Kühleborn, Hugo and Undine
Without him/you laughs no happiness
For where rancour dwells in the heart,
Peace can never hold sway!

For Lortzing the scenario is altogether more developed with radical implication. The interruption of the wedding banquet takes place at the witching hour of decision and destiny, as midnight strikes. While most moments of supercharged change or transformation traditionally occur at midnight, there is also a diurnal magic at work at midday.

This is particularly noticeable in Czech folklore, in the sinister figure of the Noonday Witch, whose frightening and destructive incursions into the ordinary lives of humans brings death. The librettist of *Rusalka*, Jaruslav Kvapil (1868-1950), himself wrote a literary ballad on the subject, which Antonin Dvořák (1841-1904) used as the subject for one of his most gripping tone poems. The topos of the fated midday hour also occurs in Debussy's *Pelléas et Mélisande*, and in many other operas. Nietzsche, after he had his idea of the eternal recurrence, standing by the pyramidal shaped boulder at Lake Silvaplana, wrote in his notebook: "And in every ring of human existence altogether there is always an hour when—first for one, then for many, then for all—the most powerful thought surfaces, the thought of the eternal recurrence of all things: each time it is for humanity the hour of midday", it is a time when the sun stands and enjoys time with his animals.[4]

The elements in pathetic fallacy reflect an eschatological participation of all nature in the events which carry judgment: the irruption of the supernatural into the ordinary world of reality. We have introduced these themes in Chapter 9. The deathly kiss is handled differently: in Lortzing, Undine says nothing; all is in her actions, raising her veil, extending her arms, implanting the kiss. Hugo's words in dying are more adumbrated than Huldbrand's, but also interestingly point to a sense of blessedness in embracing her (*die Seligkeit dich liebend zu umfangen,* "embracing blessedness in loving you"). The whole world seems to crumble in apocalypse at the kiss, and the scene beneath the waters in the river, an extended lullaby, expressing similar sentiments to Hoffmann, but with a most definite infusion of sentimentality.

Schwanengesang, Schwanenklang imply a state of magical blessedness, since the song of the swan is heard near the hour of death, and betokens movement into an alternate realm of benediction. It is no mistake that Lohengrin is the Swan Knight, a saviour figure from a spiritual kingdom of light and beatitude, brought by a swan and eventually able to free the swan from enchantment.[5]

Kühleborn's address to the transmogrified/transfigured Hugo reflects a traditional punishment-and-reward scenario: Undine's innocent suffering has enabled the forgiveness of Hugo's reprehensible betrayal. Those with souls are warned that this is how the soulless are avenged. Any ambiguity is drastically reduced and sentimentalized, this underscored by the comparative facility of the rhyme scheme. The music of this extended *Liebestod* is rocking, beautiful and soporific, and despite the intense symbolism of the swansong, the semantic ambiguity of Lortzing's lines is weakened in the prospect of *ew'ge Friede*. This could mean peacefulness or eternal oblivion: the nature of the music, however, does not suggest the darker implication. [**Fig.** Undine]

RUSALKA (1856)

Opera in four acts by Alexander Dargomyzhsky (1813-1869). Librettist: the composer, after the dramatic poem by Alexander Pushkin (1832). First performance: St Petersburg, Circus Theatre, 16 May 1856.[6]

The Slavonic version of the Undine myth is taken up by Dargomyzhsky in his opera written between 1848-55.

Natasha, the daughter of a Miller, is betrayed by the Prince and drowns herself in the stream. She becomes a Watersprite who lures men to their deaths. When the Prince marries, he hears her cries every time he approaches his bride. Wandering by the stream, to which he is compulsively drawn, and filled with sad remembrance, he meets a child who reveals herself as his daughter. The Miller hurls the Prince into the stream, where he joins Natasha, now the Rusalka, and their child.

Tsaritsa [= Natasha]
Now listen, daughter, and attend—this once
I put my trust in you. Tonight a man will come
Down to the river's bank. You will watch out
For him and go to meet him. He is kin
To us—your father, child.

Water-Babe [= Daughter]
The same one who
Abandoned you to wed a mortal woman?

Tsaritsa
The same: greet him with tenderness, and tell
Him all you know from me about your birth,
And tell him too what has become of me.
If he should ask you if I've forgot him
Or not, you may say I remember him
And love him, and invite him to my home.
Now—have you understood me daughter?

Water Babe
I understand.

Tsartitsa
Then go.

(Alone)

Since that fell hour
When, crazed with grief, I leapt into the water,
A desperate rejected, simple girl,
And woke again beneath the Dnieper River,
A Watersprite, cold of heart and potent,
Full seven long years have passed …
And every day I scheme and plan for vengeance,
At last, today, it seems my hour is come.

The Bank

Prince
Unwillingly to these sad banks I come
Drawn by some unknown power—I know not why …
For me each stick and stone speak of the past,
Retells the sad but well-beloved tale
Of my young days, my fair and carefree youth.
Here, once upon a time, love waited for me—
Freehearted, ardent love—ah, what a madman
I was to let such joy slip through my fingers,
Renounce such happiness—for I was happy …
How sorrowful, how sorrowful these thoughts,
That meeting yesterday has brought them back
The poor mad father! He is terrible.
Perhaps it may be that today I meet him
Again and he'll consent to leave the forest
And come to live with us …

(The Water-Babe emerges on the river bank)

What's this I see!
Say pretty child, whence came you?

In the underwater realm the earthly love of Natasha and the Prince is realized (through death?), and celebrated with song and ballet, similar to the final conception of Lortzing's *Undine*. The vengeance is completed, but the correlative is difficult to understand unless it is synonymous with (blissful?) oblivion. The composer as uses music to deepen the poet's insights.[7]

RUSALKA (1901)

Opera in three acts by Antonin Dvořák. Librettist: Jaroslav Kvapil, based on Friedrich de la Motte's novella, with additions from Hans Christian Andersen's *The Little Mermaid* and suggestions from Gerard Hauptmann's *The Sunken Bell*. First performance: Prague, National Theatre, 31 March 1901.[8]

The Watersprite Rusalka falls in love with Prince, and, with the help of a witch Jezibaba, becomes human so as to marry him. But she must remain silent, and when the Prince tires of her and proves unfaithful, a condition of her becoming human is violated. She dies together with the remorseful Prince who has sought her out.

The 1st Wood Nymph
I feel a tear in my eye;
a chill has suddenly blown on me.

The 2nd Wood Nymph
In gray clouds
the moon has hidden.

The 3rd Wood Nymph
Darkness is pressing into me.
Sisters, let us flee!

The Prince
My white doe!
My fairy tale! My mute vision!
Will my lamenting, my constant rush
never end?
From day to day driven by longing,
I seek you, panting, in the woods.
If night approaches, I sense you in it,
I grasp at you in the moonlit mist,
I seek you all over the earth!
My fairy tale, come back to me!
This is the place,
speak, silent woods!
Sweet vision, my love, where are you?
My white doe! Where are you?
By everything I have in my dead heart,
I invoke both heaven and earth,
I invoke both God and the demons!
Show yourself! Where are you,
my love?

Rusalka
My love, do you recognize me, do you?
My love, do you still remember?

The Prince
Kiss me, give me peace!
I don't want to return to the merriment of the world!
Keep kissing me till I die!

Rusalka
And you, my love, gave me so much,
why, my love, did you deceive me?
Do you know, my Prince, do you know?
That from my arms you will never return?
That with this annihilation
in my arms you will pay for it?

The Prince
I want to give you everything,
kiss me a thousand times!
I don't want to return, I will gladly die,
kiss me, kiss me, give me peace.
I don't want to return, I will gladly die,
I've no thought of returning!

Rusalka
My love will freeze all your senses!
I must destroy you,
I must take you into my icy embrace!

The Prince
Kiss me, give me peace!
Your kisses will absolve me of my sin.
I die happy
in your embrace!

The Watersprite
In vain he will die in your arms,
futile are all sacrifices,
Oh poor, pale Rusalka! Alas!

Rusalka
For your love, for that beauty of yours,
for your inconstant human passion,
for everything by which my fate is cursed,
human soul, God have mercy on you!

The finale of Dvořák's opera dispenses with all sentiment. The desolated Prince seeks out the betrayed Rusalka with real anguish. She, unlike Undine in both Hoffmann and Lortzing, who remain elementally detached and inviolable, has been harmed by her encounter with humanity and its untrustworthiness, feels herself defiled, and now, like Undine, must serve as an agent of natural retribution. But the Prince seeks union with her even though she warns him that he will never return from her arms: there will be only annihilation. He seeks out her kiss even though she warns him that her love will freeze him and destroy him. As in Lortzing, the Prince sees the kiss as bringing peace and absolution of his sin. But the Watersprite, that choral commentator throughout the opera, warns of the futility of all sacrifices.

The plight of the Undine figure is tragically underscored by the agency of the witch who imparts a limited humanity enabling Rusalka to seek human love. In both Dargomyzhsky and Dvořák, the hero is moulded in the manner of Huldbrand/Hugo, but is given a far more developed character in the betrayal of the soulless Watersprite. In both cases he becomes a kind of suffering young hero, and his regret leads him back to the water. The retribution of the Undine/Rusalka follows through remorse and enticement: in the Russian work through the child of his union who leads him on, and he is dispatched by the demented father of Natasha; in the Czech through his own growing remorse and increasingly urgent aspiration to death. The link between tragic love and death is common to all the manifestations of the Undine/Rusalka myths. The topos can perhaps be related to concepts of patriarchy and a mode of integrating folk memories of incest and infant mortality.[9]

A major variant on the Undine/Rusalka motif occurs in Nikolai Rimsky-Korsakov's (1844-1908) fairy operas *May Night* (1880) and *Sadko* (1898). In each instance the forlorn hero is helped, his life transformed, by the appearance of the water spirits in the form of swans—either in the moonlight, or by the banks of a sacred river—who lead him on to find his true destiny. In Tchaikovsky's *Swan Lake* (1877), the swan becomes a vector of demonic enchantment, and a symbol of good and evil, synonymous with the divided self (the self-sacrificing White Swan Odette, and the deceiving, destructive Black Swan Odile). As with the Undine operas, water is the medium of magic and transformation: at the cataclysmic finale Odette and Prince Siegfried are submerged by the rising waters of the lake, and as if in a sacramental medium of redemption, their transfigured forms are seen rising above the flooding element.

Undine's endearing laughter is mentioned twice by Huldbrand as she approaches to implant the kiss of the Love-Death. As discussed in Chapter 9, it is a motif that recurs with tragic prescience also at the conclusion of Wagner's *Siegfried,* where the overflowing emotions experienced by the young scion and the newly energized Valkyrie attain their highpoint in the seeming contradiction of *lachendes Tod.* Love and death are mystically inseparable.

The Watersprite underscores all too clearly the fundamental motif of all these stories of elemental beings and humans. In whatever disguised or transposed format, the search is for some kind of healing fulfillment we might call salvation—whether in the search of the water nymph for ensoulment, or in the pathetic dilemma of the betraying/betrayed human searching for peace or redemption. Rusalka can only emphasize the doubled-edged nature of beauty, love and human passion. The elemental annihilation is complete. The Undine myth and the symbolism of the Swan capture a yearning for some transcendent reality beyond the limitations of our sad humanity. [**Figs. 18.1-9** Undine]

Wild Swans at Coole

Upon the brimming water among the stones
Are nine-and-fifty swans.

The nineteenth autumn has come upon me
Since I first made my count;
I saw, before I had well finished,
All suddenly mount
And scatter wheeling in great broken rings
Upon their clamorous wings.

I have looked upon those brilliant creatures,
And now my heart is sore.
All's changed since I, hearing at twilight,
The first time on this shore,
The bell-beat of their wings above my head,
Trod with a lighter tread.

Unwearied still, lover by lover,
They paddle in the cold
Companionable streams or climb the air;
Their hearts have not grown old;
Passion or conquest, wander where they will,
Attend upon them still.

But now they drift on the still water,
Mysterious, beautiful;
Among what rushes will they build,
By what lake's edge or pool
Delight men's eyes when I awake some day
To find they have flown away?[10]

2. The Faust Operas

In a period of enormous questioning at the end of the 14th century and beginning of the 15th, the very composition of the world and the universe were being explored and re-defined. Questions about the nature of man, his religious dimension and affiliations, the nature of the soul and the mysteries of eternal destiny, perdition and salvation, were shaking the very foundations of belief, social and political certainties and stabilities. No single figure better represents the interrogation of human liberty and the freedom to choose than the legendary figure of Faust. In many ways he looks back to the certainties of an unbroken orthodoxy, while becoming the voice of the new Renaissance man. It is uncanny that this figure is associated with the invention of printing, and also with the dangerous questioning of theological truths, both of which were foundational to the whole new intellectual world order that was being proposed.[11]

Faust was derived from the account of a wandering conjuror who lived in Germany (c.1480-c.1540). The character is ostensibly based on Johann Georg Faust (c.1488-1521), a magician and alchemist probably from Knittlingen, Württemberg, who obtained a degree in divinity from Heidelberg University in 1509. He was a magician and astrologer about whom stories began to circulate, crediting him with supernatural gifts and evil-living. The legend was first set down and published in the *Historia von D. Johann Fausten* (compiled in Frankfurt by Johann Spiess, 1587). It immediately became popular and was re-edited and borrowed from throughout the 16th century. It was soon translated into English (as *The History of Dr Faustus, the Notorious Magician and Master of the Black Art*), into French, and other languages.

The basis of the Faust story is that he sold his soul to the Devil in return for 24 years of further life, during which he was to have every pleasure and all knowledge at his command. The climax comes when the Devil claims him for his own. The idea of making a pact with the Devil for worldly reasons is of Jewish folk origins.

Many other accounts followed, and the Faust theme was developed by authors, artists and musicians over the years. It inspired various writers, most especially Christopher Marlowe (1564-1593) in his *Tragical History of Dr Faustus* (1587-93) and Johann Wolfgang von Goethe in his *Faust* (Part 1, 1808; Part 2, 1832). It was Goethe who was responsible, however, for transforming the necromancer into a personification of the struggle between higher and lower natures in man.

LA DAMNATION DE FAUST (The Damnation of Faust) (1846*)*

Légende dramatique in four parts by Hector Berlioz; text by the composer after Goethe's *Faust* Part 1 (1773-1808) in Gérard de Nerval's (1808-1855) translation (1828 and 1840), including contributions from Almire Gandonnière (1814-1863). It was completed in 1846, incorporating *Huit Scènes de Faust* (1828) in altered form. First performance: Paris, Opéra-Comique, 6 December 1846.[12]

The plot makes use of episodes from Goethe, with Mephistopheles tempting Faust to bargain his soul in exchange for renewed youth. The pleasures provided include military glory (Berlioz introduced his version of the Hungarian Rakoczy March) (Part 1); drunken companionship in Auerbach's Cellar; powerful memory and prayer at Easter; mystical engagement with Nature (Part 2); love for Marguerite (Part 3); the betrayal of Marguerite.

Unlike Goethe's poem, Faust makes his pact with Mephistopheles after he has abandoned Marguerite and she has been sentenced to death, in the hope of rescuing her. In the final tableau her soul is received in heaven. Faust in the midst of nature, grown sated with these activities, confronted with Marguerite's ruin and death, is finally carried off to Hell in a tumultuous Ride to the Abyss, while Marguerite is saved (Part 4), with an epilogue in heaven.

The work being a sequence of scenes as in Goethe's *Faust* Part 1, tells the story episodically, but nonetheless with a sense of dramatic coherence. Marguerite remains a curiously static figure, and the tragedy of infanticide and matricide, with her implied execution, are narrated. The epilogues on earth and in heaven (touching on the final scenes of *Faust* Part 2) nonetheless provide the assurance of her transcendental translation and beatification. The episodes are underpinned by use of specific genres, like the march, the students' song, the Easter hymn, the ballad, the serenade, delicate dance pieces, the love duet, the emotionally weighted soliloquy (for both Faust and Marguerite). She has a precognition of her lover:

Scene 8

(Marguerite enters carrying a lamp. Faust hides.)

Marguerite
How stifling the air is!
I am afraid like a child.
It is my dream of yesterday that had troubled me …
In my dream I saw him … him … my future lover.
How handsome he was!
God! I was loved so passionately!
And how I loved him!
Shall we ever meet …

Scene 13

Marguerite
In my dream I saw you as I see you again now.

Faust
In a dream! … You saw me?

Marguerite
I recognise your voice,
Your features, your sweet speech …

Faust
And you loved me?

Marguerite
I waited for you.

Faust
Adored Marguerite!

Marguerite
My tenderness was already
Destined for you.
In this life …
Folly!

The Princes of Darkness
Are you forever, Mephisto,
Master and conqueror of this soul so proud?

Mephistopheles
I am Master of It forever.

The Princes of Darkness
Faust has then freely
Signed the fatal act that delivers him up to our flames?

Mephistopheles
He signed freely.

Epilogue on Earth

The Damned and Demons
Hell is silent now.
The terrible boiling of these great lakes of flames,
The gnashing of teeth and the tormenters of souls
No more are heard; and in its depths
A mystery of horror is accomplished.
O terrors!

In Heaven

Celestial Spirits
Praise! Praise! Praise! Hosanna! Hosanna!
She has greatly loved, O Lord!
Marguerite!

Marguerite's Glorification

Celestial Spirits
Mount again to Heaven, simple soul
That love misled;
Come wear again your pristine beauty
That an error changed.
Come, the divine virgins,
Your sisters the Seraphim,
Will know how to dry your tears
That earthly sorrows wring from you still.
Keep hope
And smile at happiness.
Come, Marguerite, come!

FAUST (1859)

Opera in five acts by Charles Gounod; text by Jules Barbier & Michel Carre, after Goethe's poem (1859). First performance: Paris, Théâtre

Lyrique, 19 March 1859.[13]

Faust makes a bargain with Mephistopheles: in return for eternal youth and the beautiful Marguerite, he promises Mephistopheles his soul. Valentine, Marguerite's brother, entrusts the care of his sister to the faithful youth Siebel while he goes off to the wars. Faust, aided by Mephistopheles, distracts the crowds at the kermis, enabling Faust, bearing a casket of jewels, to approach Marguerite. Mephistopheles leaves them in the garden. His distraction with the neighbour Marthe, enables Faust to bedazzle Marguerite and seduce her. Marguerite loses her peace of mind, and cannot find consolation even in church. On his return, Valentine finds Marguerite has been betrayed by Faust, challenges him to a duel, and is killed. He dies cursing his sister. Marguerite, in prison for killing her baby, refuses to escape with Faust and Mephistopheles. As she dies, she is taken by angles, assumed into heaven to the Easter hymn, while Faust is dragged down to Hell by Mephistopheles.

ACT 3 *Marguerite's garden*

Faust
Yes, believe this flower, blooming under your feet!
Let your heart hear it as the voice of heaven itself!
He loves you!
Do you understand this sweet and sublime word?
To love! To carry in our hearts
A constantly renewed flame!
To be forever drunk with eternal bliss!

Faust and Marguerite
Eternal!

Faust
O night of love, radiant sky,
O sweet transports!
Silent bliss
Instils heaven
Into both our souls!

Marguerite
I want to love and worship you!
Speak again!
I am yours!

Marguerite *(to Faust)*
I adore you!
I would die for you!

ACT 4 *The church*

Mephistopheles
No! God no longer forgives you!
The sky no longer dawns for you!
No! No!

Marguerite
Ah, this chant stifles and chokes me!
I am clamped in an iron band!

Mephistopheles
Farewell, nights of love and days of rapture!
A curse on you! Hell awaits you!

Mephistopheles
Marguerite!
Be accursed!
Hell awaits you!
(*Marguerite gives a shriek and falls senseless on the flagstones.*

ACT 5 *the prison cell*

Mephistopheles
Damnation!

Marguerite
Pure and radiant angels,
Carry my soul up to heaven!

Faust
Marguerite!

Marguerite
Why are those hands red with blood?
Go away! You fill me with horror!
(*She falls senseless.*)

Faust
Ah!

Mephistopheles
Judged!

Angelic Choir
Saved: Christ the Redeemer lives!
Te Deum laudamus!
Peace and good will on earth!
Gloria in excelsis!
Christ rose unto Heaven.

(*The prison walls open. The bells of Easter sound and a chorus of angels sings "Christ is risen!". Marguerite's soul rises to heaven. In despair Faust follows it with his eyes; he falls to his knees and prays. Mephistopheles is turned away by the shining sword of the archangel.)*

In both Berlioz's and Gounod's versions of the story, Marguerite sings the famous spinning song.

Marguerite
(she sings while plaiting her hair)

Once there was a king of Thule
Who was faithful until death,
Received, on his fair one's death,
A carved cup of gold.
As it never left him,
In the happiest festivals
Always a light tear
Moistened his eyes.

This prince, at the end of his life,
Bequeathed his towns and his gold,
Except the dear cup
That he still kept in his hand.
He made his barons and peers
Sit at his table,
In the middle of the ancient hall
Of a castle bathed by the sea.

The drinker stands and goes
To an old gilded balcony;
He drinks, and suddenly his hand throws
The sacred vessel into the waves!
The vessel falls; the water bubbles,
Then is calm again.
The old man grows pale and shudders;

He will not drink again.
Once there was a king … of Thule
Until death .. he was faithful …
Ah!

On the surface Marguerite's simple spinning song seems to serve little purpose in the play. For Berlioz it should be sung in an even, absentminded manner since nothing is further from Marguerite's mind than the story of this ballad. But there is a vital connection, as the ballad presents a very real parallel of situation and motif. Marguerite has been disturbed by Faust's interest in her and is fearful of becoming involved in an emotion that threatens to take over her life and holds little prospect of ultimate happiness, given their social disparity. The simple song is about the undying love of a king for his dead beloved of humble rank. Gretchen/Marguerite desires as well as fears loving Faust, so the song both addresses her fears and inspires hope.

The contrast between the song and the main story heightens the tragedy of the situation. The young girl believes in "Treue bis an das Grab" (faithfulness beyond the grave) and attains this ideal in herself. The tragedy lies in the fact that she is denied the realization of such a love in life. In the song the cup that the king cherishes becomes the symbol of this undying love. As a gift from his beloved it is cherished above all his possessions throughout his life, and precedes him even into death when, having drunk a last draught from it, he casts it into the sea, as an image of both oblivion and eternity. The brevity of exposition and vividness of the narrative, the hallmarks of the ballad, contrast with the elaborate Jewel Song that follows in Gounod's score.

The moulding by Gounod of Goethe's poem into a libretto is handled in a masterly manner, investing all the characters with a dramatic vividness in soliloquy and interaction. Faust's opening ruminations on futility, his joy at his restored youth, his genuine engagement with Marguerite's innocence, are all conveyed with feeling. Marguerite is depicted with real vulnerability and emotion. Her diffidence at the kermis, her innocent charm in spinning, her entranced reaction to the jewels and her own transformation, her devoted response to Faust's wooing, her anguish in church and her demented grief in prison, are all brought to life in the score. Her sudden death in the dungeon is an autonomic reaction to the sheer burden of sorrow, while her translation to heaven, again approximating the final scene of *Faust* Part 2, is pure transcendence to transfiguration. This sound and mood are sustained throughout the opera by the subtext of religious feeling using hymns (Act 1), chorales (Act 2), chants (Act 4), and the celestial chorus (Act 5), as well as modal allusions throughout in the harmonic textures.

MEFISTOFELE (1868)

Opera in three acts with prologue and epilogue by Arrigo Boito (1842-1918). Libretto by the composer, after both parts of Goethe's poem. First performance: Milan, La Scala, 5 March 1868.[14]

Unlike Gounod, Boito, himself a poet, based his opera on both parts of Goethe's work, and endeavoured to give it a substratum of philosophy. The five-part Prologue depicts the angelic choruses in Heaven, and the arrogant challenge of Mefistofele to corrupt Faust, concluding with a chorus of cherubs and psalmody of penitents and spirits. Both Faust and Mefistofele are strongly characterized: Faust in his respective paean to nature; and Mefistofele in his self-identification as "the Spirit who always denies".

The love between Faust and Margherita is consistently realized in a rapturous lyricism. The scene of Margherita's imprisonment for infanticide is extended and deeply affecting: the narration of the drowning of her baby, the exquisite duet for her and Faust as they contemplate flight, her perception of the reality of the situation, her prayer for mercy and her death, as she dies in Faust's arms. The voices of the celestial choir sing softly of her salvation.

Margherita, Faust
Far, far, far away, far away,
On the waves of a wide ocean,
Among the dewy mists of the sea,
Between the seaweed, among the flowers,
Among the palm trees,
The heaven of intimate tranquility,
The blue islet appears to me.
It appears to me on the clear sky
Ringed by the rainbow
Reflecting the smile of the sun.
The flight of the hopeful free lovers,
Migrant, radiant,
Is directed to that island,
The fight of free lovers,
Far, far, far away.

Mefistofele *(appearing at the grille)*
The day is dawning!

Margherita
Ah! Satan is roaring!

Faust
Ah! Hurry,
Time is fleeting!

Margherita *(to Faust)*
Ah! No!
Do not abandon me!

Mefistofele
From outside these gates
The fanfare of death sounds.

Margherita
Alas! Almighty God
(breaking away from Faust)
Deliver me from temptation!
They are rending my limbs
With harsh shackles.
O God, help me,
They are leading me to death,
Oh Heaven!
Ah! Already above my head,
Above my head the axe glints,
The axe, I saw it flash!

Mefistofele *(next to Faust)*
Cease, stop these vain words,
From the eastern sky
The sun is already rising,
The neighing of the black steeds
Can already be heard,
We must flee.

Faust
Calm, maiden,
Your distraught spirit;
Let me see that pale face composed;
Control the spate
Of useless sighing.
We must flee,

Ah! Yes, we must flee.

Margherita
Ah! Would that I had never been born!

Mefistophele
Well?

Margherita
Who is rising up there?
Who looms from the earth?
It is the Evil One!
Mercy!
In this holy refuge,
What does the accursed one want?
Oh! Drive him away,
Perhaps it is me he wants!

Faust
Oh! Come with me and live!
Live, live, Margherita.

Mefistofele
You follow me, or
I will leave you both for the axe

(Dawnlight. Mefistofele goes to watch at the gate. Margherita, heartbroken, agonizes in Faust's arms)

Margherita
Look, the pale dawn ...
The last day is here …
It should have been ...
The joyful day ...
Of our wedding ...
All is ended
In life.

Faust
O cruel heartbreak!

Margherita
Hush ... tell no one
That you loved Margherita
And that I gave you my heart.
Oh Lord, Thou will forgive ...

Lord ... you will forgive.
Holy Father ... Save me ...
And you, heavenly ones protect
This suppliant who turns to you.

Mefistofele
She is judged.

Faust
O torment!

Margherita
Heinrich ...
(she falls)
You disgust me!

Celestial Host
She is saved! ...

Mefistofele
Away with me, Faust!

(Faust and Mefistofele disappear. In the back of the cell the executioner appears, surrounded by his assistants.)

After Margherita's death comes the scene depicting the Night of the Classical Sabbath, introducing Helen of Troy. Boito provides this exquisite description of Helen as the embodiment of the *Ewig-Weibliche*.

Faust (*bowing before Elena*)
Forma ideal purissima
Della bellezza eterna!
Un uom ti si prosterna
Innamorato al suolo.
Volgi vêr me,
Di tua pupilla bruna,
Vaga come la luna,
Ardente come il sole.
Un uom ti si prosterna,
Innamorato, innamorato al suol.

Ideal pure form
Of eternal beauty!
A man prostrates himself
In love with the ground.
Turn to me,

Of your brown pupil,
It wanders like the moon,
Fiery as the sun.
A man prostrates himself,
In love, in love with the soil.

The Epilogue, with its beautiful choruses, shows Faust as an old man back in his study, strengthened by the angelic voices, holding to the Holy Book, falling to his knees and effectively overcoming the satanic temptation. He dies amidst a shower of rose petals, while the triumphant song of the celestial choir celebrates his salvation. Mefistofele has lost his wager and holy influences have prevailed.

EPILOGUE

Faust
God gracious, I walk away
From the demon my mocker,
Do not lead me into temptation!
(*transported in the ecstasy of vision*)
Fly the burning song
Of the celestial squad!

Celestial Host
From eternal harmony
The Universe
In the glaucous immersed space …

Mefistofele
The angelic host already is crying.
Let us catch that soul in flight.
Already the path of evil destroys
God with his foolish pardon
Destroys the work of evil!

Faust
Sacred fleeting moment,
Linger yet, you are so fair!
Grant me eternity!
(*dies*)

Heavenly Host
…. emits a chant,
Of supreme love …
(*A shower of roses descends on Faust's body*)

DOKTOR FAUST (1925)

Opera in two prologues, an interlude and three scenes by Ferruccio Busoni (1866-1924) (completed by his pupil Philipp Jarnach) (1925). The text was by the composer after the legend of Faust and Marlowe's treatment of it in *Dr Faustus* (1589/1604), and the 18th-century German puppet show. First performance; Dresden, 21 May 1925.[15]

Faust invokes the help of Mephistopheles to help him regain his lost youth. An off-stage *Credo* for the chorus, followed by the *Gloria*, marks the signing of Faust's pact with the Devil. Faust elopes with the Duchess of Parma shortly after her marriage to the Duke. The scenes at the ducal court are mostly built on dance rhythms in contrast to the exciting descriptive passages. At an inn in Wittenberg, Faust discusses philosophy with the students; this soon becomes an impressive quarrel chorus between the Catholic and Protestants, in which the fanatical Protestants intone Luther's hymn *Ein' feste Burg* against the *Te Deum laudamus* of the Catholics.

Faust hears of the Duchess's death, and he too longs to die. He meets a beggar woman, who turns out to be the Duchess, clasping the corpse of a child. Faust offers his own life if the child can be revivified. The last scene is characterized by desperate invocation. Faust lays the body of the dead child on the ground and covers it with his cloak. He throws his legendary girdle on the ground, and steps within the circle. He exerts his will in a final effort to project his personality into the body of the child: may his faults be rectified in this child, and may it accomplish what he has failed to do. He expires, and as the Night-Watchman (really Mephistopheles) announces midnight, a naked youth with arms uplifted and bearing a green twig in his hand, rises from the Faust's body and walks through the snow.

> **Faust**
> My evil spirits they play their game.
> A higher man shall banish you.
> Now stand by me, God!
>
> (*He wants to enter into the Church, which suddenly appears brightly illuminated from within. From the church door, a harassed brother steps and fights the entrance.*)
>
> **Choir**
> God will not always
> Be the Lord of Mercy
> And forgiving grace,

But sometimes of vengeance,
Of retribution and punishment.
And as that God you shall know him.
He will not hear your prayer.

Faust
You too! Let me, let me!

(The ghostly soldier stretches out the sword against him.)

Faust
Away, I have to pray!
Be gone, you hellish fiend, I still am master!

Choir
No, no!

(The apparition fades. Faust with the child in his arm, drags himself to the steps of the crucifix.)

Faust
Oh, pray, pray! Where to find the words?
They dance through the brain like magic formulas.
I want to look up to you as before.

(The Night-Watchman, sneaking in from behind, lifts his lantern. In its light, the Crucified One transforms into Helena.)

Damnation! Is there no mercy?
Is there no atonement?

(The Night-Watchman goes away. Faust rises to his feet.)

So let the work be accomplished.
Come aid … me,
Original witness,
Compelling
Fulfilling power,
I call you to the highest task.

(Faust puts the dead child on the ground, covers him with his coat, loosens his girdle, and steps into the circle.)

Blood of my blood,
limb of my limbs,
Unawakened,

Pure in spirit,
Still outside all circles
And to me in this circle
Intimately bound,
I bequeath you my life:
It goes forth
From the earth-embedded root
Of my departing day …
Into the joyous opening blossom
Of what you shall become.
Thus may I still work through you,
In what you beget
And ever deeper and deeper digging
The track of my being
Until at the end of its impulse.
Where I built crookedly,
You shall make straight,
What I failed to do,
You shall complete,
This is how I stand
Above the law then,
Shall I stand,
At once embracing
The ages
And unite myself
With last generations:
I, Faust,
One eternal will!

(He dies.)

Voice of the Night-Watchman (Mephistopheles)

You men and women, let us say,
The weather tonight is changing,
Frost is in the air.
The clock is striking the midnight hour.

(At the place where the dead child lay, a naked, half-grown youth rises, with a flowering branch in his right hand. With his arms uplifted, he walks over the snow into the night and into the city. The Night-Watchman [Mephistopheles] appears and with his lantern lights up the outstretched Faust.)

Mephistopheles
This man it would seem has met with some misfortune.

As is clear, the various composers emphasise different aspects of the legend. Faust is damned by Berlioz; Marguerite is central to Gounod (as in Schubert's song settings); Faust is more centrally handled by Boito, and especially by Busoni.

Gounod develops a much more religious approach than Berlioz: Valentine's sword-hilt forms the sign of the cross; there is the extended Church Scene of Marguerite's fraught conscience; but both composers present the chorus of angels with Marguerite's 'salvation', although Gounod develops the eschatological dimension of this much more forcefully. In neither work does Faust get his woman, just as Hoffmann fails to win Olympia, Giulietta or Antonia—the Stella of all his passions. Faust's own salvation is left unresolved.

Boito presents the whole of the Faust story: his striving and realization of the perfect moment are presented, as is the genuineness of his love for Margherita. In the end he is showered with the rose petals of salvation.

Busoni, using the old legend and Marlowe, sees Faust as daring and experimental. His final extraordinary act is a type of self-sacrifice, an immolation of his selfishness/selfhood to bring about a new life, with images of renewal and resurrection. While there is no apotheosis in the style of the other composers, the intensity of imagination and symbolism function with extraordinary power. [**Figs. 18.10-21** Faust]

The *Faust* operas present a recurring aspect of 19th-century opera where there is dichotomy between the body and the soul, the religious and the irreligious, the failings of the secular realm and the pull of the transcendent or spiritual. One sees this tendency from Mozart's *Die Zauberflöte,* with Sarastro's priestly processions and hymns, and the chorale for the Two Armed Men (1791). The Anabaptists are identified by their chorale in Meyerbeer's *Le Prophète*, where the rituals of coronation are enacted with organ and chants in Münster Cathedral (1849). In the famous *Miserere* from Verdi's *Il Trovatore*, Leonora sings of her grief as distant monks chant the great penitential Psalm 51, while Manrico voices his valedictory love ecstatically from his tower prison (1853). The piety of Lutheran Nuremberg and the spirit of the Reformation are heard in Wagner's *Die Meistersinger von Nürnberg,* from the opening church chorale to the final greeting of Hans Sachs (1868); while in *Parsifal* the Holy Grail is unveiled to the sound of mystical voices from the Dome (heaven) (1882). In Massenet's *Manon*, the heroine's renewed seduction of Des Grieux takes place against the religious background of the seminary at St Sulpice, conjured up in ecclesiastical music (1884). Celebrations of victory over Napoleon, with the great organ and choral *Te Deum*, form the backdrop in San Andrea della Valle to Scarpia's plotting in Puccini's *Tosca* (1900).

In Meyerbeer's *Robert le Diable* we hear the subterranean voices of Hell; in the finale the celestial choirs with harps consecrate the hero's salvation in marriage to Princess Isabelle. The chant of monks and the sacred tones of the organ emanate from Palermo Cathedral (1831). In *Les Huguenots,* we have entered secular history and politics, and religion is played out in terms of the musical genres and motifs of the religious parties, whether as Lutheran hymnody and Calvinist battle songs, or Catholic litanies and unctuous ecclesial blessings. Partisan confrontation results in ensembles of counterpointed anthems. The whole issue of martyrdom is realized musically in rhapsodic transformation of the chorale (1836). The same is true of Halévy's *La Juive*, where the corporate Christian celebration is depicted in the *Te Deum* from the Cathedral, and forbidden Jewish worship in the secret prayers and rituals of the Seder/Passover (1835).

These matters are more overtly and thoroughly explored in the *Faust* works than elsewhere, where we are given glimpses into the mystical realms of the celestial hosts, the very hierarchy of heaven itself. The *Faust* operas, in the manner of Goethe's great drama, allow a return to the spheres of heaven and hell, with the earth between. This is seen in the Berlioz dramatic poem, where the celebration of Easter is heard, where the climax of the work takes us on a shattering Journey to Pandemonium itself, and Marguerite's tacit pardon and elevation to heavenly beatitude is depicted in an angelic chorus with harps. The musical language of Gounod's opera is impregnated in its very harmonic textures with the music of faith: the modality and cadences of hymns and chorales can be heard throughout, with overt usage in the offstage Easter hymn, the chorale to ward off satanic evil at the tavern, the organ and spiritual music of Marguerite's visit to church, the chorale at the death of Valentine, the triumphant Easter paean marking the death and salvation of Marguerite, the triumph of good. These patterns characterize the works of Boito and Busoni as well. Indeed, *Mefistofele,* in its Prologue and Epilogue, recreate the spiritual realms of Goethe's poem its flights of angels and the music of the heavenly hosts. The finale includes a sublime setting of the Marian hymn *Salve Regina* to approximate Goethe's mystical perception of the Dantean spheres of Paradise and the mystery of the *Ewig-Weibliche.* Busoni's narrative remains terrestrial, and does not have the rhapsody of Marguerite's assumption. But his score is suffused with generic and modal citations that characterize the different stages of Faust's career: the great professions of faith in the *Gloria* and *Credo*, the *Te Deum* and Luther's chorale in the partisan confrontation (in the manner of *Les Huguenots*). Examples of specifically Easter music only emphasise this topos: most famously of all the Easter Hymn in Mascagni's *Cavalleria Rusticana* (1890).

PELLÉAS ET MÉLISANDE (1902)

Drame lyrique in five acts with music by Claude Debussy. The French libretto is a slight adaptation of Maurice Maeterlinck's Symbolist play *Pelléas et Mélisande* (1892). Translation: Henry Grafton Chapman. First performance: Paris Opéra-Comique, 30 April 1902.[16]

Synopsis

Act 1

Scene 1. A forest

The opera takes place near water. It is dark, with the sun rising over the sea. Prince Golaud, grandson of King Arkel of Allemonde, has become lost while hunting in the forest. He discovers an attractive, frightened, weeping girl sitting by a spring in which a submerged crown is visible. She is Mélisande, who is very distressed. But nothing is known about her origins, except that she has a history of abuse. She refuses to allow Golaud to retrieve her crown, which has fallen in the water, observing that she would rather die than recover it. Golaud persuades her to return home with him before the forest becomes dark.

> **Golaud**:
> I was watching your eyes
> Do you never close your eyes?
>
> **Mélisande**
> Yes, yes, I close them at night.

Scene 2. A room in the castle

Six months have passed. Geneviève, the mother of the princes Golaud and Pelléas, reads a letter to the aged and nearly blind King Arkel. It has been sent by Golaud to his brother Pelléas. Golaud reveals that he has married Mélisande, but that he knows no more about her than when they first met. Golaud fears that Arkel will be angry with him and tells Pelléas to observe how he reacts to the news. If the old man is favorable then Pelléas should light a lamp from the tower facing the sea on the third day; if Golaud does not see the lamp shining, he will sail on and never return home.

Arkel had planned to marry the widowed Golaud to Princess Ursule in the hope of ending "long wars and ancient hatreds", but he accepts Golaud's marriage to Mélisande. Pelléas arrives, he has received a letter from his friend Marcellus, who is dying. He wants to travel to say goodbye to him. Arkel thinks Pelléas should wait for the return of Golaud, and reminds Pelléas of his own father, who is lying sick in bed in the castle. Geneviève tells Pelléas not to forget to light the lamp for Golaud.

Scene 3: Before the castle

Geneviève and Mélisande walk in the castle grounds. Mélisande remarks how dark the surrounding gardens and forest are ("places so thick that the sun can never be seen"). Pelléas arrives. They look out to sea and notice a large ship leaving and a lighthouse shining, Mélisande predicts that it is destined to sink. Night falls.

Geneviève goes off to look after Golaud's young son by a previous marriage. Pelléas tells Mélisande that he might have to go away the next day. Mélisande asks him why.

Act 2

Scene 1: A well in the park)

Pelléas takes Mélisande to the "Blind Men's Well". People used to believe it had miraculous powers to cure blindness, but since the old king's eyesight started to fail, they no longer venture there. Mélisande lies on the edge of the well and tries to see to the bottom. Her hair loosens and falls into the water. Pelléas notices how extraordinarily long it is. He remembers that Golaud first met Mélisande beside a spring. He asks if Golaud tried to kiss her at that time but she does not answer. Mélisande plays with the ring Golaud gave her, throwing it up into the air until it slips from her fingers into the well. Pelléas notices that the clock was striking twelve noon as the ring dropped into the water. Mélisande asks him what she should tell Golaud. He replies, "The truth".

Scene 2: A room in the castle

Golaud is lying in bed, with Mélisande at the bedside. He is wounded, having fallen from his horse while hunting. The horse suddenly bolted for no reason as the clock struck twelve. Mélisande begins to weep and says she feels ill and unhappy in the castle. She wants to go away with Golaud, but refuses to say why she is unhappy. She complains about the gloominess of the castle: today was the first time she saw the sky. She observes that "I feel that I shall never live very long". Golaud takes her hands to comfort her and notices the wedding ring is missing. Golaud becomes angry, and Mélisande claims she dropped it in a cave by the sea where she went to collect shells with little Yniold. Golaud orders her to go and search for it before the tide comes in, even though night has fallen. Mélisande says that she is afraid to go alone, Golaud tells her to take Pelléas along with her.

Scene 3: Before a cave

Pelléas and Mélisande make their way down to the cave in darkness. Mélisande is frightened to enter, but Pelléas tells her she will need to describe the place to Golaud to prove she has been there. The moon comes out lighting the cave and reveals three beggars sleeping in the cave. Pelléas explains there is a famine in the land. He decides they should come back another day.

Act 3

Scene 1: One of the towers of the castle

Mélisande is at the tower window, singing a song (*Mes longs cheveux*) as she combs her hair. She says to Pelléas:

> My long, long hair it reaches to the foot of the tower;
> My hair is waiting for you down the tower all the way.

Pelléas appears and asks her to lean out of the window so he can kiss her hand as he is going away the next day. He cannot reach her hand, but her long hair falls down from the window and he kisses and caresses it. He winds her hair around his neck. He says he has never seen hair like hers and that "it covers me even around my heart". Pelléas then says "I suffer no more in the tangles of your hair". Pelléas playfully ties Mélisande's hair to a willow tree. A flock of doves takes flight. Mélisande becomes agitated when she hears Golaud approaching. Golaud dismisses Pelléas and Mélisande as a pair of children and leads Pelléas away.

Scene 2: The vaults of the castle

Golaud leads Pelléas down to the castle vaults, which contain the dungeons and a stagnant pool which has "the scent of death". He tells Pelléas to lean over and look into the chasm while he holds him safely. Pelléas is very uncomfortable.

Scene 3: A terrace at the entrance of the vaults

Pelléas is relieved to breathe fresh air again. It is noon. He sees Geneviève and Mélisande at a window in the tower. Golaud tells Pelléas that there must be no repeat of last night's "childish game" between him and Mélisande as she is pregnant, and the least shock might disturb her health. Pelléas should avoid her as much as possible.

Scene 4: Before the castle

In the darkness before dawn Golaud is sitting with Yniold, and questions him about Pelléas and Mélisande. The boy, too innocent to understand what he is asking, says that Pelléas and Mélisande often quarrel about the door. He admits that he once saw Pelléas and Mélisande kiss "when it was raining". Golaud lifts his son on his shoulders to spy on Pelléas and Mélisande through the window but Yniold says that they are doing nothing other than looking at the light. He threatens to scream unless Golaud lets him down again. Golaud leads him away.

Act 4

Scene 1: A room in the castle

Pelléas tells Mélisande that his father is getting better and has asked him to leave on his travels. He arranges a last meeting with Mélisande by the Blind Men's Well in the park.

Scene 2: The same

Arkel tells Mélisande how he felt sorry for her when she first came to the castle "with the strange, bewildered look of someone constantly awaiting a calamity". But now that is going to change and Mélisande will "open the door to a new era that I foresee, for you are too young and too lovely to live day and night in the shadow of death..." He asks her to kiss him.

Golaud bursts in with blood on his forehead—saying it was caused by a thorn hedge. Mélisande tries to wipe away the blood, he orders her not to touch him. Golaud asks for his sword, and notices Mélisande is trembling. He reassures her he is not going to kill her. He mocks her "great innocence", saying he sees in Mélisande's 'great eyes' "How proud they are of their beauty". He commands her to close them or "I will shut them for a long time". He tells Mélisande that she disgusts him and seizes her by the hair. Mélisande tells Arkel that Golaud does not love her anymore. Arkel comments: "If I were God, I would have pity on the hearts of men".

Scene 4: The same

Pelléas arrives alone at the well, worried about his deep involvement with Mélisande: he fears the consequences. He knows he must leave but first, he wants to see Mélisande one last time and tell her things he has kept to himself. Mélisande arrives. Pelléas admits his love for her and sings "I can hear your heart beating as though it were my own…". Mélisande confesses that she has loved him since she first saw him.

> **Pelléas**:
> I hardly heard what you said!
> We have broken the ice with red hot irons!
> And you say that in a voice that comes from the world's end!
> I was hardly able to hear…
> (*It is dark, and he cannot see her eyes*),
> your eyes were elsewhere ….
>
> **Melisande**
> I saw you somewhere else …
> *(Pelléas hears the servants shutting the castle gates for the night).*

The two kiss, Mélisande hears something moving in the shadows. It is Golaud, who has been watching the couple from behind a tree. Golaud strikes down Pelléas with his sword and kills him. Mélisande is also wounded, but she flees into the woods, saying to a dying Pelléas that she does not have courage.

Act 5

A bedroom in the castle

Mélisande sleeps in a sick bed after giving birth to her child. The doctor assures Golaud that despite her wound, her condition is not serious.

Overcome with guilt, Golaud claims he has killed for no reason, as Pelléas and Mélisande merely kissed like a brother and sister.

Arkel
Mark the way in which she sleeps …
Very slowly, very slowly …
One would think her soul were cold forever …

Mélisande wakes and asks for a window to be opened so she can see the sunset. She is somewhat confused, not knowing what she has said. Arkel thinks she is delirious. Golaud asks the doctor and Arkel to leave the room so he can speak with Mélisande alone. He blames himself for everything and begs Melisande's forgiveness, but she barely recognizes him. Golaud presses Mélisande to confess her forbidden love for Pelléas. She maintains her innocence in spite of Golaud's pleas to her to tell the truth.

Golaud
Lie no more like this at the moment of death.

Melisande
Who is it that is to die?

Golaud replies it is she, but then he will die after her. Arkel and the doctor return. Arkel tells Golaud to stop before he kills Mélisande. Golaud replies, "I have already killed her". She fades away, asking why it is so cold, and if winter is coming. Arkel hands Mélisande her newborn baby girl but she is too weak to lift the child in her arms and remarks that the baby does not cry and that she will live a sad existence. The room fills with serving women, although no one can tell who has summoned them. Mélisande quietly dies. At the moment of death, the serving women fall to their knees.

Arkel
She is going to sleep. She has closed her eyes …
Her eyes are full of tears.
But now it is her soul that is weeping …
We must not disturb her anymore …
For the soul is silent…
And departs on its own …
So quickly, so quickly …
She is gone without a word …

COMMENT

Pelléas et Mélisande is the only opera Debussy ever completed. It is considered a landmark in 20th century music. "The drama of *Pelléas* which, despite its dream-like atmosphere, contains far more humanity than those so-called 'real-life documents', seemed to suit my intentions admirably. In it there is an evocative language whose sensitivity could be extended into music and into the orchestral backcloth".[17] This opera contains virtually all of the topoi that we are concerned with.

The centre of the action is a love triangle: Pelléas, Mélisande and Prince Golaud. The latter becomes suspicious and jealous, and goes to excessive lengths to find out the truth about Pelléas and Mélisande's relationship. He even uses his own child, Yniold, to spy on the couple.

Mélisande is found by Golaud in a forest, near a fountain, she and he are lost in their own ways. She is beautiful, frightened, and clearly has been badly abused by someone in the past. She has large, startling revealing eyes. Golaud is a sad widower, and Mélisande is married into a family with a dark heritage, living in a cold, dark, sunless castle.

There are two deaths: Pelléas, killed by Golaud, and Mélisande, who expires. Golaud runs Pelléas through with his sword, a symbol of his recovered and outraged masculinity. Melisande seems to have a post-natal injury and succumbs to a type of fever. She has given birth and is a bit delirious before her death, but she drifts into a sleep, as her sad soul is silent, and gone without a word. Golaud will never know the truth.

Romantic influences are found in Maeterlinck's drama, with a Wagnerian symbolism of forests, the sea, light and dark, a sword and time. There is a tissue of other images: water (well, and pool); gold (crown, ring); tower; castle, cave, dungeon, prison; hair; and many reference to eyes, blindness, seeing and watching.

Eyes especially are powerfully and mysteriously evoked.

Golaud to Mélisande
I was watching your eyes.
Do you never close your eyes?
(Act 1, scene 1)

In Act 4 scene 2 Golaud's growing feeling of violent jealousy is explored in terms of what he sees and what he presumes she sees and construes.

> What do you look at me as though I were a beggar? ...
> Do you hope to see something in my eyes without my seeing something in yours?

Arkel sees Mélisande's eyes as symbol of innocence, but to Golaud they are tainted and the windows onto a guilty soul.

> Shut them or I'll shut them for a long time.

Flowing hair (especially derived from the Pre-Raphaelite paintings of Dante Gabriel Rossetti is a central symbolic, sexual motif.

As the opera proceeds Melisande expresses her unhappiness, often weeping and predicts her own death: "Je senses que je ne vivrais plus longtemps". Being too young to live in the shadow of death, she is desired by all the men around her. Her unloosened hair falls at the well into the water; it flows in a delicate tumbling orchestral cascade from the window of the tower, down to the importuning Pelléas. This is just as in the Grimms' fairy tale (No.12 of the *Kinder- und Hausmärchen*): Rapunzel's long hair is a way to scaling the imprisoning phallic tower giving access to the Prince, and is the means to mature erotic fulfilment. Mélisande's hair in Act 3 scene 1 becomes a fetish for Pelléas, a transferred metaphor of the erotic charge between them: he kisses her hair, placing its locks on his heart, and winding them around his neck. But conversely it is used to hurt her by Golaud, who seizes her by her hair and humiliates her (Act 4 scene 2), using the tresses as a symbol of pain in his parody of the Sign of the Cross. He alludes to Absolom, the rebellious son of King David, famed for this long flowing hair. This source of admiration and beauty becomes the agent of Absolom's death when in flight the hair is caught in the low-lying branches of a tree, and he is left "hanging between heaven and earth", and killed by the avenging general Joab (2 Samuel 18:9-15). Golaud sees Melisande's hair, her beauty and desirability as the source of her undoing, a symbol of her fatedness.

When Mélisande meets with Pelléas on their last night together (Act 4 scene 3), he hears her heart beating; we know that she is frail, pale, and delicate, and sensitive to emotions. The setting of the opera, with its Gothic flourishes, has a female protagonist reminiscent of Poe's waifs. The castle becomes a louring gloomy presence, and time a recurring motif. We find out that Mélisande was born at noon; she loses her ring and Golaud falls from his horse at midday. The couple are locked out of the castle at midnight—the midnight chimes emphasise fate. Coincidences, cosmic order, high noon and midnight, engulf Mélisande, a poor mysterious little being, "C'etait un pauvre petit être mysterieux…" in settings that are emotionally too much for her weak frame.

Mélisande's death can be explained by life and social circumstances in a world where she was lost and no-one understood her, with swift currents of underlying violence. It seems rather like that of Isolde after the death of Tristan. She is trapped in a crepuscular half-world, and dies as the sun goes down towards the depths of the sea.

The music is characterized by the almost complete absence of arias or set pieces. There are only two reasonably lengthy passages for soloists: Geneviève's reading of the letter in Act 1 and Mélisande's song from the tower in Act 3. Debussy set the text of the play one note to a syllable in a speech rhythm, as a continuous, fluid lyrical line, a *cantilena*, hovering between chant and recitative. The orchestral textures are shimmering, vaporous, harmonically shifting, in a typical Impressionistic manner. The musical style reveals Debussy's deeply ambivalent attitude to Wagner. Despite his avowal to escape from German influences, Wagner's theory of music drama is evident in the handling of the orchestra and the declamatory vocal line. There is a blurring of the usual distinction between aria and recitative, with the orchestra providing a ceaseless commentary, with some *Leitmotifs*. There are musical themes readily identified with three main identities: they appear on the first page of the score (Golaud, Mélisande and the castle in Allemonde).

The Tower Scene constitutes an extended love duet, with occasional overtones of the nocturnal tryst of the lovers in *Tristan und Isolde*, and a harmonic language familiar from Puccini, especially in the undulating horn passages for Pelléas that intensify his feelings. The second love scene in Act 4 scene 4 (A Fountain in the Park) follows the model of the *Tristan* duet even more closely: it opens with loud agitation as Pelléas musters his courage to bid farewell to Mélisande; sinks into a reverie-like middle section with overt declarations of love, the harp making a subtle appearance at these dreamy moments; becomes agitated again with the closure of the gates (depicted orchestrally), their exclusion, and the fears that they are watched; becomes disturbed again as they kiss and discern the presence of Golaud (announced by bassoons); and reaches a brief climax with his murder of Pelléas (in the manner of Tristan's fight with Melot) as Mélisande looks on helplessly.

The whole of the last act is in some ways a variant on the Love-Death. As Mélisande lies dying, she is relentlessly questioned by Golaud, with all held within the oracular serenity of Arkel's observations (all-wise, all-knowing, like Gurnemanz in *Parsifal*). Golaud ranges through the gamut of emotions as he tries to extract the truth from his frail and dying wife, she always comforted by Arkel. Her responses hover around the same subdued monotones, but represent in their own elusiveness themselves a type of testimony to her love-until-death for Pelléas. Arkel's words assume the nature of a lullaby. The harp becomes discreetly ever-present in the last moments, a sure sign of Mélisande's ascendant love, even though her voice is now mute in death. The hushed orchestral epilogue establishes a transcendent peace.

Arkel
She was a quiet little thing, so timid and silent. She was a poor mysterious little person like anyone.

[**Figs. 18.22-25** Pelléas and Mélisande]

Notes

[1] Hoffmann, *Undine,* with Rita Streich and Raimund Grumbach, Symphonieorchester des Bayerischen Rundfunks, cond. Jan Koetsier (Memories); with Heidrun Plesch and Johannes Beck, Jugendorchester Bamberg, cond. Hermann Dechant (Bayer Records).
[2] The passage reads literally: "That means to get salvation, /To whom such a farewell is granted."
[3] Lortzing, *Undine* with Anneliese Rothenberger and Nicolai Gedda, Radio-Symphonie Orchester Berlin, cond. Robert Heger; with Monilka Krause and Josef Protschka, Kölner Rundfunkorchester, cond. Kurt Eichhorn (Capriccio).
[4] Prideaux, S. *I am Dynamite! A Life of Friedrich Nietzsche* (London, Faber & Faber, 2018), 191, 230.
[5] The swan is a symbol of great complexity. "The dedication of the swan to Apollo, as the god of music, arose out of the mystic belief that it would sing sweetly when on the point of death… [It] has even deeper significance: hermaphroditism, since in its movement and certainly in its long phallic neck it is masculine yet in its rounded, silky body it is feminine. In sum, then, the swan always points to the complete satisfaction of a desire, the swan-song being a particular allusion to desire which brings about its own death." See J. E. Cirlot, *A Dictionary of Symbols* (New York: Philosophical Library, 1962), 306-307. Because Zeus took the form of a swan to seduce Leda, it came to symbolize love and the gods. But it is also stands for solitude, music and poetry, while its whiteness represents sincerity. See David Fontana, *The Secret Language of Symbols: A Visual Key to Symbols and Their Meanings* (London: Pavilion Books Ltd., 1993), 87.

[6] Dargomyzhsky, *Rusalka*, with Natalia Mikhailova, Konstantin Pluzhnikov and Alexander Vedernikov, Symphony Orchestra of Moscow Radio, cond. Vladimir Fedoseyev (Relief).

[7] Emerson, Caryl. "To What End Rusalka? Pushkin's Folk Tragedy and Dargomyzhskii's Opera". *The Slavonic and East European Review*, 97:1; *1917 and Beyond: Continuity, Rupture and Memory in Russian Music* (January 2019), 169-200. "Only an opera libretto can accommodate and transcribe the simultaneities of ensemble thinking. The astonishing potency of the libretto as a narrative-musical form, competent to extend dramatic poetry into psychological realms it could not otherwise go, is everywhere apparent in this opera, especially in its trios, quartets and peasant choruses. The composer's compact, multi-layered librettistic solutions to unfolding dramatic conflict—alternately helpless, repressed, horrific—in turn fed into further dramaturgical attempts to complete Pushkin's play. Dargomyzhskii's gift for psychological musical portraiture had ripened in a 'realistic' decade very distant from the aristocratic Age of Pushkin a mere quarter-century earlier. But the composer, like the poet and also like his predecessor Mikhail Glinka, had been raised and trained in West European idioms" (172).

[8] Dvořák, *Rusalka*, with Gabriela Benaackov-Capova and Wieslaw Ochman, Czech Philharmonic Orchestra, cond. Vaclav Neuman; with René Flemming and Ben Heppner, Czech Philharmonic Orchestra, cond. Charles Mackerras (Decca).

[9] See Cooke, Brett. "Constraining the Other in Kvapil and Dvorak's *Rusalka,*" in Brett Cooke, George E. Slusser, and Jaume Marti-Olivella (eds), *The Fantastic Other: An Interface of Perespectives* (Amsterdam: Editions Rodopi, 1998), 121-142. "One hint is the daughter's defiance of her father, her seeming lack of loyalty. Dvorak's Rusalka, early in Act I, desires to be 'set free' from the Vodnik's domain. In Dargomyzhsky's opera, the Daughter is openly abusive of her father. Also conspicuously missing from these texts is the mermaid's mother, indeed, she is very rarely mentioned in mermaid tales, if at all, while the father or an uncle is always present. Nor are there aunts or stepmothers. Why should the single father be such a regular feature of mermaid tales? To some extent this may reflect the harsh conditions of life both when folklore flourished and when the literary mermaid made her appearance, especially in terms of childbirth mortality. But this pattern of single fathers, nevertheless, seems to be too regular."

[10] Yeats, William Butler. "Wild Swans at Coole'. *Collected Poems of W. B. Yeats* (London: Macmillan and Co Ltd, 1955), 147.

[11] *The Faustian Century: German Literature and Culture in the Age of Luther and Faustus*. Ed. J. M. van der Laan and Andrew Weeks (London: Camden House, 2013).

[12] Berlioz, *La Damnation de Faust*, with Nicolai Gedda, Janet Baker and Gabriel Bacquier, Orchestre de Paris, cond. Georges Prêtre (EMI).

[13] Gounod, *Faust*, with Nicolai Gedda, Victoria de los Angeles and Boris Christoff, Orchestre du Théâtre National de l'Opéra de Paris, cond. André Cluytens (EMI).

[14] Boito, *Mefistofele*, with Placido Domingo, Monserrat Caballé and Norman Treigle, London Symphony Orchestra, cond. Julius Rudel (EMI).

[15] Busoni, *Doktor Faust*, with Dietrich Fischer-Diskau, Hildegard Hillebrecht and William Cochrane, Symphonie-Orchester des Bayerischen Rundfunks, cond. Ferdinand Leitner (DG).
[16] Debussy, *Pelléas et Mélisande*, with Claude Dormoy and Michele Command, Orchestre de Lyon, cond. Serge Baudo (Eurodisc).
[17] Langham Smith: "Pelléas et Mélisande", *Grove Music Online* ed. L. Macy.

CHAPTER 19

THE OPERAS OF RICHARD WAGNER

Part 1

The transcendental makes the beautiful more beautiful, the moving more moving, and the profound more profound.

(Roger Scruton, *Music as an Art*)[1]

Introduction

Chapter 18 drew attention to operas which we had referred to as being special cases. In the overall context of this book, they were selected for further detailed investigation because they represented narratives which concerned *Unexpected Death*, in several areas with some hint of the autonomic fluctuations. They are presented separately because the operatic deaths are linked to dramatic events such as metamorphosis, transfiguration and other transcendental epiphanies. Further they relate to special themes, such as the Faust series. The composers who appear most on the list are Wagner (six operas), and Donizetti (three). We discuss these findings later. In this chapter we wish to shift our focus to the librettos and music of Wagner.

We include an analysis of two of his first three operas, *Die Feen* and *Rienzi.* These works include many thematic and musical ideas which are to be found in the later operas and which are of importance for the central themes of our book. *Das Liebesverbot,* and *Die Meistersinger von Nürnberg,* while of considerable relevance for the theoretical and dramatic development of Wagner's works, shy away from *Unexpected Death.* Because of the importance of *Der Ring des Nibelungen*, we have treated this separately from the point of view of the chronology. The librettos of all the following operas were composed by Wagner. [**Figs. 19.1-28**]

> When we look at Wagner, we are gazing into a magnifying mirror of the soul of the human species.
>
> (Alex Ross, *Wagnerism: Art and Politics in the Shadow of Music*)[2]

THE EARLY OPERAS

DIE FEEN (The Fairies) (1834/1888)

Romantic Opera; the libretto was written by the composer after Carlo Gozzi's *La donna serpente* (1762). First performance: Munich, 29 June 1888.[3]

Pursuing a doe, Prince Arindal plunges into a river and awakes in the castle of the fairy Ada. They fall in love and marry on condition that for eight years he shall not ask her name.

Act 1

Ada, half-fairy, half-mortal, marries Arindal, King of Tramond. He is living in fairyland, in joy and splendour, but must not ask her name. Morald and Gunther have been sent to find out what has happened to Arindal. Ada persuades him to return to help save his country, under heavy attack. He asks her name, so Ada and the magic kingdom disappear. Arindal sings and grieves at the loss of Ada.

Arindal
Where do I find you, where do I find comfort?
You have gone, and with you all my happiness!
My searching eyes
have looked for you everywhere;
every valley and every hill
echoes the sigh of passionate longing.
Oh, grief! All my efforts are vain.
The wilderness resounds with her name;
the echo mocks my torment
and only calls "Ada, Ada!"
But no answering voice calls "Arindal!"
Your eyes light up for me no more,
your bosom does not warm me;
no kiss quenches the thirst of my lips,
your arms do not enfold me,
only deathly cold breathes on me.
Oh, grief!
Was it all but a dream?

My sun is where you are,
life is only where you are,
to be far from you is death
and terrifying night.
Ah, let me find life,
free me from fear of death.
Where are you, ah, where are you,
where do you abide, far from me?
Oh, end my torment
and take me to yourself again.

Maddened by the ordeals set up by Ada's jealous companions, he thinks Ada is a sorceress who has abandoned him and that he should return to his kingdom.

(As he turns to go, he is overcome by sudden fatigue and slowly sinks down on a stone.)

But what power restrains my limbs?
I would go, but my feet refuse to stir.
My eyelids droop. Is this the approach of sleep?
I feel it. Farewell, my love,
your husband leaves you thus. Adieu!
(He falls asleep).

In the fairy garden, Ada sings of how she is willing to sacrifice her immortality for Arindal ("Wie muss ich doch beklagen"). Her father has died, and she is now Queen. Arindal awakens and declares his joy at seeing Ada again, but she announces that he will abandon her the next day.

Act 2

There is a prophecy that the kingdom will not fall if Arindal returns, and he does. Ada's companions hope that she will renounce Arindal and remain immortal. She fears that Arindal will be cursed with madness and die, she being turned to a statue, but then expresses hope that Arindal's love will prove strong.

Ada
In perpetual beauty I could flourish,
immortal, ever young.
Betrayed, wretched woman!
What is immortality to you?
An endless and eternal death,
but every day with him

is new, eternal life!
So be it! The choice is made,
for such a life I give up all.
My Arindal!
For even if he should succumb
and the rock encase me,
love will make the stone itself
weep tears of longing.

Ada wins the battle against Arindal's foes who have attacked his kingdom since his father died. Ada appears with her two children by Arindal. She seems to throw them into a fiery abyss. Arindal curses her. She sorrowfully explains that the Fairy King had required, as a condition of her renouncing her immortality, that she conceal her fairy background from Arindal for eight years and on the last day torment him as best she can. If he cursed her, she would remain immortal and be turned to stone for a hundred years while he would go mad and die.

Act 3

All pray for the curse to be lifted. Arindal is hallucinating that he is hunting a doe. As it is killed, he realizes it is his wife. He continues to experience visions before falling asleep.

Go to it! I'll shoot my arrow!
See, how it flies—I aimed well!
Haha, it went straight into her heart!
Oh, look, the creature can weep!
Tears glisten in her eyes.
How the dying eyes look at me!
How fair she is!
Horror! It is no animal!
See here, it is my wife!

The voice of the petrified but weeping Ada is heard calling for him.

Ada
(off-stage)
Cold stone encloses
my burning love.
Only tears can soften
the harsh prison.
Through every barrier
my love comes to you,
and if you hear my lament,

hasten to me!

Arindal awakens, and he expresses his willingness to die for her. A sword, shield and lyre are presented to Arindal by the benevolent magician Groma: these will win him victory. Arindal is led to a portal guarded by earth spirits. He is about to be defeated when he is reminded of the shield: the earth spirits disappear when he holds it up. Next, they encounter bronze men who guard a holy sanctuary. Here the shield fails. Arindal holds up the sword: the bronze men vanish. He then reaches a grotto where Ada has been turned to stone. Arindal is threatened that failure will mean that he too is turned to stone. He is urged to play the lyre. When he does so ("O ihr, des Busens Hochgefühle"), Ada is freed from the stone. The Fairy King has decided to grant Arindal immortality. Ada invites him to rule her fairyland with her, and she remains immortal.

Fairy King
You, mortal, penetrated our realm
and the infinite power of love
lent you that sublime strength which only
is given to immortals.
Know then: through your human failings,
Ada remains immortal as she was,
but he who wrested her from us with divine strength
is more than mortal—be immortal, as she is!

COMMENT

Die Feen was Wagner's first completed opera but remained unperformed in his lifetime. The fantasy plot anticipates many themes that are recognised embryonic features of his later compositions. These include a forbidden action and, importantly, redemption.

The Weberian overture in E major, the key in which the opera begins and ends, includes many of the principal themes. The work as a whole does not have the complex chains of melody and chromatic harmony that distinguished the composer's mature works. There is, however, already a tendency in the score to move away from strict generic numbers to long through-composed passages. Recurring themes or simple *Leitmotifs,* associated with characters and situations, already reveal a tendency that Wagner would later use in a far more sophisticated manner in his mature works. The orchestra often carries the tune while vocal parts are declamatory, another anticipation of the later style. There are some set arias, especially Ada's big scene in Act 2, which requires dramatic soprano singing. There are also melodies re-used in, and ensembles anticipating, *Rienzi*, *Tannhäuser* and *Lohengrin*. On the lighter side is a buffo duet for Gernot and Drolla that looks forward to *Das Liebesverbot*, notable for its unassuming tunefulness.

RIENZI, DER LETZTE DER TRIBUNEN **(Rienzi, The Last of the Tribunes) (1842)**

Grand Opera in five acts. The libretto is based on Mary Russell Mitford's drama (1828) and Edward Bulwer-Lytton's novel (1835) of the same name. First performance: Dresden, Hofoper, 20 October 1842.[4]

Act 1

Outside Rienzi's house

Orsini (a Roman patrician) attempts to kidnap Rienzi's sister Irene. The Roman people support Rienzi's condemnation of the tyrannical nobles. Irene and Adriano Colonna (son of one of the noble families) realise their mutual attraction. A crowd, inspired by Rienzi's speeches, offers him the crown, but he insists that he wishes only to be a Tribune of the Roman people.

Act 2

The patricians plot the death of Rienzi; Adriano is horrified when he learns of this. Rienzi greets a group of ambassadors for whom a ballet entertainment is laid on, Orsini attempts to stab Rienzi, who however is protected by a vest of chain mail. Adriano pleads with Rienzi for mercy for the nobles, which Rienzi grants.

Act 3

The patricians have recruited an army to march on Rome. The people upbraid Rienzi for showing clemency to the treacherous nobles. Adriano is filled with conflicting feelings. The people are alarmed. Rienzi rouses the people and leads them out to battle as the women pray for success. Rienzi scores victory over the nobles; in the course the battle Adriano's father, Stefano Colonna, is killed. Adriano swears revenge, but Rienzi dismisses him.

> **Irene**
> Hapless man! Look, it is too late!
> Would you go to your death senselessly?

Adriano
Almighty God! Yes! It is too late!
My senses are fading!

Irene
See, I clasp you tightly,
and only with my life will I let go!

Adriano
Double death and agony of love!
O Heaven, finish my torment!
(*Irene draws Adriano to her, kneeling*)

Act 4

Adriano's intention to kill Rienzi wavers when Rienzi arrives together with Irene. The Pope has laid a papal interdict on Rienzi and he and his associates are excommunicated. Despite Adriano's urgings, Irene resolves to stay with Rienzi.

Adriano
He is accursed and cast out
but I shall save you, flee his presence!

Irene
My brother!
(*tearing herself from Adriano's grasp*)
Go away, wretched man!
(*gazes at Rienzi*)
Rienzi! Rienzi! Oh, my brother!
(*She throws herself upon Rienzi's breast*)

Adriano
(*raging*)
Madwoman! Go with him to your ruin!
(*He hurries off*)

Rienzi
(*coming out of a dazed state, feels Irene on his breast, lifts her up and looks into her eyes with a troubled gaze*)
Irene, you? There is still a Rome!
(*They stand in a long embrace*)

Act 5

Scene 1. Rienzi's room

Rienzi prays for guidance. He suggests to Irene that she seeks safety with Adriano, but she demurs.

(*Rienzi, and Irene embrace ardently*)

Rienzi
If I am abandoned by the Church, for the glory
of which I started my work, and by the people
to which I first gave the name of a people,
if I am abandoned by every friend who brought me
happiness, then two things shall be faithful to me for ever:
Heaven itself and my sister!

Irene
The sister of Rienzi defies death!
Put me to death—
(*falling at Rienzi's feet*)
I shall never leave you!

Rienzi
(*clasps her powerfully to his breast*)
Come, proud maiden, to my breast!

An apologetic Adriano enters and tells the pair that the Capitol is to be burnt and they are at risk.

Scene 2. The Capitol

Rienzi's attempts to speak are met with insults from the fickle crowd. Adriano, in trying to rescue Rienzi and Irene, is killed with them as the building collapses.

(*The Capitol is ablaze; Rienzi and Irene can be seen, clasped in each other's arms and surrounded by flames, on the balcony; the people throw stones at them*)

People
Come on, honour the Church's high command!
Death and destruction to them!
(*Adriano enters with others, set about attacking the people.*)

Adriano *(seeing Irene)*
Irene! Irene! Come, through the flames!

(*As Adriano rushes towards the Capitol, the tower in which Rienzi and Irene are standing collapses with a dreadful noise, burying Adriano as well as Rienzi and Irene.*)

COMMENT

Rienzi is Wagner's third completed opera and is mostly written in the style of grand opera: depictions of crowds, acts ending with an extended finale ensemble, ballet sequence, and replete with solos, duets, trios and crowd scenes. The liberal ethos associated with the hero and the political intervention of a reactionary clergy recall *La Vestale, Les Huguenots*, and *La Juive*. *Rienzi* prefigures themes (brother/sister relationships, social order and revolution) to which Wagner was often to return in his later works. *Rienzi* remained one of Wagner's most successful operas until the early 20th century.

In this opera, with its historical subject, the transcendental issues are not so obviously amenable to motivic treatment. Nonetheless, the opera operates on a skein of 35 themes, some of which recur with the pointed thematic purposefulness of *Leitmotif.* The arresting trumpet call that begins the overture represents the liberating call to arms that features so much in the action, and this is substantiated in Rienzi's battle hymn "Santo Spirito, cavalieri". But the chief theme is that of Rienzi's prayer, which begins in the 18th bar of the overture and unfolds in a broad serene cantabile (*molto legato e espressivo*) in D major, deeply contemplative and moving, which is then repeated in great augmentation to rhapsodic effect, the emotion intensified by upward rushing string triplets. The cataclysmic finale is prefaced by the extended scene of Rienzi's lonely confrontation with his failed enterprise, and his sense of a heavenly mission captured in his prayer: this becomes the medium of his transcendent death with Irene when the prayer is heard in variant, with huge chords dominating the tumult. The transcendent theme is also reflected in the beautiful Peace Motif which accompanies the entry of the emissaries in Act 2, and in more secular manner in the exultation of the military triumph at the end of Act 3.

THE MIDDLE-PERIOD OPERAS

DER FLIEGENDE HOLLÄNDER (The Flying Dutchman) (1843)

Romantic Opera; the libretto is based on legend as recounted in Ch. 6 of a tale by Heinrich Heine, *Aus den Memoiren des Herrn von Schnabelewopski* (1831). First performance: Dresden, Hofoper, 2 January 1843.[5]

Act 1

Captain Daland's ship is pushed off course by a storm, and after dropping anchor, a mysterious ship appears next to his. The helmsman has fallen asleep singing a ballad about his sweetheart. The Flying Dutchman (Vanderdecken) laments his fate. Once every seven years he may leave his ship in search of a woman whose perfect love will redeem him from his deathless wandering, otherwise he must sail on until the Day of Judgement. Daland wakes and speaks with the Dutchman, who offers him a large sum of money for the night's lodging. He learns that Daland has a daughter. Daland, mesmerized by the Dutchman's wealth, agrees to allow his daughter Senta to marry the Dutchman.

Dutchman
How often into the sea's deepest maw
have I longingly hurled myself,
yet death, ah, I found it not!...
You I ask, blessed angel of God,
who won for me the terms of my salvation,
was I the sorry plaything of your scorn,
when you showed me the way to redemption?
Vain hope! Terrible, futile folly!
There is no eternal fidelity on earth!

Act 2

Senta, is with a group of local women, singing and making sails. They tease her about her boyfriend, Erik the Huntsman, but she is in a trance staring at a picture of the fabled Dutchman.

Senta (*in the armchair*)
Yohohoe! Yohohohoe! Yohohoe! Yohoe!
Have you seen the ship upon the ocean
with blood-red sails and black masts?
On its bridge a pallid man,
the ship's master, watches incessantly.
Whee! How the wind howls!

Yohoheh! Whee!
How it whistles in the rigging!
Yohoheh! Whee!
Like an arrow he flies on,
without aim, without end, without rest!
Yet there could be redemption one day for that pale man
if he found a wife on earth who'd be true to him till death!
Ah, when, pale seaman, will you find her?
Pray heaven that soon a wife will keep faith with him!
(*Towards the end of the stanza Senta turns to the picture.*)

She vows to be faithful and rescue him. Erik arrives and is shocked. He tells her of his dream of the advent of a strange man with her father, who carries her out to sea. Daland arrives with the mysterious guest. She tells the Dutchman that she will remain truthful and faithful to him until she dies. They both sing of redemption (*Erlösung*), Daland happily blesses their union.

(Senta sitting down at the beginning of Erik's tale of his dream, falls into a trance-like sleep, so that she appears to be dreaming the dream being related to her).

He seeks me out! For him I look!
With him surely I will share his doom!…

(The Dutchman has just entered and Senta's gaze moves from the portrait to the Dutchman. She utters a cry of astonishment and remains as if spellbound, without taking her eyes off him. The Dutchman, his eyes fixed on Senta, comes slowly forward)

Dutchman
(deeply moved)
As from the mist of times long gone
this girl's image speaks to me:
as I dreamt of her for restless ages,
I see her now before my eyes.
The dull glow I feel burning here,
can I in my misery call it love?
Ah, no! It is a yearning for redemption:
would that through such an angel it came true!

Senta
Am I deep in a wonderful dream?
What I see, is it mere fancy?
As I have often seen him, here he stands.
The pain that burns within my breast,

ah, this longing, how shall I name it?
What you yearn for, salvation,
would it came true, poor man, through me!
(Towards the end of the stanza Senta turns to the picture.)
It is I who will save you with my true love!
May God's angel show me to you!
Through me you shall find grace!
One has my oath. Here do I pledge him:
My honour until death!

Act 3

The women of the village and Daland's's crew celebrate the impending marriage and vainly try to engage with the silent and terrifying Dutch sialors. Erik pleads with Senta to remain faithful to him. The Dutchman overhears Erik's plea, and believes Senta is faithless. He tells her that since she has not proclaimed her vows before God, she will escape the eternal damnation of those who betray him. The Dutchman and his ghostly crew quickly depart, their ghostly forms prompting screams from the villagers who run to the shore. Senta makes her way to a tall cliff overlooking the bay. Remembering her vow of faithfulness unto death to the Dutchman, in the glow of the setting sun, she throws herself off the cliff. The heavens open and the Dutchman and Senta are seen in firm embrace as they rise to heaven.

Erik
Senta! Do you want to perish?
Come to me! You are in Satan's clutches!

Dutchman
I am condemned to the most ghastly fate.
From the curse a woman alone can free me,
a woman true to me till death.
You did vow to be true but
not solemnly before God: this saves you!
For know, poor girl, the fate awaiting
those who break faith with me:
eternal damnation is their lot!
But you shall be saved.
Farewell!

Senta
(checking the Dutchman)
I know you well! And well I know your fate!
I knew you when first I saw you!

The end of your torment is near! I am she
by whose true love you shall find salvation!

Dutchman
(to Senta)
You do not know me, cannot guess who I am!!
(He points to his ship its red sails spread)
the *Flying Dutchman* I am called!

Senta
(*From the rock overhanging the sea, calls after the departing Dutchman*)
Praise your angel and his decree!
Here I stand, true to you unto death!

(She leaps into the sea; at once the Dutch ship sinks with all her crew. The sea heaves and falls in a whirlpool. In the glow of the rising sun, the transfigured forms of the Dutchman and Senta, clasped in each other's arms, are seen rising over the wreck and soaring into the sky)

COMMENT

The story was derived from Wilhelm Hauff's *Das Gespensterschiff*, used in Heine's collection of stories *Salon* and his novel *Memoiren des Herren von Schnabelewopski.* Based on the universal myth from Odysseus to the Wandering Jew, the story captured Wagner's attention, providing him with the idea of the redemption of the Ahasuerus of the ocean, redeemed by the love of a pure woman. Wagner described the story, with elements of the Romantic, Byronic and Gothic, as "the first folk-poem [*Volksgedicht*] which forced its way into my heart".[6] It was a turning point in his artistic career: he determined to be become a poet, and "no more a concoctor of opera-texts".

With Senta's ballad Wagner claimed to have grasped the idea of the *Leitmotif* for opera. The elemental nature of the folk genre with its recurring themes of love, loss and death, and its literary adaptation by Goethe in *Faust,* was introduced into operatic currency by Scribe, initially in his libretti for Auber (*Le Maçon*) and Boieldieu (*La Dame blanche*) (both 1825). Elemental cries are heard ("Hohoje! Hohoje!" of the helmsmen; "Hui! Johoh" of Senta) which rebound in *The Ring:* we have dreams, curses, predestination, prefiguration of images long buried in the psyche (the Hollander, the pallid man of the portrait). There is recognition when eyes first meet; Senta's cry when they come face to face (the moment which brings an old dream into recall, the same for him—a dream from eternity); the sea, storms, the mystery ship with a black mast and blood red sails; and, most significantly, redemption through a woman's self-sacrifice. Such themes will recur in one form or another in future Wagner operas. There is a *Liebestod* in Act 2.

Wagner turned the female protagonist of the story in the *Memoirs of Herr von Schnabelewopski* from that of a middle-class girl to that of a redemptrix of man. Her vow is not to bourgeois law, but to the everlasting law of eternity.

The score of this opera is characterized by the sustained evocation of natural imagery, as the sea is depicted on nearly every page. The character of the accursed seafarer emerges from the restless chromatic surging of the waves and in the solemn portentous theme for the Dutchman, from D4 to D5 (taken from his great monologue in Act 1), expressing a great yearning for rest (in D minor). The peace comes in the juxtaposed theme associated with Senta and her ballad, with its concluding assertion: an aspiration to bring release and relief to the doomed wanderer. It emerges in the overture immediately after the frenzy of the Dutchman's desolation, in the relative F major key, in serene and perfect harmony, designed to represent the redeeming principle, the sacrificial love of the Eternal Feminine, the *Ewig-Weibliche*, typified by Senta. This theme, growing out of her ballad, also consecrates the moving duet with the Dutchman, when on their first meeting their elective affinity, already established in an eternal predestination, is expressed in a type of Love-Death. It returns at the conclusion of the opera, marking the sacrificial leap of Senta to her death for and with the Dutchman. Here it becomes a glorious redemptive apotheosis, emphasized by Wagner's revision of the closing bars of the opera in 1847.

TANNHÄUSER UND DER SÄNGERKRIEG AUF WARTBURG (Tannhäuser and the Singing Contest at the Wartburg) **(1845)**

Grand Romantic Opera in three acts after two medieval legends and their contemporary re-telling by E.T.A. Hoffman, Heinrich Heine and Ludwig Tieck. First performance: Drsden, Hofoper, 19 October 1845.[7]

Place: near Eisenach, Thuringia, early 13th century.

Act 1

The Venusberg, the Hörselberg of "Frau Holda".

Tannhäuser is held captive through his love for Venus in her erotic realm under the Hörstelberg. He longs for freedom, spring and the sound of church bells. He pleads to be allowed to depart.

> **Tannhäuser**
> Fortunate forever the man who has tarried with you!
> Forever envied he who, with ardent passion,
> has shared the godlike glow in your embrace!
> Entrancing are the marvels of your kingdom,
> I breathe the magic of all pleasure here;
> no country on the broad earth offers the like,
> what they possess you can easily spare.
> But amid these rosy perfumes
> I long for the woodland breezes,
> for the clear blue of our skies,
> for the fresh green of our meadows,
> for the sweet song of our little birds,
> for the dear sound of our bells.
> From your kingdom I must fly—
> O queen, goddess, let me go!

Resisting the charms of Venus, he declares: "My salvation rests in Mary", the mother of God. We then find Tannhäuser just below the Wartburg. It is springtime; a young shepherd pipes an ode to vernal season. Pilgrims heading for Rome in procession pass as Tannhäuser stands motionless, and he sinks to his knees. He is discovered by the Landgrave and his companions, Wolfram, Walther, Biterolf, Reinmar and Heinrich, who welcome him. (In the past Tannhäuser had fled from the Court after he had fallen out with his fellow knights in the prize-singing contest, one particular song having aroused erotic feelings, which also caught the ear of Venus). Wolfram informs him that his singing gained for him the heart of Elisabeth who has never been the same since he left. He joins the Landgrave and the singers to the Wartburg.

> **Wolfram** (*in the valley of the Wartburg*)
> Was it by magic or by your might
> that you achieved the miracle
> or captivating the most virtuous of maids
> by your singing filled with joy and sorrow?
> For, when, in haughtiness, you left us,
> her heart closed to our song;

we saw her cheeks grow pale,
she ever shunned our circle.
Oh, return, you valiant Singer,
let not your song be far from ours.
Let her no longer be absent from our festivals,
let her star shine on us once more!

Tannhäuser
The heavens look down upon me,
the meadows sparkle, richly-decked!
The spring, the spring
with a thousand lovely sounds
has entered into my soul, rejoicing!
In sweet impetuous urgency
my heart cries aloud:
To her, to her!
Lead me to her!

Act 2

The hall of Wartburg in Eisenach

Since Tannhäuser's disappearance, Elisabeth has been disconsolate. Now she is filled with elation at his return. There is to be a song competition and she joyfully agrees to be present. The Landgrave and Elisabeth receive the noblemen of the neighbourhood, who appear in rich attire. The Landgrave announces that the contestants' song is to be about "love's awakening". Elisabeth will grant the victor one wish, whatever it may be. In the contest, Wolfram declares that love is like a pure stream, which should never be troubled. Tannhäuser replies that the highest love is found only in the pleasure of the senses. He answers Wolfram and the palled songs othe others with a paean to Venus. The women, with the exception of Elisabeth, leave the hall in horror, and the knights draw swords against Tannhäuser, but Elisabeth protects him. Tannhäuser expresses his penitence for his blasphemous sentiments, and the Landgrave allows him to join the band of pilgrims bound for Rome, where he may perhaps obtain forgiveness and redemption from the Pope.

Tannhäuser (*at the contest*)
Oh Wolfram, you who have sung thus,
have woefully misrepresented love!
If you languish so fearfully,
the world would come to an end, forsooth!
To praise God in the sublime and lofty distance,
raise your eyes to heaven, look up to His stars!

But that which inclines to touch,
lies near the heart and senses,
that which, conceived of the selfsame stuff
in weaker mould, nestles to one —
I do boldly approach that fount of delight
with which no fear is ever mixed,
for the fount is inexhaustible,
as my longing is unquenchable!
That my desire may ever burn
I will ever refresh myself at the source!
To thee, goddess of love, shall my song ring out!
Now let thy praise be sung aloud by me!
Thy honeyed fascination is fount of all beauty,
and every sweet wonder stems from thee!
That man who has held thee locked in passionate embrace,
knows what love is, and he alone.
Poor creatures, who have never enjoyed her love,
fare hence, fare hence into the Venusberg! ...

Elisabeth
I pray for him, I pray for his life!
May the spirit of belief be granted him anew
since for him, too, the Saviour suffered once!
For him only will I pray,
may my life be prayer;
grant that he may see Thy light,
before he is lost in night!
In joyful trepidation,
let a sacrifice be dedicated to Thee!
Take, oh, take my life:
I no longer call it mine!

Act 3

The valley of the Wartburg, in autumn

The pilgrims return, and Elisabeth, accompanied by Wolfram, falls on her knees in prayer. She asks the returning pilgrims for news of Tannhäuser, but in vain.

Elisabeth
He has not returned
(*sinks to her knees*)
Almighty Virgin, hear my prayer!
I cry to thee, All-glorious!
Let me perish in the dust before thee,

oh, take me from this earth!
Make me, pure and angel-like,
enter into my blessed realm!
If ever, engrossed in foolish fancies,
my heart did stray from thee,
if ever a sinful longing,
a worldly yearning did spring up within me,
I wrestled then beneath a thousand smarts,
to kill it in my heart!
But if I could not atone for every fault—
yet receive me of thy grace,
that, as a worthy maid, I may
draw near thee in humble greeting,
only to implore the richest favour
of thy mercy for his sin!
(*She remains for a long time in devout rapture*)

Wolfram, who loves her with faithful devotion, has a presentiment of her death ("Song to the Evening Star"). He percieves a woebegone returned Tannhäuser. The Pope refused his plea for absolution and declared that Tannhäuser had no more chance of being forgiven than the Pope's staff had of sprouting leaves.

Tannhäuser (*Rome Narration*)
Hear me, Wolfram, hear me!
Then I saw him through whom God proclaims himself—
all men prostrated themself before him in the dust.
And he bestowed grace on thousands, pardoned,
thousands he commanded joyfully to rise.
Then I, too, drew near— my head bowed down to earth—
with sorely grieving mien, myself I accused
of the sinful delights my senses had experienced,
of the longings no penance had yet cooled;
and for deliverance from the searing bonds,
shot through with savage pain, I implored him.
And he whom I so begged began:
"If you have enjoyed such sinful delights
and enflamed your passions at the fires of hell,
if you have sojourned in the Venusberg,
then, now from henceforth, you are eternally damned!
As this staff in my hand
no longer bedecks itself in fresh green,
so from the burning brands of hell
deliverance can never blossom for you!"
Then I sank, annihilated, speechless, to the ground;

I fainted clean away.
When I woke, night brooded over the deserted square.
From the false sound of promise
which, icy-cold, pierced my soul,
shuddering horror forced me away with wildly staggering step!
It drove me there where I had enjoyed so much delight
and pleasure on her warm breast!
To you, dame Venus, do I return,
into thy magic's sweet night;
to your court do I descend,
where your alluring charm will smile upon me now for always!
(*The funeral procession*)

Tannhäuser despairingly seeks the way back to the Venusberg. Venus appears and beckoning him back to her abode. Wolfram notices a funeral procession, mourners bearing the corpse of Elisabeth on a bier. Tannhäuser races to her side and collapses dead upon her body, invoking her holy name. The younger pilgrims are heard approaching. They enter and announce that the Pope's staff has miraculously sprouted young green leaves, a sign that Tannhäuser has obtained God's forgiveness.

Men
Holy the pure maid who, now united with
the heavenly host, stands in the presence of the Everlasting!
Blessed the sinner for whom she wept,
for whom she implores salvation of heaven!
(*Wolfram leads Tannhäuser to the bier bearing Elisabeth's body. He sinks slowly to the ground*)

Tannhäuser
Holy Elisabeth, pray for me!
(*He dies. The scene is flooded by the red glow of dawn*)

Younger Pilgrims (*bearing a staff covered in fresh green leaves*)
Hail! Hail! Hail to the miracle of mercy!
Redemption is conferred upon the world.
It came to pass in the holy hour of night,
the Lord manifested Himself in a miracle.
The barren staff in a priest's hand
He decked with fresh green:
for the sinner in the fires of hell
redemption shall blossom thus afresh!
Throughout all lands let it be proclaimed to him
who has found forgiveness by this miracle!
High above all the world is God,

and His Mercy is no mockery!

Walther, Heinrich der Schreiber, Wolfram, Biterolf, Reinmar, Landgrave, Knights and Older Pilgrims
The salvation of grace is the penitent's reward,
now he attains the peace of the blessed!

Younger Pilgrims
Alleluia! Alleluia! Alleluia!

COMMENT

Unlike the text of *Der fliegende Holländer,* there is some autonomic instability within this opera, and both Tannhäuser and Elisabeth reveal this in the settings of extreme emotion, with trances and collapses. This opera takes us back to the legends and history of the Middle Ages, with several sources, including *Der Kampf der Sänger* by E.T.A. Hoffmann, *Der Getreue Eckart und der Tannenhäuser* by Ludwig Tieck, and Heinrich Heine's poem *Der Tannhaüser* (also from *Salon*).

Pagan sensuality and Christian virtue, Venus and Elizabeth, are polarities which are within the triangular context of Tannhäuser's tensions between the Langrave's Court, the purity of Elisabeth (as Madonna—"Mein Heil liegt in Maria" ("My salvation lies in Maria")—and Venus in her mountain [**Fig. 19. 4-6**]. Tannhäuser is damned like Faust to Hell. There is a death wish sung by both Tannhäuser and Elisabeth, and she experiences previously unfelt emotions, "Longings I have never known" ("Verlangen, das ich nie gekannt"). The eroticism embedded within the opera is plangent (a possible early title having being *Der Venusberg*, The Mount of Venus).

The main pull is between two worlds, the subterranean realm of eroticism ruled by the pagan Goddess of Love, and the bright terrestrial sphere of the Wartburg, where the clear rule of order and orthodoxy is presided over by the saintly Elisabeth. The overture delineates these realms, beginning with the stately diatonic walking progress of the E major Pilgrims' Chorus, which is interrupted by the shimmering chromatic iridescence of the Venusberg music in its enchanting sensuousness. This eventually bursts into Tannhäuser's praise of voluptuous sensuality (which in its raptures and melodic contour is strikingly similar to an accelerated version of Rienzi's prayer). These tonal and emotional areas dominate the narrative structure of the whole opera, with Elisabeth's Greeting to the Hall of Song, her chaste love duet with Tannhäuser, her impassioned pleading for mercy for the errant minstrel, her deeply-felt and extended prayer to the Virgin, embodying various gradations of transcendence.

Wolfram, a character of virtue and self-sacrifice, shares in this transcendent modality in his special enunciations of Elisabeth's name at key moments; in his attempts to save Tannhäuser; and especially in his beautiful Hymn to the Evening Star. Elements of a transcendent faith also come into Tannhäuser's fraught Rome Narration where he recounts to Wolfram the bitter suffering on his own pilgrimage to find forgiveness. The evocation of the medieval city of Rome and his meeting with the Pope sustains the orthodox and sublime imagery. The pull to the transcendent is fully realized in the choruses at the end of the opera: in the grief of Elisabeth's bier carried in procession, and then sublimely in the ecstasy of the returning pilgrims (in E-flat) who bring with them the Pope's miraculously flowering staff, the symbol of the divine grace of forgiveness bestowed.

LOHENGRIN (1850)

Romantic Opera, the libretto after the myth of Perseus and the anonymous Middle High German epic. First performance: Weimar, Court Theatre, 28 August 1850.[8]

Place: Antwerp, on the Scheldt, 10th century.

Act 1

King Henry the Fowler has arrived in Brabant to address the German tribes to expel the Hungarians from his dominions. There is also a dispute involving the disappearance of the child Duke Gottfried of Brabant. The Duke's guardian, Count Friedrich von Telramund, has accused the Duke's sister, Elsa, of murdering her brother in order to become the Duchess of Brabant. He calls upon the King to punish Elsa and to make him, Telramund, the new Duke of Brabant. He is the next of kin to the late Duke, and having renounced to his right to marry Elsa.he has chosen the pagan Ortrud instead.

Elsa has to answer Telramund's accusation. The King declares that he cannot resolve the matter and defers it to God's judgment through ordeal by combat. Telramund agrees. Elsa takes as her champion, a knight she has beheld in her dreams.

The King
Speak Elsa! What have you to confide to me?

Elsa
(quietly transfigured, staring ahead of her)
Lonely, in troubled days
I prayed to the Lord,
my most heartfelt grief
I poured out in prayer.
And from my groans
there issued a plaintive sound
that grew into a mighty roar
as it echoed through the skies:
I listened as it receded into the distance
until my ear could scarce hear it;
my eyes closed
and I fell into a deep sleep.

All the Men
How extraordinary! Is she dreaming? Is she enraptured?
(Elsa's expression goes from one of dream-like detachment to one of frenzied transfiguration)

Elsa
In splendid, shining armour
a knight approached,
a man of such pure virtue
as I had never seen before:
a golden horn at his side,
leaning on a sword—
thus he appeared to me
from nowhere, this warrior true;
with kindly gestures
he gave me comfort;
I will wait for the knight,
he shall be my champion!
(It is midday, all wait for Elsa's champion to appear)

Twice the Herald sounds the horn in summons, without response. Elsa sinks to her knees and prays to God. A boat drawn by a swan appears, in it a knight in shining armour. Elsa places her honour in his keeping. He asks but one thing in return for his service: she is never to ask him his name or where he has come from. Elsa agrees to this.

Elsa
(All the time she has been looking at Lohengrin, she has been unable to move,

> *as if spellbound; but as soon as he addresses her she seems to wake up and falls down at his feet, overcome with joy*)
> My knight, my saviour!
> Take me to you;
> I give to you all that I am!

Ortrud privately expresses confidence that Telramund will win. The unknown knight defeats Telramund but spares his life. Taking Elsa by the hand, he declares her innocent and asks for her hand in marriage.

> **Elsa**
> Would that I could find tunes of jubilation
> equal to your glory, worthily to laud you,
> tunes rich in the highest praise!
> In you I must melt away,
> before you I fade into nothingness;
> that I may be blissfully happy,
> take all that I am!
> *(She falls upon Lohengrin's breast)*

Act 2

Night in the courtyard outside the cathedral

Telramund and Ortrud are banished. Ortrud reveals that she is a pagan witch (daughter of Radbod Duke of Frisia) and assures him that her people (and he) are destined to rule the kingdom again.

Elsa hears Ortrud lamenting and pities her. Ortrud prays to her pagan gods, Wodan and Freia, to deceive Elsa and restore pagan rule. She warns Elsa that her champion is unknown, and he could leave her at any time.

The King announces that the Knight will lead the people to glorious new conquests, and will marry Elsa. Ortrud further unsettles Elsa. Telramund pleads to the King that his defeat was invalid because the Knight did not give his name; trial by combat is traditionally open only to established citizens, he then accuses him of sorcery. Elsa is shaken. The Knight consoles her, then they enter the church.

> **Telramund** (*to the King*)
> How carelessly you proceeded with the trial
> that yet robbed me of my honour,
> for you spared him one question
> when he came to do battle!
> You shall not prevent that question now,

for I shall put it to him:
His name, station and honour
I ask him to reveal before all here present!
(*The crowd stirs, deeply shocked*)
Now shall he answer the charge;
if he can, then I received just punishment -
but if he cannot, you shall see
that his innocence is not what it seems!

Lohengrin
Only one person must I answer:
Elsa—
(*Lohengrin stops in consternation, he turns to Elsa and notices that with heaving breast, she is torn by an inward struggle*)

Lohengrin
Elsa! How she trembles!
I see her brooding wildly!
Has the lying tongue of hatred beguiled her?
O Heaven, protect her heart from danger!
May the innocent one never be plagued with doubts!

Elsa
(oblivious to what is happening around her, staring ahead)
The secret he conceals would bring him disaster
if he revealed it here to all present;
how woefully ungrateful I should be to betray my saviour
by forcing him to reveal it.
If I knew his lot, I would keep it secret!
Yet my breast is torn with doubt!
(deeply agitated and in a state of confused embarrassment)
My deliverer, who brought me salvation!
My knight, in whom I must melt away!
High above the force of all doubt
shall my love stand.
(She sinks upon his breast: Organ music is heard from the minster)

Lohengrin
Hail to you, Elsa! Let us now go before God!

Act 3

Scene 1. The bridal chamber

Bridal Song
Faithfully guarded, remain behind

where the blessing of love shall preserve you!
Triumphant courage, love and happiness
join you in faith as the happiest of couples.
Champion of youth, remain here!
Jewel of youth, remain here!
Flee now the splendours of the wedding feast,
may the delights of the heart be yours!
This sweet-smelling room, decked for love,
has now taken you, away from the splendour.

The couple express their love for each other. Ortrud's words, however, are impressed upon Elsa, and, despite his warning, she asks her husband the fatal questions.

Lohengrin
If, O fair one, you are able to say you are happy,
then you fill me too with heavenly joy!
As I feel my heart go out to you,
I breathe delights that God alone bestows;
How noble is the nature of our love!
Though we never met, we sensed each other;
I was chosen to be your champion,
love paved my way to you:
your eyes told me that you were free of guilt—
your countenance compelled me to serve your grace.

Elsa
But I had already seen you,
for you had come to me in a wondrous dream;
when in waking hours I saw you standing before me,
I knew that you had come following God's counsel.
I wanted to dissolve before your gaze,
like a stream I wanted to wind around your feet,
like a sweet-smelling flower in the meadow
to incline enraptured towards the fall of your feet.
Is this merely love? What shall I call
this word, inexpressibly divine
as your name—that I, alas, may never know,
that I may never use to address my most revered!

Lohengrin
Elsa

Elsa
How sweet the sound of my name from your lips!
Will you not grant me the fair sound of yours?

Only when we are led to the stillness of love
shall you allow my lips to pronounce it.

Lohengrin
(sternly and gravely, moving back a few steps)
You have already to thank me for the highest confidence,
since I gladly believed the oath you made;
if you never falter before my command,
I shall consider you to be above all other women!
(She starts in agitation ... and pauses, as if listening)

Elsa
Did you hear nothing? Did you not hear anyone approaching?
Ah, no!
(staring ahead of her)
Yes, there—the swan—the swan!
There he comes, swimming across the water –
you call him—he brings the boat!

Lohengrin
Elsa! Stop! Calm your madness!

Elsa
Nothing can bring me peace,
nothing can tear me from my madness,
save—even if it should cost me my life—
knowing who you are!

Lohengrin
Elsa, what do you venture to say?

Elsa
Ill-fatedly noble man!
Hear the question I must ask you!
Tell me your name!

Lohengrin
Stop!

Elsa
Whence did you come?

Lohengrin
Woe unto you!

Elsa
What is your origin?

Lohengrin
Woe unto us, what have you done?

Telramund and his followers rush into the room to attack him. The Knight kills Telramund. Then he sorrowfully turns to Elsa and asks her to follow him to the King, to whom he will now reveal his mystery.

(*Elsa, who has thrown herself at Lohengrin's breast, sinks slowly to the floor in a faint*)

Lohengrin
Woe, now all our happiness is gone!

Scene 2. On the banks of the Scheldt as in Act 1

(*Elsa appears*)

Men
Behold, Elsa the virtuous approaches! How pale and melancholy she looks!

The King
How sad you look!

The Knight discloses his identity by telling the story of the Holy Grail, on Monsalvat, and reveals himself as Lohengrin, Knight of the Holy Grail and son of Parsifal, sent to protect an unjustly accused woman. The rules of the Holy Grail determine that Knights of the Grail must remain anonymous, and if their identity is revealed, they must return from whence they came.

Lohengrin
You all heard her promise me
that she would never ask who I am!
But now she has broken her solemn vow,
her heart has succumbed to perfidious counsel!
That the wild questionings of her doubt
might now be stilled,
I shall no longer withhold the answer:
now I must reveal my name and origin.
Hear how I reward the forbidden question!
I was sent to you by the Grail:
my father Parzival wears its crown,

I it's knight—am called Lohengrin.

Elsa *(devastated)*
I swoon! What dreadful darkness!
I gasp! I gasp for air, wretch that I am!
(She is about to collapse when Lohengrin catches her in his arms)

Elsa
If you are truly as divine as I believed,
banish not God's mercy from your heart!
If this most wretched of women expiates her great sin in misery,
let not your gracious presence flee her!
Do not repudiate me, however great my crime,
Do not leave me, oh do not leave me, most wretched of women!

Lohengrin
There is but one punishment for your crime!
Alas! I, as you, feel its cruel pain!
We must be parted, separated:
this must be the punishment, this the atonement!
(*Elsa falls)*

The swan reappears. Lohengrin tells Elsa that if she had maintained her oath, she could have recovered her lost brother, and gives her his sword, horn and ring, for her brother is to become the future leader of Brabant.

Elsa
(*Waking up from her dazed state, she sits up in the seat and looks over towards the river bank*)
O horror! Ah, the swan!
(*She remains sitting up in the same position as if frozen*)

Lohengrin
The Grail sends for the one who is late in returning!

Ortrud appears. She tells Elsa that the swan was actually Gottfried, Elsa's brother, on whom she put a curse by transforming him into a swan. Lohengrin prays and the swan sinks into the water, to emerge as young Gottfried. A dove descends from heaven and taking the place of the swan at the head of the boat, leads Lohengrin away, back to the castle of the Holy Grail. Elsa, transfigured in a last moment of joy, looks at Gottfried, who moves forwards and bows before the king. Gottfried then rushes into Elsa's arms. Elsa, after a brief moment of rapture, quickly turns her gaze to the river bank, but she can no longer see Lohengrin.

Elsa
My husband! My husband!

(Elsa, whom Gottfried is holding in his arms, slowly sinks lifeless to the ground, dead. Lohengrin is seen receding into the distance).

COMMENT

In this opera, referred to by Wagner as 'romantic',[9] we find prescience with both Elsa and Lohengrin. Unlike *The Flying Dutchman*, here it is the female that needs a rescuer. Here he is an ambassador for Christ; he refuses the title of Duke. She has met him, quietly transfigured, in her dreams, and he 'knows' her, their eyes confirm the immediacy of their love. At her first appearance, Elsa appears dream-like, and can recount her vision of her handsome champion. There are several references to her autonomic instability, with trances, falls, faints, heaving heart, trembling, at the end swooning and gasping.

The pressures and loyalties, social, romantic, legally bounded, and constrictive (as to the forbidden question, and by the fears instilled and embedded by Ortrud who, like Iago, poisons the love relationship) are so great, and give all reasons, with her delicate nervousness, that lead to her Sudden Death. This is not only an example of *Unexpected Death*, but with its heavy Christian overtones, is transcendental. The powerful love duet has everything as passionate and sorrowful as that from *Tristan und Isolde*, although does not reach the pitch of a *Liebestod*.

This opera also operates on a dual world of inversions, symbolized by the keys of A major and the relative minor of F sharp: the world of light and redemption, and that of darkness, confusion and ruination. The first is the realm of the Holy Grail represented by Lohengrin, who is sent to save the innocent, reveal the truth and bring back those who have been lost. The second is the dark pagan realm of the sorceress Ortrud, who seeks to obfuscate the truth, sow dissention, breed violence, and triumph over the ruin of many. Elsa and Telramund are the victims caught up in the struggle between these forces that share in an eschatological symbolism of salvation and damnation. The opera is really dominated by the ethereal tones of the prelude with its radiant four- to eight-part string harmonies conjuring up the airy spaces of heaven, and representing a vision of angels bringing the Grail down from celestial realms and then returning. This represents the knight's errand, sent from the Grail to save Elsa. The translucent Lohengrin harmonies infuse all the music associated with him, especially the wrenching Farewell to the Swan, the declaration of love for

Elsa, the first part of their love duet (especially Lohengrin's elated self-declaration "Höchstes Vertauen") and finally the spiritual harmonies that return fully in his Narration in the last act which distil his transcendental associations.

THE LATER OPERAS

TRISTAN UND ISOLDE (1865)

Music Drama after Gottfried von Strassburg's epic *Tristan* (c. 1210), in turn based on Thomas of Britain's *Tristan* (c. 1150), in turn based on a lost earlier version of the legend. First performance: Munich, Court Opera, 10 June 1865,[10]

Act 1

Isolde, promised to King Marke in marriage, and her handmaid, Brangäne, are aboard Tristan's ship being transported to the King's lands in Cornwall. Isolde is enraged with Tristan, the knight responsible for taking her to Marke. She commands Tristan to appear before her. He refuses, but his henchman, Kurwenal, answers, saying that Isolde is in no position to command Tristan. He reminds Brangäne that Isolde's previous fiancé, Morold, was killed by Tristan.

Brangäne
Let me know what is tormenting you!
My lady Isolde.

Isolde
Air! Air!
My heart is stifled!
Open up! Open wide there!
(whose gaze falls on Tristan, and remains coldly fixed on him)
Chosen for me,
lost to me,
great and strong,
bold and cowardly!
Death-devoted head!
Death-devoted heart!

Isolde tells her of how, following the death of Morold, she happened upon a stranger who called himself Tantris, mortally wounded, whom she cured used her healing powers. She then discovered Tantris was Tristan. She attempted to kill the man with his own sword. However, Tristan looked into her eyes ('Er sah' mir in die Augen') and she could not harm him.

Isolde
He who with sly cunning called himself 'Tantris'
Isolde soon recognised
as Tristan since in his sword, as he lay there,
she perceived a notch into which,
there fitted exactly a splinter
which once, in the head of the Irish knight,
had been sent back to mock her.
Then a cry awoke from the depths of my heart!
With the gleaming sword I stood before him,
ready to avenge on him, Lord Morold's death.
From his bed, he looked up—not at the sword, not at my hand—
his eyes met mine directly, and his misery troubled my heart.

Tristan left with the promise never to come back, but has returned to take her to King Marke. Isolde, furious at Tristan's betrayal, insists that he drink atonement to her and asks Brangäne to fetch the drink. She is shocked to see that it is a death potion.

Isolde (*angry at Tristan's deception*)
Curse you, vile creature, a curse upon your head!
Vengeance! Death! Death for us both!

Tristan arrives, and Isolde tells him that he owes her his life: Tristan offers her his sword. She sings they must drink atonement. Brangäne brings in the potion that will seal their pardon; Tristan knows that it may kill him. Tristan drinks half, the Isolde takes the rest. As the sailors hail the arrival of King Marke, Brangäne tells that it was not poison, but a love potion.

Isolde
Put up your sword! When I wielded it before,
when vengeance tore at my breast,
when your measuring gaze stole my likeness,
the sword—I let it sink.
Let us now drink reconciliation!
(*Isolde advances with the cup to Tristan, who gazes fixedly in her eyes*).

Tristan
I shall take the goblet that I may be fully cleansed.
And witness to the oath of reconciliation
which I take, in gratitude to you.
Tristan's misery, keenest defiance!

Heart's deceit, wishful dreaming!
The only consolation in eternal mourning.
Beneficent draught of forgetfulness,
I drain you unwaveringly!

(*She drinks, and both are seized with shuddering, gazing at each other with deepest emotion. Developing into a passion as they clutch their hearts*)

Tristan
Isolde!

Isolde
Tristan!

Together
How our hearts are borne aloft!
How all our senses pulsate with bliss!
Longing devotion's burgeoning blossoms,
yearning love's blessed glow!
My breast bursting with exultant delight!
Isolde! Tristan!
Broken free of the world, won for me!
You my only awareness, utmost rapture of love!

Isolde
Where am I? Am I alive?
Ah! What was that draught?

Brangäne
(in despair)
The love potion.

Isolde
Must I live on?
(She falls on his breast, unconscious)

Act 2

King Marke is out hunting, Isolde and Brangäne alone in the castle. Isolde, listening to the hunting horns, believes the hunting party is far enough away to warrant extinguishing the brazier—the prearranged signal for Tristan to join her.

Brangäne
You are misled by your impetuous desires
into hearing what you imagine.

Brangäne warns Isolde about the King's knight Melot, who suspects that something is going on between Tristan and Isolde. The flame is extinguished and Brangäne retires to the ramparts to keep watch as Tristan arrives. They are lost in the bliss of each others' person, and disregard the warnings uttered by Brangäne.

Isolde
Do you not know the Love Spirit,
not know her magic's power?
The Queen of boldest courage,
Regent of the world's course?
Love and Death are subject to her,
she weaves them out of bliss and sorrow,
transmuting envy into love.
Beckon the night!
My heart is filled with rapturous awe
Oh now let the light be quenched
Put out its frightening glare
Let my beloved in.
(The lovers declare their passion.)

Tristan
Is it I? Is it you? No illusion?

Together
Not a dream? O heart's rapture,
o sweet, most sublime,
Eternal!

Isolde
You in darkness,
I in light!

Tristan
The light! The light! Oh, this light,
how long before it was extinguished!
The sun set, day ran its course
but it would not stifle its spite:
lighting its dread signal.

Isolde
From the light of Day
which made you appear to me
a traitor
I wished to flee
into Night,

to take you with me,
where my heart would bid me
end all deception,
where the vain premonition
of treachery might be dispelled,
there to pledge to you
eternal love,
to consecrate you to Death
in company with myself.

Tristan
Oh, now we are night consecrated.
dedicated to Night!
Caught in the sunbeam!
Amid the vain fancy of day
he still harbours one desire—the yearning
for sacred Night where, all-eternal, true alone,
love's bliss smiles on him!

Isolde
heart on your heart, mouth on mouth!

Tristan
the single bond of a single breath

Together
Eyes are dazed by blinding rapture
the world pales away
with its blinding radiance:

Tristan
All earthly illusion, here I defy

Together
then am I myself the world;
floating in sublime bliss, life of love most sacred,
the sweetly conscious undeluded wish
never again to waken.

Tristan
Never waken!

Isolde
Must Day then waken Tristan?

Tristan
Let Day give way before death!

Tristan
Our love
Tristan's love? Yours and mine,
Isolde's love?
Were I to give my life to that
for which I would so gladly die,
how could love die with me,
how could the ever-living perish?
And if his love never were to die
could Tristan die yoked to love eternal?

Isolde
Yet may we name it, this love as Tristan and Isolde?
This little word: and, would death not destroy
the bonds of love which it entwines
if Tristan were to die?

Tristan
But should we die, we would not part, joined forever,
Without end, never waking never fearing,
Nameless there is love enfolded,
each to each belonging,
with love alone our life source
What could die but that which troubles us,
preventing Tristan from ever loving Isolde,
forever loving only her?

Isolde
Yet this little word: and, were it destroyed,
how else but together with Isolde's own life
would death be given to Tristan?
Let the night never end.

Together
O eternal Night, sweet Night!
Gloriously sublime Night of love!
Those whom you have embraced,
upon whom you have smiled,
how could they ever waken without fear?
Now banish dread, sweet death,
yearned for, longed for death-in-love [*Liebestod*]!
Free from languishing
enclosed in sweet darkness.

No evasion no parting,
just we alone, ever home,
in unmeasured realms of ecstatic dreams.

The day breaks as Melot leads King Marke and his men to find Tristan and Isolde in each other's arms. Marke is heartbroken, not only because of his nephew's betrayal but also because Melot chose to betray his friend Tristan to Marke.

When questioned, Tristan is unable to give the reason of his betrayal, since Marke would not understand. Tristan announces that Melot has also fallen in love with Isolde. Melot and Tristan fight, but Tristan throws his sword aside and Melot wounds him.

Act 3

Tristan's home at his castle at Kareol in Brittany

A shepherd pipes a mournful tune and asks if Tristan is awake. Kurwenal replies that only Isolde's arrival can save Tristan. The shepherd offers to keep watch and he will pipe a joyful tune at the arrival of any ship. Tristan awakes and laments his fate—to be, once again, in the false realm of daylight. Kurwenal tells him that Isolde is on her way. Tristan asks if her ship is in sight, a sorrowful tune from the shepherd's pipe is heard.

Tristan rails once again against his desires and the fateful love potion. Exhausted, he collapses in delirium. The shepherd is heard piping the arrival of Isolde's ship, but Tristan tears the bandages from his wounds. Isolde arrives as Tristan dies with her name on his lips.

Tristan
She who my wound will finally heal,
like a hero approaches,
she comes for my salvation!

Isolde
Tristan! Beloved!

Tristan
(in extreme agitation)
What? Is it the light I hear?
The torch, ah!
The torch is extinguished!
To her! To her!

Tristan, hardly conscious, totters towards her. She takes him in her arms. Tristan sinks slowly to the ground in her arms)

Isolde
Tristan!

Tristan
(dying, looks up at her)
Isolde!
(He dies)

Isolde collapses beside him. Marke and Brangäne arrive. Kurwenal attacks Melot: Both are killed. Marke, grieves over the body of Tristan, his "truest friend", Brangäne has revealed the secret of the love potion and that he had come not to part the lovers, but to unite them. Isolde in a final aria describes her vision of Tristan risen again (the *Liebestod*), and dies.

Isolde
In the heaving swell,
in the resounding echoes,
in the universal stream
of the world-breath—
to drown,
to founder—
unconscious—
utmost rapture!
(Isolde sinks gently, as if transfigured, in Brangäne's arms, on to Tristan's body.)

COMMENT

Night and day, dark and light, night and death frame the actions and tragedy of this opera. Tristan is guilty to be alive, and like Parsifal, his father is dead, and his mother died in childbirth. In Wagner's operas *Tristan*, *Parsifal* and *Siegfried* the protagonists are all motherless, and give their own lives, or complete a life of self-denial, to resolve these situations. This opera explores the fulfilment and the agonies of love: its destruction imbricated by the political and social settings of the lovers, it is full of Schopenhauerian leanings regarding yearnings.

Isolde has episodes of autonomic reactivity. Even at the very beginning she seems to be quite depressed. Brangäne tells us that she is pale and silent, not eating, not sleeping, numb and wretched, wild and distraught.

Tristan and Isolde do not share death together. The final *Liebestod* will be further discussed later, but there is the precognitive unconscious awareness of their love for each other, the revealing of this with a potion, the significance of the eye glances, and the powerful eroticism within the libretto and the music.

This work revolves around the single obsessional internal world of the two lovers and the unwithstandable compulsion of their passion, given as external correlative in the love potion (intended as a deathly poison). The A-minor prelude, with its innovative chromatics and unresolved harmonies, represents the secret inner world of this passion and its engrossing convolutions—rapturous but also unsettling. This extends into the worlds of light and darkness, day and night, with the Act 2 love tryst in the night, the focus of the extended love duet where the moods of palpitating nocturne and aspiration into transcendence unfold, the two voices slowly merging into the unison dream of their rapture. This is interspersed with the slow, rapt warnings of Brangäne which are not heeded, and indeed hardly differ in texture and sound from the part of Isolde. The warnings are the embodied voice of the Isolde's superego, disregarded and overwhelmed by the id of the subject's all-consuming passion.

The tone painting of the grey ocean bounding Tristan's Breton home sets the mood for the desolate ruminations of the wounded Tristan, the mournful shepherd's pipe etching the loneliness of the dying man and eliciting memories of his lost childhood. This creates the scene for Isolde's arrival, the self-inflicted death of Tristan, and the transformatory finale of Isolde's own *Unexpected Death* at the end of her famous rhapsody "Mild und leise". Known as her *Liebestod*, this soliloquy works on the tonal system of the tonic keys of B (B minor–B major). The translucent harmonies for winds, horns and strings, with the arpeggiated harp and then string figurations sustaining the incrementally rising vocal line, has the effect of merging consciousness and life into unconsciousness and death in a musical process of wish-fulfilment. "The monologue is ultimately concerned with the metamorphosis of the lovers' story into myth, into art. Tellingly, even as the language dissolves into fragments, the music achieves radiant simplicity, the vocal line becoming fixed almost exclusively on the notes of the B-major scale, with the elemental interval of the octave rising at the end. You have the sense that Isolde is not so much dying as disappearing back into the music that set the tale in motion."[11]

PARSIFAL (1882)

Bühnenweihfestspiel (Religious Stage Festival)/Music Drama loosely based on Wolfram von Eschenbach's *Parzifal* (early 13th century), and Chrétien de Troyes's *Perceval, the Story of the Grail.* First performance: Bayreuth, Festspielhaus, 26 July 1882.[12]

Place: *the castle of Monsalvat in the mountains in the north of Spain, and Klingsor's magic palace in the south of Spain.*

Act 1

In the realm of the Grail Knights

Scene 1

In a forest near the castle of Monsalvat, home of the Grail and its Knights, Gurnemanz, eldest Knight of the Grail, sees Amfortas, King of the Grail Knights, approaching. Amfortas has been injured by the Holy Spear and the wound will not heal.

A wild dishevelled woman—Kundry—rushes in. She is in wild garb, her skirts tucked up, her black hair loose, her complexion deep ruddy-brown. Her eyes are dark and piercing, sometimes flashing wildly, more often lifeless and staring. She seems to have flown through the air. She gives Gurnemanz a vial of balsam, brought from Arabia, to ease the King's pain and then collapses, exhausted.

Amfortas arrives, borne on a stretcher by Knights of the Grail. He accepts the potion from Gurnemanz and tries to thank Kundry. The squires eye Kundry with mistrust. Gurnemanz tells them she has often helped the Grail Knights. The squires think she is a witch and asks why will she not find the Holy Spear for them. This spear, which pierced the side of the Redeemer on the Cross, and the Holy Grail, which caught the blood flowing from his side, had come to Monsalvat to be guarded by the Knights of the Grail under the rule of Titurel, the father of Amfortas. Gurnemanz reveals that this deed is destined for someone else. He says Amfortas was given guardianship of the Spear, but lost it when he was seduced by a beautiful woman in Klingsor's domain. Klingsor grabbed the Spear and stabbed Amfortas: this wound causes Amfortas both suffering and shame and will never heal on its own.

Gurnemanz's explains how Klingsor had yearned to join the knights but, unable to keep impure thoughts from his mind, resorted to self-castration. Expelled from the order Klingsor, with dark arts, he filled the valley below with beautiful flower-maidens to seduce wayward Grail Knights. Klingsor schemes also to obtain the Grail. Gurnemanz tells how Amfortas later had a holy vision which told him to wait for a "pure fool, enlightened by compassion" ("Durch Mitleid wissend, der reine Tor") who would finally heal the wound.

Suddenly the Knights cry out in anguish ("Weh! Weh!"): a flying swan has been shot by a young man, who is brought forward. Gurnemanz says this is a holy place, and asks what harm the swan had done?

Gurnemanz
What did the birds sing to you from the branches?
What harm did that faithful swan do you?
Seeking his mate, he flew up
to circle with it over the lake
and gloriously to hallow the bath.
What is he now to you? Here—look! here you struck him,
the blood still congealing, the wings drooping lifeless,
the snowy plumage stained dark,
the eyes glazed—do you see his look?
(Parsifal breaks his bow and throws his arrows away.)
How could you commit this crime?

Parsifal
I don't know.

Gurnemanz asks him questions, to each the boy replies, "I don't know". He has a mother, Herzeleide, and Kundry tells them that his father was Gamuret, a knight killed in battle. The youth recalls that upon seeing knights pass through his forest, he left his home and mother to follow them. Kundry recounts that, as she rode by, she saw Herzeleide die of grief. Hearing this, he then collapses. Kundry, weary for sleep, wishes that she would never again waken, and disappears into the undergrowth.

Kundry
As I rode by, I saw her dying:
she bade me greet you, fool.
(Parsifal springs at Kundry. Gurnemanz restrains him.)

Gurnemanz
Insane youth! Again violent?

(Parsifal stands as if dazed.)
What has the woman done to you? She spoke the truth;
for Kundry never lies, though she has seen much.

Parsifal *(seized with violent trembling)*
I am fainting!
(Kundry brings water and sprinkles Parsifal with it and then gives it to him to drink.)

Gurnemanz
Good, according to the Grail's mercy:
they vanquish evil who requite it with good.

Kundry
I never do good; I long only for rest,
only rest in my weariness,
To sleep! O that no one would wake me!
(starting in fear)
No! Not sleep! Horror seizes me!
(She falls, violently trembling, then lets her arms and head drop wearily and totters away.)
In vain to resist! The time has come.

Gurnemanz says that in this realm, time becomes space ("Zum Raum wird hier die Zeit"). He invites the boy to observe the Grail ritual.

They arrive at the Hall of the Grail, where the Knights are assembling to receive Holy Communion ("Zum letzten Liebesmahle"). The voice of Titurel is heard, telling his son, Amfortas, now in great shame, to uncover the Grail. He cries out for forgiveness ("Erbarmen!) but hears only the promise of future redemption by the pure fool.

Amfortas
All-merciful one, have mercy on me!
Take back my inheritance, heal my wound,
that I may die holy, pure and whole for Thee!
(He sinks back as if unconscious.)

(*The boy, on hearing Amfortas loud cry of agony makes a violent movement towards his heart, which he clutched convulsively for a long time: now he again stands motionless, as if petrified.)*

Gurnemanz
Why are you still standing there?
Do you know what you have seen?
(*Parsifal pressing his heart shakes his head*)

So you are only a fool then!
Off with you, and go on your way!
But heed Gurnemanz …

A Voice
The pure fool, enlightened by compassion.

Act 2

Scene 1. In Klingsor's magic castle

Klingsor, staring into his magic mirror, conjures up and awakens Kundry, calling her many names: First Sorceress, Hell's Rose, Herodias, Gundryggia and, lastly, Kundry. She wakes with a scream.

Kundry
Oh! Oh! Blackest night!
Frenzy! O rage! O misery!
Sleep … sleep …
deep sleep! Death!

She is transformed into an alluring woman. Klingsor observes that Parsifal is approaching, and summons his enchanted knights to fight him. They are vanquished. Klingsor instructs the Flower-Maidens and Kundry to seduce Parsifal.

Scene 2

Surrounded by seductive Flower-Maidens, a voice calls out, "Parsifal!" He now recalls this name is what his mother used when appearing in his dreams. The Flower-Maidens call him a fool.Kundry knows his name, she learned it from his mother. She reveals much of Parsifal's history to him and he, remorseful, blames himself for his mother's death. Kundry says this realization is a first sign of understanding and that, with a kiss, she can help him understand his mother's love. As they kiss Parsifal suddenly recoils in pain and cries out the name Amfortas. He feels the wounded king's pain burning in his own side ("Amfortas! Die Wunde! Die Wunde!"). He rejects Kundry's advances.

Kundry
When her arms clasped you tight
did you perhaps fear her kisses?
But you did not consider her woe,
her desperate grief,
when you finally did not return

and left no trace behind!
She waited night and day
till her laments grew faint,
grief consumed her pain
and she craved for death's release:
her sorrow broke her heart,
and Herzeleide [Heart's Sorrow] died.
She who once gave you life and being,
to subdue death and folly sends you this day,
as a last token of a mother's blessing,
the first kiss of love.
(She has bent her head completely over his and gives him a long kiss on the lips. Parsifal suddenly starts up with a gesture of the utmost terror: his demeanour expresses some fearful change; he presses his hands hard against his heart as if to master an agonizing pain.)

Parsifal
Amfortas! The wound! The wound!
It burns within my heart!
O sorrow, sorrow! Fearful sorrow!
From the depths of my heart it cries aloud.
Most wretched! Most pitiable!
I saw the wound bleeding: now it bleeds in me!
Here—here! No, no! It is not the wound.
Flow in streams, my blood, from it!
Here! Here in my heart is the flame!
The longing, the terrible longing
which seizes and grips all my senses!
O torment of love!
How everything trembles, quakes and quivers
in sinful desire!
(As Kundry stares at Parsifal in fear and astonishment, he falls into a complete trance.)
My dull gaze is fixed on the sacred vessel;
the holy blood flows: the bliss of redemption, divinely mild,
trembles within every soul around:
only here, in my heart, will the pangs not be stilled.
The Saviour's lament I hear there,
"Redeem Me, rescue Me
from hands defiled by sin!"
Thus rang the divine lament
in terrible clarity in my soul.
And I fool, coward, fled hither to wild childish deeds!
(He flings himself in despair on his knees.)
Redeemer! Saviour! Lord of grace!
How can I, a sinner, purge my guilt?

Kundry tells Parsifal that if he can feel compassion for Amfortas, then he should be able to feel compassion for her as well. She has been cursed for centuries, unable to rest, because she saw the Saviour on the cross and laughed at His pains. Now she can never weep, only laugh, and she is enslaved to Klingsor. Parsifal asks her to lead him to Amfortas. She begs him to stay with her for just one hour, he refuses, and she curses him to wander without ever finding the Kingdom of the Grail.

Klingsor appears and throws the Spear at Parsifal but it stops in mid-air, above his head. Parsifal takes it and makes the sign of the Cross. Klingsor's castle crumbles.

Act 3

Scene 1. The grounds of the Grail Castle

Many years later. Gurnemanz is now aged and bent. Kundry lies unconscious in the brush,

> **Gurnemanz**
> Awake! Awake to the Spring!
> Cold and stiff!
> This time she may well be dead:
> yet it was her groaning I heard.

He revives her using water from the Holy Spring, but she will only speak the word "serve" ("Dienen"). An armed stranger in a helmet and black armour approaches, Gurnemanz recognizes the youth who shot the swan, and then recognizes that the Holy Spear is now returned. Parsifal expresses his desire to return to Amfortas ("Zu ihm, des tiefe Klagen").

Gurnemanz tells him that in his absence Amfortas has never revealed the Grail, and that Titurel has died. Gurnemanz tells him that today is the day of Titurel's funeral rites, and that Parsifal has a great duty to perform. Kundry washes Parsifal's feet and Gurnemanz anoints him with water from the Holy Spring, recognizing him as the Pure Fool, now enlightened by compassion, and as the new King of the Knights of the Grail. Today is Good Friday, when all the world is renewed. Parsifal now baptizes the weeping Kundry. Tolling bells are heard afar off; Gurnemanz says "Midday: the hour has come. My lord, permit your servant to guide you!" and they set off for the Castle of the Grail.

Scene 2. Within the Castle of the Grail

Amfortas is brought before the Grail shrine, and Titurel's coffin. He wishes to join his father in death ("Mein Vater! Hochgesegneter der Helden!"*).*

Amfortas (*in wild despair*)
Already I feel the darkness of death enshroud me,
and must I yet again return to life?
Madmen! Who would force me to live?
Could you but grant me death!
(He tears open his garment.)
Here I am, here is the open wound!
Here flows my blood, that poisons me.
Draw your weapons! Plunge your swords
in deep, deep, up to the hilt!
Up, you heroes!
Slay the sinner with his agony,
then once more the Grail shall shine clear on you!

The Knights of Grail urge Amfortas to uncover the Grail again, but Amfortas refuses, commanding the Knights to kill him and end his suffering and the shame he has brought on the Knighthood. Parsifal steps forth with the only weapon that can heal the wound ("Nur eine Waffe taugt"): with the Spear he touches Amfortas's side, and both heals and absolves him. He commands the revealing of the Grail. As all present kneel, Kundry, released from her curse, sinks lifeless to the ground as a white dove hovers over the head of Parsifal.

Parsifal
But one weapon serves: only the Spear that smote you
can heal your wound.

(Amfortas's features light up in holy ecstasy: he seems to stagger under overpowering emotion: Gurnemanz supports him.)

Be whole, absolved and atoned!
O blessed be your suffering, that gave pity's mighty power
and purest wisdom's might to the timorous fool!
I bring back to you
the holy Spear!

(All gaze in supreme rapture at the uplifted Spear, to whose point Parsifal raises his eyes. Parsifal mounts the altar steps, he takes the Grail from the

shrine already opened. The Grail gradually glows with a soft light. Increasing darkness below and growing illumination from above.)

Boys, Youths, Knights
Miracle of supreme salvation!
Our Redeemer redeemed!

(*A beam of light: the Grail glows at its brightest. From the dome a white dove descends and hovers over Parsifal's head. Kundry slowly sinks lifeless to the ground in front of Parsifal, her eyes uplifted to him. Amfortas and Gurnemanz kneel in homage to Parsifal, who waves the Grail in blessing over the worshipping brotherhood of knights.*)

COMMENT

Wagner wanted this opera to be a *Bühnenweihfestspiel,* a religious stage festival, only to be performed at Bayreuth. The religious symbolism is profound, and the features of so many of Wagner's ideas are within the libretto and the music. Kundry is clearly an instance of *Unexpected Death*, showing signs of autonomic dysfunction from her early appearance. Although he does not die, Parsifal is quite unstable, but significantly has powerful embodied feelings, with trances, and episodes that are literally heartfelt occasioned by such significant events—as learning about the death of his mother, his guilt about that, and Kundry's kiss. He actually physically feels the wound of Amfortas. Mothers (Herzeleide dies of a broken heart), sex versus purity, light versus dark, bells and midday, and the significance of Good Friday all coalesce within the opera: ideas of compassion and redemption make up the central theses. There are bells in both the first and last acts. Bells, hanging between earth and heaven, are vectors of transcendence. The motifs of looking (eyes) and feeling (heart) are integral to Parsifal's experience: at the first sight of the agony of Amfortas, Parsifal convulsively touches his heart.

The ecstatic dominates the score of *Parsifal* and underlines the situations of sin and despair that characterize the ruined lives of the ageless Kundry, the fallible Amfortas, the despairing Titurel and the longsuffering Gurnemanz, who are all open to redemption in the messianic figure and actions of the hero. He is the guileless fool, the innocent who learns the depth of the divine compassion. He follows and represents Jesus, renewing the ceremonial of the Last Supper, exposing the Holy Grail afresh, and healing the various forms of suffering, especially that of Amfortas. Parsifal touches him with the Sacred Spear of the Passion, lost by the negligent Amfortas and recovered from the evil Klingsor.

The slow, monolithic prelude, operating on F minor and the relative major of A flat (as with *Lohengrin*) gradually instils a mood of sublime stateliness in which the drama unfolds. The imperturbable plasticity of the motifs of Faith (the Last Supper), Hope, and Love (the Eucharist or Grail) generates an impassioned austerity which infuses the whole work, even in the darkest moments. Changes of key, figures of rustling strings and modulations moving up by thirds, all sustain and increase the power of the instrumental recital. In the statement of the Eucharistic Motif it rises up and up, trailing off into ethereal whisperings. The same music and motifs dominate the two Last Supper Scenes, with their ritualistic choruses. This sound picture recurs in all the scenes representing sacramental prototypes—in both the expositions of the Grail, and in the serene but similarly sublime baptism of Kundry on Good Friday, where the whole of nature in pathetic fallacy is drawn into the spiritual mysteries. These reach their controlled transcendent fulfilment in the closing scene of the opera, where the Faith Motif is restated, held in ecstatic harmonies.

Part 2

How terrible this darkness was, how bewildering,
And yet how mysteriously beautiful.

(Stephan Zweig, *The Burning Secret*)[13]

We shall not cease from exploration
And the end of all our exploring
Will be to arrive where we started
And know the place for the first time.
Through the unknown, unremembered gate
When the last of earth left to discover
Is that which was the beginning;
At the source of the longest river
The voice of the hidden waterfall
And the children in the apple-tree
Not known, because not looked for
But heard, half-heard, in the stillness
Between two waves of the sea.
Quick now, here, now, always—
A condition of complete simplicity
(Costing not less than everything)
And all shall be well and
All manner of thing shall be well
When the tongues of flame are in-folded

Into the crowned knot of fire
And the fire and the rose are one.

(T.S. Elliot, *The Four Quartets*)[14]

DER RING DES NIBELUNGEN (The Ring of the Nibelung) (1876)

Ein Bühnenfestspeil für drei Tage und einen Vorabend (A Stage Festival Play in three parts with a preliminary evening (a prelude), sometimes called a tetralogy), based on the Middle High German Nibelung Saga and other Nordic and Teutonic myths. Translation: Stuart Spencer and Barry Millington. First performance: Bayreuth, Festspielhaus, 13, 14, 16, 17 August 1876.[15]

DAS RHEINGOLD (The Rhinegold) (1869)

At the bottom of the Rhine the three Rhinemaidens play together. Alberich, a Nibelung dwarf, tries to woo them. They maidens mock him. A sudden ray of sunshine pierces the depths, revealing the Rhinegold. They explain that the gold, which their father has ordered them to guard, can be made into a magic ring which gives power to rule the world if its bearer first renounces love. Alberich curses love and seizes the gold.

Wotan's fortress of Valhalla

Wotan, ruler of the gods, is asleep and woken by his wife, Fricka, She reminds him of his promise to the giants Fasolt and Fafner, who built the castle, that he would give them Fricka's sister Freia, the goddess of youth and beauty, as payment. Wotan is confident that Loge, the demigod of fire, will find an alternative payment.

Loge
I see full well:
in the whole wide world
there's naught so rare
as to strike mankind as a worthy ransom
for women's delights and worth.
Wherever there's life and breath
what might man
deem mightier
than woman's delights and worth?
in water, earth and air
none will relinquish
love and women,
except the Nibelung Night-Alberich.

Loge tells of Alberich, who has renounced love, stolen the Rhinegold and made a powerful magic ring from it. The giants will accept the Nibelung's treasure in payment, instead of Freia. Wotan resolves to travel with Loge to Alberich's subterranean kingdom to obtain the gold.

In Nibelheim, Alberich has enslaved the rest of the Nibelung dwarves. He has forced his brother Mime, a skillful smith, to create a magic helmet, the Tarnhelm, which can make him change form. Loge asks how he can protect himself against theft. Alberich replies with the Tarnhelm. Loge asks him first to transform himself into a giant snake and then into a toad. When he is in this small form, Wotan and Loge capture him.

Wotan and Loge force Alberich to exchange his wealth for his freedom. The Nibelungen bring up the hoard of gold. Alberich then asks for the return of the Tarnhelm, but Loge says that it is part of his ransom. Wotan tears the Ring from Alberich's hand and puts it on his own finger. Alberich lays a curse on the Ring: the one who owns it will eventually be robbed of it and killed.

> **Alberich**
> As my freedom came to me through a curse,
> so shall this ring be accursed in turn!
> so shall its spell now deal
> death to whoever shall wear it!
> may he who owns it
> be wracked by care,
> and he who does not
> be ravaged by envy!
> Doomed to die,
> my curse you'll not escape!

The Giants demand both the Tarhelm and the Ring as part of the Nibelung treasure. Loge protests that the Ring belongs to the Rhinemaidens. The giants seize Freia and start to leave. Erda, the earth goddess, appears and warns Wotan of impending doom. He surrenders the Ring. Fafner kills Fasolt in a fight over Freia and the gold.

Wotan leads the gods across a rainbow bridge to the castle, which he names Valhalla. The Rhine maidens mourn the loss of their gold.

COMMENTS

In *Das Rheingold* a tacit dichotomy is presented in the polarities of nature and love, and the world of power and lovelessness. Loge the god of fire plays a crucial role in *Rheingold*, and his narrative about nature and the universality and costliness of love is central, dramaturgically and thematically. His Narration, recounting how in his journeying throughout the world he has never found anything more precious than love, is a key statement. The fluid serenity of the music brings the Rhine music to the fore again. The counterpart is the dark world of Alberich, generated in the depths of the Rhine in his renunciation of love to enable him to steal the Rhinegold and to acquire the power to alter nature by being able to forge the Ring. Its elusive chromatic harmonies capture the unsettling intangibility of this elemental power. Alberich's loss of the Ring to Wotan generates the Curse that fashions the whole drama. Its strong, declamatory nature, while harmonically related to the Ring, is heard in its intransigent power, always keeping its melodic contour but relentlessly in its force and intensity, embodying the destructive hatred it represents.

DIE WALKÜRE (The Valkyrie) (1870)

Act 1

An exhausted Siegmund finds shelter from his enemies in a dwelling built around a massive ash-tree, in which is a sword buried to the hilt. Sieglinde, the wife of Hunding, offers him rest here until her husband's return. They look at each other with growing interest and emotion.

Hunding enter, and questions Siegmund's presence. Calling himself Wehwalt ("woeful"), Siegmund explains that he grew up in the forest with his parents and twin sister. Their home burned down, his mother was killed and his sister gone. Hunding identifies Siegmund as the enemy of his clan, says Siegmund may stay, but they must fight in the morning. Sieglinde returns, having drugged Hunding's drink.

She is longing for the hero who will draw the sword from the tree, put there by an old man in grey with a hat hung so low that one of his eyes was hidden. They realize their kinship (with much interactive eye-gazing) and Siegmund draws the sword from the tree. She is Sieglinde, his twin sister. Siegmund names the sword *Nothung* ("needful") and declares that it will be her protection. The two sing of their passionate love for each other, and, as the act ends, they rush joyfully out of the hut into the spring night.

Siegmund
Whatever I longed for
I saw in you;
in you I found

whatever I lacked!
I laugh aloud in hallowed delight
as I hold you within my embrace
and feel your beating heart!
(The main door flies open.)

Sieglinde
Ha, who's left? Or who's come in?

Siegmund
No one left—
but someone came:
see how Spring
smiles into the hall!
Winter's storms have waned.
Far from here
follow me now,
away to springtime's
smiling home:
there Nothung the sword shall shield you,
when Siegmund succumbs to your love.

Sieglinde
If you are Siegmund
whom I see here—
Sieglinde am I,
Who has longed for you:
your own true sister
you've won for yourself with the sword.

Siegmund
Bride and sister
you are to your brother—
so let the blood of the Wälsungs blossom!
(He draws her towards him with furious passion; the curtain falls quickly.)

In *Die Walküre* the passion of nature and its expression in sexual union is represented in the first encounter between Siegmund and Sieglinde. Their extended love duet, revealed in the incipient lyricism of the lines, vocal and orchestral, is given correlative in the elemental power of the burgeoning spring night, and the growing mutual rapture of the lovers. We forget the incestuous nature of this coupling, and the GSA (Genetic Sexual Attraction) it explores, in the fervour of the music.

Act 2

In a rocky landscape, Wotan instructs Brünnhilde, his Valkyrie daughter, to protect Siegmund in his forthcoming battle with Hunding. Fricka arrives and insists that Siegmund and Sieglinde be punished for their adultery and incest which contravene the natural law and the contract of marriage. Wotan reluctantly agrees that he will not protect Siegmund. After Fricka leaves, the troubled Wotan rescinds his earlier instruction; he explains how he is inextricably bound and trapped in the history of his own contracts and covenants, and instructs Brünnhilde to give the victory to Hunding.

Brünnhilde announces to Siegmund his impending death; when she tells him Sieglinde cannot accompany him to Valhalla, he refuses to follow Brünnhilde. Siegmund threatens to kill both Sieglinde and himself. Brünnhilde, with compassion, decides to defy her father and grant victory to Siegmund. Sieglinde, in terrified and altered state of mind, collapses into Siegmund's arms.

Siegmund begins to overpower Hunding, but Wotan appears and shatters Siegmund's sword with his spear. Hunding stabs him to death. Brünnhilde gathers up the fragments of the sword and flees. With a contemptuous wave of the hand Wotan strikes Hunding dead, and swears that Brünnhilde will be punished.

Sieglinde
You fall—
the sword is shivered in shards·
the ash-tree topples—
the trunk is riven!
Brother! My brother!
Siegmund—ha!
(She sinks senseless into Siegmund's arms.)

Siegmund
Sister! Beloved! …

Wotan
(to Hunding)
Be gone, slave!
Kneel before Fricka:
tell her that Wotan's spear
has avenged what brought her disgrace.
Go! Go!
(At a contemptuous wave of his hand, Hunding falls to the ground, dead.)

The power of love and nature is turned on its head in Act 2 in the Volsungs' terrified flight and the faltering faintness of Sieglinde, who is almost rendered demented by her physical and emotional debilitation. The fraught stillness that prevails after Wotan's intervention in the clash between Siegmund and the avenging husband is etched even more strongly in the almost whispered scorn of Wotan's words to Hunding, and his mortal gesture that sees Hunding fall dead at his feet. This is one of the very few *Unexpected Deaths* in *The Ring*.

Act 3

The Valkyries congregate on the mountain-top, each carrying a dead hero. Brünnhilde arrives with Sieglinde, and begs her sisters for help, but they dare not defy Wotan. Brünnhilde gives the fragments of the sword Nothung to Sieglinde, who resolves to save her child.

> **Brünnhilde** (*to Sieglinde*)
> Then hurry away
> and head for the east!
> Know this alone
> and ward it always:
> the world's noblest hero,
> O woman, you harbour
> within your sheltering womb!
> *(She hands the fragments of Siegmund's sword to Sieglinde)*
> For him keep safe
> the sword's stout fragments;
> from his father's field
> I haply took them:
> let him who'll wield
> the newly forged sword
> receive his name from me—
> may 'Siegfried' rejoice in victory!
>
> **Sieglinde**
> Sublimest wonder!
> Glorious maid!
> You true-hearted woman,
> I thank you for sacred solace!

As she departs, Wotan comes with great wrath. Brünnhilde is to be stripped of her demi-divinity as a Valkyrie and become a mortal woman, to be held in defenseless sleep on the mountain, prey to any man who finds her. But Wotan consents to her fervent request that he surround her resting place with a circle of fire that will protect her from all but the bravest of heroes. Wotan bids her a loving farewell, lays her sleeping form down on a rock and summons Loge, the demigod of fire, who places a circle of flames around her.

Wotan
Farewell [*Leb'wohl*], you valiant,
glorious child!
You, my heart's
most hallowed pride,
Farewell! Farewell! Farewell!
if I must lose
you whom I loved,
you laughing delight of my eye:
a bridal fire
shall burn for you
such as never blazed for a bride!
Fiery flames shall
encircle the fell;
with withering fears
let them fright the faint-hearted;
For one man alone shall woo the bride,
one freer than I, the god!
That radiant pair of eyes
which I often caressed with a smile,
this glittering pair of eyes
which oft glistened on me in the storm
amidst wildly weaving fears:
for one last time
let them joy me today
with this valediction's
final kiss!
On a happier man
their stars shall shine:
on the hapless immortal
they must close in parting!
And so—the god
turns away from you:
so he kisses your godhead away.
(He lingeringly kisses both her eyes. She sinks back, with eyes closed, into his arms, as consciousness gently slips away.

COMMENT

The power of love and compassion is manifested in the brief scene of Brünnhilde's rescue of Sieglinde from Wotan's wrath, where Sieglinde's almost deathly grief and hopelessness are turned into sudden ecstasy at the news that she is bearing the heroic fruit of her love with Siegmund. The rising rapture of the promise of redemption through advent of the fearless hero floods the music with movement and light as Sieglinde is now galvanized with hope and able to confront the future for all its uncertainty and danger. The introduction of what becomes the Redemption *Leitmotif* briefly appears here.

Brünnhilde's Pleading, as well as Wotan's Farewell, distils the various themes of the story, with its tragic exploration of doomed plans and stricken love, of disobedience and reconciliation, of defiance and forgiveness. Wotan at this stage shows a marked strength in anger that is subjected to slow transformation through his love for Brünnhilde. Through the latter's prophecy of Siegfried to the encroaching tone of Wotan's softening anger, the intensity of emotion is sustained in its slowly encompassing tenderness, finally symbolized in Wotan's kissing of Brünnhilde's eyes. This is captured in the aching yearning of Wotan's caressing farewell kiss. Nature and love see the transformation sustained in the Sleep Motif and the Magic Fire Music with its crystalline purity, the serene and restful beauty from the orchestra drawing one into the mystery of the enchanted slumber enfolded in Loge's fire. Even the concluding and portentous enunciation of the Fate Motif shares in the sustained orchestral mellifluence.

SIEGFRIED (1876)

Act 1

In a cave in dense forest the dwarf Mime, the brother of Alberich, is trying to weld a broken sword. He has raised Siegfried as a foster-child to kill Fafner, who has since transformed himself from a giant to a dragon, and hoards the treasures. Siegfried wants to know his parentage. Mime explains that Siegfried's mother (Sieglinde) died when she was in labour. He shows Siegfried the broken pieces of Nothung. Mime is unable to reforge the sword.

Wotan in disguise arrives as the Wanderer. There follows a ritual series of three questions and answers that Wotan and Mime put to each other. The third from Wotan asks: who can repair the sword? Mime has no problem with the first two questions but cannot answer the last one. Wotan replies that only "he who does not know fear" can reforge Nothung.

Mime realizes that Siegfried is "the one who does not know fear" and that unless he can instil fear in him, Siegfried will kill him as the Wanderer foretold. Mime promises to teach Siegfried fear by taking him to Fafner. Siegfried ecstatically files down and re-forges Nothung. Mime brews a poisoned drink for Siegfried after he has defeated the dragon.

Act 2

Siegfried and Mime arrive at Fafner's cave. Siegfried speculates that Mime cannot be his father, and on what his mother might have looked like (with bright-shining eyes of a roe-deer, only fairer): why did she have to die giving birth to him? Siegfried hears a forest bird sing as he waits for the dragon to appear. He then plays a tune on a reed and then on his horn, Fafner comes out, they fight; Siegfried stabs Fafner but when he withdraws Nothung, he tastes the dragon's blood. Siegfried complains to Mime that he has still not learned the meaning of fear. Mime offers him the poisoned drink; however, the magic power of the dragon's blood allows Siegfried to read Mime's treacherous thoughts, and he stabs him to death. He also finds he can now understand the Woodbird's song: it sings of a woman sleeping on a rock surrounded by magic fire. He follows the bird towards the rock.

Voice of the Woodbird
Hey! Siegfried's now slain
the evil dwarf!
Now I know
the most glorious wife for him.
High on a fell she sleeps,
fire burns round her hall:
if he passed through the blaze
and wakened the bride,
Brünnhilde then would be his!

Siegfried
O welcome song!
Sweetest breath!
Its meaning burns my breast
with searing heat!
How it thrills my heart
with its kindling desire!
What courses so swiftly
through heart and senses!
Tell me the answer, sweet friend!

Voice of the Woodbird
Delighting in sorrow,
I sing of love;
blissful I weave
my lay from woe:
lovers alone can know its meaning!

Siegfried
Can I awaken the bride?

Voice of the Woodbird
He who wins the bride
and awakens Brünnhilde
shall never be a coward:
only he who knows not fear!

Siegfried
The foolish boy
who knows not fear,
my woodbird, that is I!
Now I burn with longing
to learn it from Brünnhild'.

Act 3

Wotan awakens Erde the Eternal Woman (*Ewiges Weib*) for advice but she cannot help him. Siegfried meets the Wanderer but he does not recognize his grandfather. The Wanderer blocks his path, Siegfried mocks him, and breaks his spear (the symbol of Wotan's authority) with a blow from Nothung.

Siegfried passes through the fire, finds Brünnhilde, removes her armour, and at the sight of the first woman he has ever seen, Siegfried experiences tumultuous emotion. He kisses Brünnhilde, waking her from her magic sleep. Together, they hail "Light-bringing love, and laughing death".

Siegfried
(At Brünnhilde's rock, at first thinking he sees a sleeping man, he removes the helmet and long curling hair breaks free, he removes the breastplate)
This is no man!
Burning enchantment
charms my heart;
fiery terror
transfixes my eyes:
my senses stagger and swoon!

To save me, whom shall I
call on to help me?
Mother! Mother!
Be mindful of me!
How shall I waken the maid
so that she opens her eyes for me?
Opens her eyes for me?
What if the sight should yet blind me?
Might my bravery dare it?
Could I bear their light?
Around me everything floats
and sways and swims;
searing desire consumes my senses:
on my quaking heart
my hand is trembling!
Is this what it is to fear?
A woman lies asleep:
She has taught him the meaning of fear.

Brünnhilde
Hail to you, sun
Hail to you, light
Hail to you, light-bringing day!
If only you knew, you joy of the world,
How I have always loved you!

Siegfried
Ha! As the blood in our veins ignites,
as our flashing glances consume one another,
as our arms clasp each other in ardour—
my courage returns,
and the fear, ah!
the fear that I never learnt –!

Brünnhilde
Laughing I must love you;
laughing I must grow blind;
laughing let us perish—
laughing go to our doom!
Be gone, Valhalla's
light-bringing world!
May your proud-standing stronghold
moulder to dust!
Fare well, resplendent
pomp of the gods!
End in rapture,

you endless race!
Rend, you Norns,
the rope of runes!
Dusk of the gods,
let your darkness arise!
Night of destruction,
let your mists roll in!
Siegfried's star
now shines upon me;
he's mine forever,
always mine,
my heritage and own,
my one and all:
light-bringing love
and laughing death!

Siegfried
Laughing you wake
in gladness to me:
Brünnhilde lives!
Brünnhilde laughs!
Hail to the day
that sheds light all around us!
Hail to the sun
that shines upon us!
Hail to the light
that emerges from night!
Hail to the world
for which Brünnhilde lives!
She wakes! She lives!
She smiles upon me!
Brünnhilde's star
shines resplendent upon me!
She's mine forever,
always mine,
my heritage and own,
my one and all:
light-bringing love
and laughing death!
(Brünnhilde throws herself into Siegfried's arms.)

COMMENT

The beauty and transforming power of nature, proposed and evident in both *Das Rheingold* and *Die Walküre*, is almost the principal focus of *Siegfried* in depicting the hero's early life and path to maturity. The forest with its animals and birds, and also the dark secret of the lurking dragon, is the key to this development, and all the related autonomic and emotional references. The dark cave of his childhood with the dwarf Mime, and the cavern containing the transmogrified Fafner, who guards the Ring and related hoard, give up their treasures to the young Siegfried. He acquires the skills to forge anew the sword Nothung and so to vanquish the Dragon. Woken by his woodland horn-calls. Siegfried in overcoming Fafner becomes custodian of the treasure. By accidentally tasting the Dragon's burning-hot blood Siegfried is given access to the secrets of nature, and can hear the meaning of birdsong. The Woodbird imparts the knowledge of the dormant Brünnhilde, sleeping on the fiery hilltop. This mysterious knowledge "burns" Siegfried's breast, and is one of a register of emotional fire affecting his heart: burning, searing, thrilling, kindling.

The idyll of the Forest Murmurs and the enchanting song of the Woodbird make present the beauty described in Loge's Narration, and lead Siegfried on to the Rock of Brünnhilde's Sleep.

Siegfried's tumultuous and rhapsodic ascent of the mountain, having unconsciously overcome the Wise Old Man in his grandfather, Wotan, also provides a wonderful sense of carefully shaped. inexorable movement. His passing through the fire is clear, exhilarating, and slowly tapers away, capturing perfectly the tumult that reduces to the solitary string line, suggesting the airy heights and the protagonist's incredulity. This depicts his amazed perceptions as he discovers a whole new world in the sleeping Brünnhilde. The glorious and extended love duet that follows in so many different parts represents various aspects of the growth to mature sexuality for both the characters. The movement through fire and air, through bemusement, fear, emotion growing through the stages of discovery, interest and passion to exultation, is described as a "burning enchantment"; Brünnhilde's awakening hails the advent of the sun that eventually penetrates the very being of both characters.

The rapturous final section, the closing section of their duet in unison to emphasize their oneness now of mind and heart, is marked by their "flashing glances" and the peak of emotion in an overflowing of the spirit as "laughing I love you". The climax presents the disturbing but enrapturing paradox of "laughing death". This is all too soon the fate visited upon the heroic couple. We are reminded of the intimate association between Love and Death, which turns this duet into a triumph of exhilarating beauty. It is significant that 'Tod' is the last sung word in *Der fliegende Holländer* and *Siegfried*, foreshadowing the end of the Ring.

The collocation of death and laughter in *Siegfried* invest the prospect of mortality with a paradoxical joyousness.

GÖTTERDÄMMERUNG (Twilight of the Gods) (1876)

Prologue

The Three Norns are weaving the Rope of Destiny. They sing of the past and the present, and of the future. They disappear, disconcerted when the Rope unaccountably snaps at the mention of the Curse.

Brünnhilde and Siegfried emerge from their cave. Brünnhilde sends Siegfried off to new adventures. As a pledge of fidelity, Siegfried gives her the Ring taken from Fafner's hoard.

> **Brünnhilde**
> Feast your eyes
> on this blessed pair!
> Parted—who would divide us?
> Divided—they'll never part!

Act 1

In the Hall of the Gibichungs, Gunther, lord of the Gibichungs, with his half-brother and chief minister, Hagen. The latter says they need to find a wife for Gunther and a husband for their sister Gutrune. He suggests Brünnhilde for Gunther's wife, and Siegfried for Gutrune's husband. Gutrune has a potion that can make Siegfried forget Brünnhilde and fall in love with Gutrune; under its influence, Siegfried will win Brünnhilde for Gunther.

Siegfried appears at Gibichung Hall, Gutrune offers him the love potion. Siegfried falls in love with Gutrune. Gunther and Siegfried swear blood-brotherhood. Meanwhile, Brünnhilde is visited by her Valkyrie sister Waltraute, Wotan has ordered branches of the World Ash tree to be piled around Valhalla; he waits in Valhalla for the end. Siegfried arrives, disguised as Gunther by means of the Tarnhelm, and claims Brünnhilde as his wife. Though Brünnhilde resists violently, Siegfried overpowers her, taking the Ring from her.

Act 2

Alberich visits Hagen in his half-sleep, and urges him to continue the plan of destruction. Siegfried arrives having left Brünnhilde with Gunther.

Hagen summons the Gibichung vassals to welcome Gunther and his bride. Gunther leads in a downcast Brünnhilde, who is astonished to see Siegfried. Noticing the Ring on Siegfried's hand, she realizes she has been betrayed—that the man who conquered her was not Gunther, but Siegfried in disguise. She denounces Siegfried and accuses him of having seduced her himself. Siegfried swears her accusations are false. Siegfried then leads Gutrune and the bystanders off to the wedding feast, leaving Brünnhilde, Hagen, and Gunther alone. Shamed by Brünnhilde's outburst, Gunther agrees to Hagen's suggestion that Siegfried must be killed. Brünnhilde, seeking revenge for Siegfried's treachery, tells Hagen that she had kept Siegfried's body protected by magic, but as he would never flee, she spared his back such protection. Hagen and Gunther invite Siegfried on a hunting-trip with a view to murdering him.

Act 3

In the woods by the bank of the Rhine, the Rhinemaidens mourn the lost Rhine gold. Siegfried happens by and they urge him to return the Ring and avoid its curse, but he laughs at them and says he prefers to die rather than bargain for his life.

Siegfried with the hunters tells them about his past adventures. Hagen gives him another potion, which restores his memory: he tells of his youth and his supreme happiness with Brünnhilde. Hagen stabs him in the back. Siegfried's body is carried away to the Gibichung Hall (Siegfried's Funeral March).

Back in the Gibichung Hall. Gutrune is devastated when Siegfried's corpse is brought in. Gunther blames Siegfried's death on Hagen, but is killed by the latter. Hagen moves to take the Ring, Siegfried's hand rises threateningly. Hagen recoils in fear. Gutrune, who earlier had been cursed by Brünnhilde falls into despair, and filed with shame, dies suddenly.

Gutrune
Accursèd Hagen!
That you counselled the poison
that robbed her of her husband!
Ah, sorrow!
How swiftly I see it now:
Brünnhilde was his own true love,
whom the philtre made him forget.

Brünnhilde arrives and orders a huge funeral pyre. Lighting the pyre with a firebrand, she sings about the dead hero, then mounts her horse Grane and rides into the flames.

A sequence of *Leitmotifs* unfold as the hall of the Gibichungs catches fire and collapses. The Rhine overflows its banks, quenching the fire, and the Rhinemaidens swim in on the crest to claim the Ring. Hagen tries to stop them but they drag him into the depths and drown him. As they celebrate the return of the Ring and its gold to the river, a red glow is seen in the sky. Fire flares up in the Hall of the Gods, who are consumed in the flames. The music is gradually overtaken by the *Erlösungsmotif*, the Motif of Redemption.

Brünnhilde
Purer than sunlight
streams the light from his eyes:
the purest of men it was
who betrayed me!
False to his wife
while true to his friend,
from her who was faithful
—she alone who was loyal—
he sundered himself with his sword!
Never were oaths
more nobly sworn;
never were treaties
kept more truly;
never did any man
love more loyally:
and yet every oath,
every treaty,
the truest love—
no one betrayed as he did!
Do you know why that was so?
(looking upward)
O you, eternal
guardians of oaths!
Direct your gaze
at my burgeoning grief:
behold your eternal guilt!
Hear my indictment,
most mighty of gods!
By the bravest of deeds,
which you dearly desired,
you doomed him who wrought it

to suffer the curse
to which you in turn succumbed:
it was I whom the purest man
had to betray,
that a woman might grow wise!
Do I now know what you need?
All things, all things,
all things I know,
all is clear to me now!
I hear the rustle
of your ravens' wings:
with anxiously longed-for tidings
I send the two of them home.
Rest now, rest now, you god!
Let the fire that consumes me
Cleanse the ring and its curse:
In the floodwaters let it dissolve.
Feel how the flames burn in my breast,
Effulgent fires seize hold of my heart:
To clasp him to me
While held in my arms
And in mightiest love
To be wedded to him!
Siegfried! Siegfried! See!
Your wife greets you in bliss!

COMMENT

In the Prologue, the Break of Dawn is very thoughtful, providing a repeat of the unsupported string line, lower this time. The emotion of the finale of *Siegfried* is rehearsed but with a weightier, more restrained tone, as though both characters have matured. The Motif of Siegfried the Hero now proclaimed is portentous, stately, even majestic; and Brünnhilde's role strong and resolute, with brilliant tessitura. The gradual build-up to the triumphant unison climax for them both embodies musically the chiasmus of parting and dividing and its double negation. Gutrune is an example of *Unexpected Death*.

The true resolution of the epic drama must see the salvific interaction of both hero and heroine, Siegfried and Brünnhilde. The famous finale of *The Ring*, Brünnhilde's Immolation, provides the feminine dimension, as the heroine pays her loving homage to the betrayed hero, bidding farewell to the era of the gods, issuing in a new age of human love in joining Siegfried in the flames. This is a sacrificial death of cosmic dimension, initiating the apocalyptic final moments. The orchestral postlude is tumultuous: the wonderful lapping waves of the rising waters of a restored nature sees the gold returned to the Rhine, providing a beautifully shaped conclusion to this unique musical-dramatic meditation on life and love—their origins, meaning and destiny.

Notes

[1] Scruton, R. *Music as an Art* (London: Bloomsbury, 2018), 84.
[2] Ross, Alex. *Wagnerism: Art and Politics in the Shadow of Music* (London: 4th Estate, 2020), 660.
[3] Wagner, *Die Feen*, with April Cantelo and John Mitchinson, BBC Northern Symphony Orchestra, cond. Edward Downes (BBC in *Wagner: Complete Operas* DG).
[4] Wagner, R. *Rienzi*, with René Kollo, Siv Wennberg and Janis Martin, Staatskapelle Dresden, cond. Heinrich Hollreiser (EMI, also in Wagner: Complete Operas DG).
[5] Wagner, R. *Der fliegende Holländer*, with Dietrich Fischer-Diskau and Marianne Schech, Berlin Deutsche Oper, cond. Franz Konwitschny (EMI).
[6] Wagner, R. *A communication to my friends* (1851). Trans. W.A. Ellis (Wokingham: Dodo Press), 33.
[7] Wagner, *Tannhäuser*, with Hans Hopf, Elisabeth Grümmer, Marianne Schech, Dietrich Fischer-Diskau and Gottlob Frick, Staatskapelle Berlin, cond. Franz Konwitschny (EMI).
[8] Wagner, R. *Lohengrin*, with Jess Thomas, Elisabeth Grümmer, Christa Ludwig, Dietrich Fischer-Diskau and Gottlob Frick, Wiener Philharmoniker, cond. Rudolph Kempe (EMI).
[9] In the first editions of the libretto and the full score.
[10] Wagner, R. *Tristan und Isolde*, with Wolfgang Windgassen, Birgit Nilsson, Christa Ludwig and Marti Talvela, Bayreuther Festspiele 1966, cond. Karl Böhm (DG).
[11] Ross, A. *Wagnerism*, Ibid. 2020, 69.
[12] Wagner, R. *Parsifal*, with René Kollo, Christa Ludwig, Gottlob Frick, Dietrich Fischer-Diskau, Wiener Philharmoniker, cond. George Solti (Decca).
[13] Zweig, S. *The Burning Secret and other stories* (London: Pushkin Collection, 1989), Chapter 4.
[14] Eliot, T.S. Ibid., 44.
[15] Wagner, R. *Der Ring des Nibelungen*. There are many recordings, but recommended is the first full stereophonic version (1959-66) with Birgit Nilsson, Wolfgang Windgassen, Hans Hotter, George London, Gottlob Frick etc., Wiener Philharmoniker, cond. George Solti (Decca).

PART 6:

ANALYSIS AND EXPLANATIONS

CHAPTER 20

LESSONS FROM THE LIBRETTI: WHY WHEN AND WHERE? AN ANALYSIS OF *UNEXPECTED DEATHS*: RESULTS AND 19TH-CENTURY CULTURE, SCIENCE AND MEDICINE

Twelve o'clock.
Along the reaches of the street
Held in a lunar synthesis,
Whispering lunar incantations
Dissolve the floors of memory
And all its clear relations,
Its divisions and precisions,
Every street lamp that I pass
Beats like a fatalistic drum,
And through the spaces of the dark
Midnight shakes the memory
As a madman shakes a dead geranium.

(T.S. Eliot, 'Rhapsody on a Windy Night')[1]

When time stops and time is never ending;
And the ground swell, that is and was from the beginning,
Clangs
The bell.

(T.S. Eliot. 'Dry Salvages': *Four Quartets)*[2]

Introduction

In the previous six chapters we have identified 50 deaths in opera librettos which we have referred to as *Unexpected Death*. We have drawn attention to those events and sections of the librettos in which we have examined references to the ANS, hinting at possible 'causes' for sudden demise. We now wish to discuss several themes which have been highlighted, central to our analysis of the understanding of opera generally, especially but not exclusively to Romantic topoi, and the underlying dialectic of love and death, subsuming the interpenetration of opposites. These include affairs of the heart and autonomic instability, abnormal states of mind from trances, hallucinations and madness, and the nature and meaning of the transcendental, which have bearings on the *Liebestod* and ideas of redemption. We have acknowledged the rather selective nature of these operas, but we have scoured the canon, used our own knowledge of the narratives and music, and provided, as far as space could allow, samples from the librettos and discussion of the music to illustrate our points.

List 1 reveals some interesting facts. Of the 50 *Unexpected Deaths*, 31 are female and 19 male. In the former group, 22 are from the 19th century, nine from the 20th century. In the latter, nine are from the 19th century, and 10 from the 20th century (we have not ventured in our survey selections of any operas from the 21st century).

Referring to the other main texts about operatic deaths, Poizat's *The Angel's Cry* and Conrad's *A Song of Love and Death*, our survey confirms certain conclusions. Poizat's selection extended up to 1973, about the same time span as ours. His "biological deaths", including those of "characters in torment … who are subject to a particular kind of suffering such as madness, unjust condemnation, self-sacrifice in response to unrequited love" are dominated by females about 2:1, and females die more frequently at the end of the operas (Poizat: 70 per cent female to 47 per cent males).[3]

With regards to the historical distribution, Poizat found in the subset of "deaths of characters in torment", 50 per cent of females died in operas composed between 1800 and 1900, compared with 43 per cent of males (80 per cent referred to as "heroes"). Male deaths that conclude operas in the 20th century accounted for 22 of 40 (55 per cent): our figures are seven out of nine. As we show, 'modern times' are more dangerous for males.

Clément tells us that "opera concerns women … they suffer, they cry, they die". Women, she controversially contends, have the most beautiful music. Her list of 'causes' is quite limited, and includes "Those who die for nothing, just like that—of fear, or fright, or sadness, or anxiety. Those who die poisoned, gently; those who are choked; those who fold in on themselves peacefully. Violent deaths, lyrical deaths, gentle deaths, talkative or silent deaths …". She lists suicides, drowning and poisoning "and a few unclassifiable, thank God for them, dying without anyone knowing how or why". We attempt to look at that last question.

Poizat's data "support the idea that opera in the Romantic Era stages the suffering and death of women—what Catherine Clément has called their 'undoing'".[4] Yet she pleads they die without explanation. Our point of departure was to delve deeper into the questions and conundrums posed by Clément, stimulated by our own observations of many operas, but with a guiding eye of historical, medical, musical and cultural backgrounds.

In music we have traced the early introduction of the transcendental with Monteverdi's *La favola d'Orfeo* (1607), the lament for the death of Dido in Purcell's opera (1689), and the rise in the 18th century of magical transformations such as seen in Handel's *Alcina* (1735), the sorceress who seduces knights changing them into anything she fancies. The formalism of such operas inhibited the development of the drama, but with Gluck's *Orfeo et Euridice* (1762) music changed, as does the portrayal of love. The spirit of individuality begins to shine through. Until the end of the 18th century the male is the predominant protagonist. Then comes the Romantic heft linked with shifting social values (Mozart's *Marriage of Figaro*), eschatological concerns (Mozart's *Don Giovanni*), and revolution with rescue operas (Beethoven's *Fidelio*). The rise of the female as pivotal in the dramas and philosophy of opera in the 19th century is noted generally. This is apparent from our analysis of the librettos. But first we must return to some medico-cultural issues.

The Rise of the Sciences—Physics and Mathematics

In Germany, the 1830s onwards witnessed the rise of experimentation in the sciences. By the 1860s and 1870s the study of physics was considered an independent discipline, and a way to explore the secrets of nature. Herman von Helmholtz (1821-1894), who had studied medicine at the University of Berlin, was an ideal person to bring physics and mathematics together at a time when Newton's mechanical views of nature were being challenged. He wanted to explore the relationship between the senses and the external world in animals and humans, seeking to unify the sciences with close attention to links with the arts. His method was that of the *Geisteswissenschaften,* bringing together physics with ideas in philosophy, history, linguistics and musicology, with a reverence for a scientific approach.

The intellectual environment at the time was full of discussion about physico-chemical forces active in an organism, but there was a movement to abolish vitalism (the idea that emerged from chemistry, that living organisms were driven by some non-physical element, an *élan vital*: life could not be reduced simply to mechanical forces). With a growing influence of empiricism, *Naturphilosophie* was under attack and Helmholtz's doctrine of the Principle of the Conservation of Energy, namely that the sum of forces remains constant in every isolated system, was powerfully argued. Forces acted together or could inhibit each other, ideas which altered understanding of nervous forces and energy, giving the young Sigmund Freud a physiological perspective to his early theories of how the mind and brain interacted.

Helmholtz was a pioneer in the scientific study of human vision and audition and argued for some unifying force of nature. Inspired by *psychophysics*, he was interested in the relationships between measurable physical stimuli and their corresponding human perceptions. For example, the amplitude of a sound wave can be varied, causing the sound to appear louder or softer, but a linear step in sound-pressure amplitude does not result in a linear step in perceived loudness.[5] In experimental studies Helmholtz examined the relationship between physical energy (physics) and its perception (psychology), with the goal in mind to develop "psychophysical laws". He wanted to know what went on within the ear itself. In 1863 Helmholtz published *Sensations of Tone*, a book which has influenced musicologists even through to today. Musical tones were like waves: he discovered the speed of nerve impulses in humans was between 24 and 38 metres a second, depending on circumstances, which was considerably slower than electricity, hence bodily action and thought were not simultaneous but serial in nature.

Helmholtz contrasted the ideological views of Kant, chasing the limits of what we could and could not know, from the empirical approach of science, and opined that Hegel's idealistic philosophy was useless. He thought the area of sensory perception was a key to potential cooperation between philosophy and science.

At this time Berlin was at the centre of a *Kulturekampf.* The materialist views such as those of Rudolf Virchov (1821-1902), the father of modern pathology, were set against the Catholic Church and anti-modernists such as the spiritual astrophysicist Johann Karl Friedrich Zöllner (1834-1882): strife between Berlin and Leipzig. Helmholtz was open-minded and was even attacked as echoing Schopenhauer.

Helmholtz undertook comprehensive analyses of acoustics. Joseph Fourier's (1768-1730) analysis of harmonic structure laid the basis for his researches into musical tones, which, like waves, could be analysed by Fourier's analysis.[6] With his *Tonempfindungen* (resonator), he found that the overtones are consigned by the structure of the ear and its nerves. The non-linear responses of the ear and the role of unconscious inferences are thus related to our interpretation of the sounds we hear: musical scales are the product of artistic creativity. He presented views on musical harmony and sympathetic vibration, examined musical consonance and dissonance, and referred to the "spiritual ear" (*geistige Ohr*), the ear as a harmonic analyser which held the power of imagination.[7]

Johannes Peter Müller (1801-1858) differentiated nervous from electrical conduction. His *Law of Specific Nerve Energies* (1835) stated that the origin of the sensation is not important: our perceptions are as dependent on any object as on the senses themselves. Perceptions were representations. We never directly experience objects themselves, only sensory effects, and it is acts of psychical activity which serve towards our knowledge of the world. Empirical pathways to knowledge cannot leave out one half of the equation.

Since the speed of nerve conduction was much slower than electricity, Helmholtz argued that visual perceptions are not simply external imprints upon the retina but required *a priori* unconscious inferences, for how else could the retina sort out all the incoming sense information into a meaningful whole? A new philosophy and physiological understanding of perception was needed, since perceptions of the present require knowledge of the past; hence his interest in the influence of beauty on human memory, and Goethe's Eternal Feminine.[8]

Helmholtz discovered the rods and the three colour receptors in the retina, with different light waves of sensitivity. Our interpretation of colour was thus dependent on our unconscious influences, rather than on the property of the viewed object, findings which had a considerable influence on the arts: for example, on Monet's Impressionism and Cézanne's imaginative reconstructions of nature. Wagner was somewhat ambivalent about Helmholtz the man but their wives were friendly and he was interested in Helmholtz's views on harmony, with associations to the *Gesamtkunstwerk,* a distance from the dry academic theories of the time.[9]

The Enlightenment project was coming adrift. 1848 saw widespread revolutions in Europe. Frederick William Herschel's (1838-1822) advanced telescopes expanded the universe, investigations of the nature of electricity and magnetism shrunk the nature of power and energy to waves or particles, and the early work of James Clarke Maxwell (1831-1879) on light and gases opened up investigations that were later to unify physics. Such a climate of discovery was set alongside Alexander von Humboldt's Romanticism, derived from the book of Nature not that of the Linnaean classifiers, giving rise to pre-Darwinian evolutionary theories crafted in a language stimulating feelings including the sublime.

Continental thought, dominated for a while by the metaphysics of Hegel's three-step dialectic, brought the fourth dimension to prominence; relations between time and space were part of the *Zeitgeist.* Frankenstein had used 'galvanism' to give the "spark of life" to his monster, while the diagnosis of *Monomania* as a mental disorder, especially in France, had led to arguments around the insanity defence in courts and to the rise of the independent discipline *médecin des aliénés* (later *alienism* and psychiatry). Coleridge had visited Kubla Kahn.[10]

Analysis of List 1

From the operas in List 1, some reference to autonomic instability is found in 14 of 31 females (45 per cent) and three of 19 males (16 per cent). The occurrence of relevant signs or symptoms in the text is much more apparent in the 19th than the 20th century, especially for the females, 11 of 22 (50 per cent) in the 19th and two of 9 (22 per cent) in the 20th, compared with males of one of 9 (11 per cent) for the 19th and two of 10 (20 per cent) for the 20th. Females account for 42 per cent and males 16 per cent with a physiological vulnerability that may at least underpin their sudden demise. Donizetti and Wagner are the most represented for female deaths, all from19th century. No composer stands out with regards to males for the 19th or 20th centuries. There were several operas in which it was difficult to decide about the precise meaning of the libretto when it came to emotional expressions, and metaphorical references to 'heart' were avoided. The situation with, for example, Salammbô and St Francis was not so clear, autonomic references are present in Giulietta, more obvious in Anna Bolena, variable in the operas related to Manon, prominent for Hoffmann's Antonia, and Elsa in *Lohengrin* is an exemplar of one who was very vulnerable.

The autonomic perturbations are central to several other operas, where sudden death is not the key feature, but nevertheless possible when the intensity of the dramatic situation is before us and intense visceral feelings are sung. Mime's description of fear as described to Siegfried, and then later the actual feelings Siegfried has when he first experiences fear in encountering Brünnhilde are arresting. Likewise, Parsifal in his encounter with the suffering of Amfortas and the wound, recrudesced by the embrace with Kundry, becomes an overwhelmingly potent physical experience. The element of grief and fear is sufficient to account for the death of Anna Bolena, and in *Das Wunder der Heliane* the signs of Heliane's autonomic arousal give her potential guilt away. Concern about guilt is related to several other deaths, for example with Boris Godunov, and the internal Dionysian struggles of Aschenbach in *Death in Venice*. Another expression of the bodily exertion is found in the two operas which end with exhaustion driven by the music, namely Elektra, and Roberto in *Le Villi,* who both dance themselves to death. The ballets *Giselle* (1841, music by Adolphe Adam) and *Espana* (1908, music by Jules Massenet) also explore the motif of dancing to death, the first by malevolent enchantment, and the second by psychological compulsion. A variant is Stravinsky's Sacrificial Dance in *The Rite of Spring* (1913), when the Chosen One, driven by wild Dionysian forces, falls dead.

The description of autonomic reactions in the 19th-century operas, much related to cardiac instability, contrasts with use of the heart as a common metaphor in poetry and opera in earlier times (and throughout history) as a love motif, especially the anguish of love. More comic variants are found in the duet for Fiordiligi and Guglielmo in *Così fan tutte* (1790) and in Johann Strauss's *Die Fledermaus* (1874), where in the duet for Rosalinde and Eisenstein the beat of the amorous heart is measured on his pocket watch.

We do not wish to infer that the autonomic references allow an explanation for the whole story. We explore the psychological predispositions and vulnerable personalities of the protagonists leading to the unexpected death, and we note many of the characters chosen have reasons to experience bodily and psychological distress. But the composers, by seeding such clues in the librettos, reflect an awareness of the medical advances noted above, especially Wagner, as we discuss later.[11]

Buried within our analyses of the libretti are many references to Romantic topoi that capture symbols, times, places and events, cynosures that are so often disregarded or passed over by productions, and whose significance is lost on the audience. Yet if one turns a blind eye or a deaf ear away from them, the meaning of the works are etiolated. We now wish to consider these. [**Figs. 20.1-12**]

Time

Zum Raum wird hier die Zeit (Here space becomes time)

(Wagner, *Parsifal*, Act 1)

We have noted the profound statement from Gurnemanz about space and time.Time is a recurring theme in myth, philosophy, everyday life and opera: it is almost impossible to conceive of time without resorting to metaphor. It is motion in space, sounded by soothsayers, town criers and bells, as "time and the hour run through the roughest day".[12] Time flows, times are located in space, the time-space metaphor of general relativity is a four-dimensional space: events occur at and in spots of time, many marked out by nature's rhythms and culturally endowed rituals. Rituals are saturated by symbols endowing spiritual and other meaning. Repetition, the past within the present, iteration with cycles of return of indefinite duration, are experienced by individuals. But as Merleau-Ponty expressed it: "My life is made up of rhythms that do not have their *reason* in what I have chosen, but rather have their *condition* in the banal milieu that surrounds me".[13]

Rituals are saturated with meaning, endowing habit with spiritual inscape. The entrained rhythms of our heart and lungs, of our diurnal responses to the living day, the key temporal points of midday and midnight, the monthly rhythms of nature and the four seasons, the origins of new life, and of death: all figure in operas, especially the ones we have been discussing. They symbolically underpin issues of Love and Death, the irresistible autonomic features of the rarefied world of distilled emotion.

The opening years of the 20th century seemed a special time of meditation on the nature of mutability. In the opening scene of Richard Strauss's *Der Rosenkavalier* (1911), one of post-coital reflection, the issue is of time, its passing and meaning. The clock ticks as the older Countess tries to make an end of her erotic involvement with a young man. Janáček's *Osud* (1904) covers early, middle and later scenes in the principal character's life, which, with the dramas of encounter, creativity, death by accident and retrospective reflection, bring the whole issue of time and mutability into consideration and sharp reflection. Schreker's *Der ferne Klang* (1912) is constructed around the passing of time, its incomprehensible and uncontrollable movement that shapes human life and tragically underpins human limitation. Greta and Fritz are never able to realize the opportunities for love, and are fatefully separated through circumstances all their lives, only to be united at the very end when time has run out on all their dreams.

Midday

The great hinge of the day, when the waxing and the waning of light are held in perfect balance, is a time of decisive, seemingly fateful, change. The mysterious water nymph Undine appears as an abandoned waif at the hour of midday when powerful forces are at play, both for good or evil. Midday is also the time of the Slavonic Midday Witch, who exercises a malign hold over Rusalka. Mélisande, like Undine, appears to Golaud in the forest at midday. In Wagner's *Lohengrin,* Elsa appears before the King to be accused of fratricide and recounts her dream of a champion who will save her; Lohengrin's swan-drawn boat appears at the all-changing hour of midday. In *Parsifal* we hear Kundry's curse but also Gurnemanz and Parsifal moving to the Grail Temple at this portentous hour. Many years later the knight Parsifal appears at Gurnemanz's cave and baptizes Kundry at this special time amidst the radiance of Good Friday. The bells fatefully toll midday as the fraught Boris Godunov is overcome with fatal remorse and psychiatric suffocation: the chiming of the hour transforms into the death knell.

Midnight

The middle of the night is the witching hour, the moment of destiny and decisive change. In Lortzing's *Undine* the interruption of the wedding happens at midnight. In Weber's *Der Freischütz* (1821) the evil tempter Kaspar inducts the gullible young huntsman Max into the dark arts of the Magic Bullets (*Freikugeln*) in the haunted Wolf's Glen. The incantation to the Dark Huntsman Samiel begins as the distant bell strikes midnight, various ghostly apparitions accompanying the casting of the bullets until the seventh unleashes the dark forces of the Wild Hunt and advent of Samiel. In the ensuing silence the bell strikes one. Faust in Busoni's version dies as the Night-Watchman calls midnight.

In *Robert le Diable* (1831), the wavering Robert, torn impossibly between the paternal and maternal paths that life sets before him, is saved by grace at the decisive hour of midnight. In Adam's ballet *Giselle* (1841), the evil enchantment of the Wilis, the ghostly spirits of abandoned brides, begins at midnight as the Queen of these nefarious spirits causes Giselle's wraith to emerge from her grave and to seek the destruction of her perfidious lover, Prince Albrecht.

In *Rigoletto* (1851), the Jester's planned assassination of his profligate employer the Duke of Mantua is supposed to take place after the chiming of eleven during a terrible storm. In the final moments Rigoletto holds a murdered body in a sack, but at the stroke of midnight hears the Duke singing in the distance. He tears open the sack to find his beloved Gilda dying, the victim of his own rebounding curse.

Conversely, in the Romantic comic opera *Martha, oder Der Markt von Richmond* (1847) by Friedrich von Flotow (1812-1883), the farmer-brothers Lionel and Plunkett bring home two 'maids', 'Martha and Julia' (really Lady Harriet her companion Nancy) hired from Richmond Fair; the two couples inadvertently, but perhaps in a way predestined, fall in love: there is a sequence of quartets, and at midnight a tacit recognition of the love growing between them.

The decisive events of the ghost story in Gounod's *La Nonne sanglante* take place at midnight, when the guilty character is slain by his son, so ending a nexus of secret crimes in the past and setting the wandering ghost of the murdered Nun free. In Rubinstein's opera *The Demon* the final and fatal colloquy between the fragile and fated Tamara and the roving spirit of her satanic tempter takes place at the same hour against the choral prayers of the nuns and the anguish of the Demon, as Tamara dies and is transported to heaven.

In *The Queen of Spades*, Lisa waits for the possessed Herman on the Nevsky Prospect Bridge; she sings of her grief and foreboding and throws herself into the icy river at midnight, appalled by Herman's possessed state. Later the Countess's Ghost comes to Herman in a dream-like vision in his barracks at midnight to impart to her obsessed victim the fatal secret of the cards.

Cinderella in Massenet's *Cendrillon* (1899) must flee the ball before the stroke of midnight when her enchanted transmogrification will end. And the vengeance planned against the gross and irresponsible Falstaff in Verdi's opera (1893) takes place in Windsor Great Park, beginning at midnight when the 'fairy' shenanigans of his disguised victims-turned-persecutors begin.

The sounding of midnight in Johann Strauss's *Die Fledermaus* (1874), heard in the overture and at the centre of the work, marks a joyous but symbolically pertinent use of the midnight hour. The reign of confusion, mistaken identities, disguise, and misrule deriving from the celebration of Bacchus is brought to its climax at the witching hour. The chimes mark the turn of the year and also the fulfilment of hidden desire and forbidden passion, obviated in a round of giddy pleasure under the influence of champagne. It will require the sober light of the morn to bring about a more rational understanding of Prince Orlovsky's party and all its Dionysian implications.

Bells

We have noted above the ominous bells for Boris Godunov's passion and passing, and how Robert the Devil, the wavering hero, is 'saved by the bell' from his existential dilemmas by the chimes of midnight, a symbol of divine grace. In *Giselle*, the erring Prince Albrecht is saved from the demonic forces of the Wilis, who are dancing him to death when the bells of six in the morning announce the break of day and end the evil spell. But in Puccini's operatic variant, *Le Villi*, the protagonist, Roberto, is not so lucky and is danced to death by the vengeful spirits.

A tremendous booming of bells heralds the miraculous denouement of Korngold's *Das Wunder der Heliane* (1927), the resurrection of the dead Stranger and the murder of the heroine, both essential to their heavenly translation.

Conversely, a bell from the Church of Saint Germain d'Auxerre tolls to announce the beginning of the Massacre of St Bartholomew's Day in *Les Huguenots* (1836). The death of Anna Bolena in Donizetti's opera (1830) is similarly heralded by the pealing bells that ironically mark the marriage of Henry with Jane Seymour and the end of Anna's reign and her life itself. And of course it is the vesper bell, the moment of evening prayer in Verdi's *Les Vêpres Siciliennes* (1855), that is the signal for the partisan massacre of the French in Palermo (as in *Les Huguenots*), including the two lovers, Henri and Hélène, who have just had their union blessed.

In the Meyerbeer's opera *Dinorah* (1859) the goat, a symbol of pastoral beatitude and conversely also of devilish alignment, wears a bell that serves as a *Leitmotif* for this animal and its association with Dinorah's alienated state of mind.

Wagner intones moments of great import with bells. After Elsa has asked Lohengrin the fatal questions, and Telramund and his men have tried to kill him, Lohengrin strikes a bell that chimes slowly and sadly, marking the end of his association with Elsa, and preparing the way for the valedictory last scene. In *Parsifal*, as Gurnemanz leads Parsifal to the Grail Temple, its bells ring out in various pitches, calling the knights to prayer, and symbolizing the divine grace, represented by the bell, suspended between earth and heaven, depicting in its conical form the vault of heaven, the source of all blessing and creativity represented in Holy Grail and its association the Holy Spirit.

In Verdi's *Il Trovatore* (1853) the bell tolling for the dead strikes relentlessly as the distant monks intone the *Miserere*, while Leonora desperately seeks to save Manrico, who sings of his love from his tower prison. In *Lakmé* the famous Bell Song characterizes the call to prayer of the Hindu priestess and celebrates her loving free spirit, further embroidered by her vocal imitation of the tintinnabulation.

There is also the famous 'Bell Chorus' in Leoncavallo's *Pagliacci* (1892): the villagers sing as they naively anticipate, with tragic irony, the joyful theatricals by the advent of a troupe of Strolling Players. These actors, however, are riven with dark secrets of deceit and adultery that culminate in murder during the performance.

In *Der ferne Klang* the mysterious mystical sound of divine creativity, so obsessively sought by the hero, is associated with the distant sound of bells and glockenspiel. They are finally heard in full at the end. Ironically, Fritz expires just as he attains his fated love and illusive artistic aspirations only in death.

> Ring out, wild bells, to the wild sky,
> The flying cloud, the frosty light:
> The year is dying in the night;
> Ring out, wild bells, and let him die.[14]

Madness and Melancholia

The theme of madness in opera has been discussed by many commentators. In List 1, six females and only one male (Boris Gudunov) develop some variant of an altered mental state sufficient to classify as either temporary insanity or prolonged derangement, with hallucinations or delusions. Other examples of disturbed characters are found in many operas. Variants are provided by the guilty Assur in Rossini's *Semiramide;* the treasure-crazed Hoël in Meyerbeer's *Dinorah;* and the stricken Peter Grimes in Britten's opera. In *Die Feen*, Arindal is hallucinating that he is hunting a doe; there is Herman's flight of fantasy in *The Queen of Spades* and Verdi's Lady Macbeth's guilt-ridden visions. Alma suffers from syphilitic madness in *Lulu*. The madness of Renata in Prokofiev's *The Fiery Angel* (1955) lead to her being burned at the stake. In some operas the dramatic alterations of the mental state occur in the central acts, as with Elvira in Bellini's *I Puritani* (1835), but many emerge at the end, as with Mélisande and Lucia di Lammermoor. Why Mila's mother goes mad and kills herself in the middle of Janáček's *Osud* (1904), provoking the tragic trail of subsequent events, is not revealed.

The mad scene of the grieving Imogene at the end of Bellini's *Il Pirata* (1827) and the final ending for Donizetti's *Anna Bolena* (1830) reflect on the increasing attention to madness and its causes in the early years of the 19th century, not only as portrayed on the stage and in literature, such as through the novels of Sir Walter Scott and others, but also within medical circles. The first performance of *Lucia di Lammermoor* was in Naples. The Asylum La Maddalena, established in 1813 at Aversa just outside Naples, was widely celebrated for its progressive treatments of insanity, and may even have allowed public visitors. The important publications on asylum reform by the renowned French alienist Jean-Étienne-Dominique Esquirol (1772-1840) were widely discussed, as were his descriptions of different forms of mental illness. His *Des maladies mentales considérées sous les rapports medical, hygiénique et medico-légal* (1838) was translated into Italian by Dr Luigi Calvetti, who was one of Donizetti's doctors when the composer was brought to Bergamo from Paris in the tertiary stages of meningovascular syphilis. Esquirol established a private asylum at Ivry with his nephew Jean Mitivié in 1824. Donizetti was first confined to this asylum under his care with the director Dr Moreau, who was the composer's personal physician while he was in Paris.[15]

In their paper *Opera and Neuroscience* Lorenzo Lorusso and colleagues looked at the various portrayals of insanity in operas of various epochs, noting, as we do, the shift in emphasis in the Romantic Era. The debt to Classical tragedy is evident in some early instances. They cite, for example, Monteverdi's *L'incoronazione di Poppea* (*The Coronation of Poppea,*1642) as a good starting point (with the suicide of Seneca), noting then the Baroque Period portraying "grotesque or terrifying madness". They discuss Mozart's interest in Mesmerism, his family knowing Mesmer in Vienna, and the use of a magnetic stone applied to the heads of Ferrando and Guglielmo (*Così fan tutti*) to assuage the arsenic they had taken. They identify the heartbeats within his musical orchestrations (e.g. in *Don Giovanni,* when Masetto listens to the chest of Zerlina).[16]

The authors discuss the *melodrama* of early 19th century operas in Paris, the comedic finales being followed by the Romantic tragedies, beginning around 1820. A similar shift was seen in the English novel, as Jane Austen's innovative explorations of the psyche (as in *Emma,* 1815) compare with those of Emily Brontë, classically in *Wuthering Heights* (discussed in Chapter 6), in a tragedy which confounds love and hate, and good and evil, as Catherine and Heathcliffe seem to be reunited by a passion transcending death.

Lorusso *et al* also have a list of female madness in operas, 40 in all, from 1589 to 1909. As with our List 1, the 19th century is the most represented (32). Interestingly, in relation to our themes, the composer most cited is Donizetti (seven); Bellini comes second (two).

Variants on this theme relate to alterations of the mental state at, let us say, the borderlands of consciousness, areas of the mind that so interested William James. Sleepwalking is the theme of Amina in Bellini's *La Sonnambula* (1831) and is important in Verdi's *Macbeth* (1847). Trances and other dream-like states are well represented. In our list they are a part of the drama in *Anna Bolena* and *Heliane*, and in Miles from *Turn of the Screw.* Wagner's operas have many: Senta, Elisabeth, and Elsa all become affected and we have the mental transformations of Tristan and Parsifal. The latter are positive developments in the traditions of education and spiritual growth. A more subtle counter-image is that to the monk Athanaël in Massenet's *Thaïs* (1894), who begins with earnest and idealistic attempts to save the soul of the courtesan heroine but is then drawn compulsively into his own emergent and torrential passion for her; she herself has experienced an inner transformation from a life of erotic blandishment to one of consecrated sanctity. Elektra in Strauss's opera (1909) undergoes a process from depressed hopelessness in anguished loss and disappointment to one of ecstatic joy at the return of her brother and the vengeance wreaked on the murderers of her father. Her triumphant dance of elation sees her sudden and unexpected death.

The paraplegic Amaranta in Pavel Haas's (1899-1944) opera *Šarlatán* (1938) is 'cured' of what turns out to be either hysterical paralysis or malingering; Hamlet feigns madness. Krzysztof Eugeniusz Penderecki's (1933-2020) *The Devils of Loudun* (1969) is about an outbreak of sexual hysteria among the nuns of St Ursula's Convent that leads to the death of

Father Urbain Grandier at the stake. Exorcism followed by death occurs in *Love and Other Demons* by Péter Eötvös (2008): the possibly rabies-infected Sierva is thought to be influenced by demons and dies possibly in a convulsion.

On the fringes also we find the role of dreams, not simple wishes or fantasies but factors that are important to the theme of the opera. Marguerite in Berlioz's opera/oratorio has dreamt of the handsome Faust; Elsa has dreamed of her knight in shining armour; Eric has seen the stranger and Senta, and she the image of The Dutchman in their dreams; Parsifal remembers his name that his mother used from a dream. In *Die Meistersinger von Nürnberg* (1868) Stoltzing's dream is creative (Wagner himself is said to have derived the beginning of *Das Rheingold* from a dream-like state). Ophelia dreamt that Hamlet was her husband in Thomas's opera (1866). Klytemnestra in Strauss's *Elektra*, guilty of her husband's murder, is plagued by nightmares and fears of retribution. In Bohuslav Martinů's *Julietta* (1938) the protagonist Michel, like most people in the opera, has no memory but recalls how he fell in love with the singing of Julietta. At the Central Bureau of Dreams, people come to ask for their fantasy dream, Michel fears being imprisoned in the dream world forever, where he chooses to remain with Julietta.[17] Aschenbach's dreams of Apollo and Dionysus in Britten's *Death in Venice* (1973) bring about some internal change, leading him to simply resign to his fate.

Hecate and the Moon

The Romans referred to epilepsy as the *morbus comitialis,* since observing an attack could spoil the assembly:

> A kind of sudden sickness 'tis …
> When members down in fatal weakness fell.
> And God himself
> Through changing faces of the unstable moon
> Proclaims conception of a man oft thus to be outstretched.[18]

A disorder, often associated with madness, the Falling Sickness as Hippocrates called it, was considered to be caused by Hecate, the Snake goddess of the night, charmed by the music of Orpheus, also a witch known as the Crone Goddess of the Waning Moon. The seizures were associated with the full moon and by some it was called "the disease of the moon", the disorder was often associated with madness. As a suffer of epilepsy Fyodor Dostoevsky's (1821-1881) experiences are well known:

> There are seconds, they come only five or six at a time, and you suddenly feel the presence of eternal harmony, fully achieved. It is nothing earthly; not that it's heavenly, but man cannot endure it in his earthly state. One must change physically or die. The feeling is clear and indisputable. As if you suddenly sense the whole of nature and suddenly say: yes, this is true. God, when he was creating the world, said at the end of each day of creation: 'Yes, this is true, this is good.' This ... this is not tender heartedness, but simply joy ... What's most frightening is that it's so terribly clear, and there's such joy. If it were longer than five seconds—the soul couldn't endure it and would vanish. In those five seconds I live my life through, and for them I would give my whole life, because it's worth it. To endure ten seconds one would have to change physically … [19]

Divine possession influenced 'lunatics' (*Mondsucht*), a term which embraced a much wider spectrum of illness than epilepsy.[20] It is not surprising the moon is such a powerful symbol in operas.

The moon is the mysterious feminine counterpart of the sun, associated with the tides, the physiology of women and lunar goddesses. The moon measures and determines terrestrial phases; it also unifies through its activities the waters and rains, the fertility of vegetation and fecundity of animals and women. Its phases are analogous to the seasons of the year and the ages of man. It is mutable and transitory, celestial and infernal, associated with Diana (Jana, Roman Selene/Hecate Greek), a guide to the secret or occult side of nature. Its painfully changing shape symbolizes either for good or bad. Endymion, who first observed the movements of the moon, was seen by Luna who was enraptured by his beauty: "A thing of beauty is a joy for ever".[21]

In opera the central sinister supernatural events in *Der Freischütz* and *Robert le Diable* take place in the moonlight. Some of the most famous arias of transformation in the Undine stories are at night in moonlight. They are to do with fleeting shades and changing identity, as in the Shadow Song in *Dinorah* and the Song to the Moon in *Rusalka*, where the heroine is caught in worlds of half-truths and shifting, yearning identities that can only be resolved in love and sacrifice.

Moonlight is the medium of passion and of transition or uncertainty. In Gounod's opera, Faust seduces Marguerite in the moonlight in an enraptured, otherworldly circumstance, an almost transcendental state, and she vows to die with him. Faust's love is genuine but also treacherous. Valentine eventually curses Faust as a seducer and betrayer.

The female symbolism takes on more corporeal nature in the role of the moon in *Salome*, where the crude passions of the heroine take place in the moonlight; the nocturnal medium where the nexus of passion and murder occur in Elektra's family will be avenged.

Conversely it is mystical medium, as in *Salammbô,* where on a clear starry night the priest invokes the goddess Ishtar, who will appear invisible, as the moon. Moonlight floods the interior of Hunding's Hall in *Die Walküre* when the door flies open, inviting Siegmund and Sieglinde to consummate their love in full tide of spring rebirth. The moonlight transfigures the neglected Paradise Garden in *A Village Romeo and Juliet,* where the soft mellow light initiates something mysterious and beautiful, touching the garden as if by enchantment, preparing the way for the Love-Death of Sali and Vrenchen. The same is true of *Daphne,* where Apollo appears to declare his love for the young woman in the mystical aura of moonlight, accompanied by the especially evocative 'moon-music', a preliminary to her metamorphosis in love. In Puccini's *Turandot* (1926) the full ambiguity of this variable symbol is revealed. The moon is associated with the terror of the icy Princess's war on men. The people bray for the blood of her latest fated suitor, and pray for the moon to rise, the time for execution. The night and the moon will be the milieu of waiting, when Turandot seeks to discover the name of her latest suitor, and Liù is sacrificed to save her prince. It will also herald the advent of true love when Calaf melts the icy heart of the Princess.[22]

The moon is the insubstantial medium of ambiguity, of yearning and aspiration, and hence the perfect medium of delusion. At the crisis at the end of Act 1 of Schreker's *Der ferne Klang* (1912) Grete, abandoned by the obsessed Fritz, escapes her dreary milieu and arranged marriage only to find herself in the labyrinthine perplexities of a wood where she is alone. The moon rises and she is impressed by the beauty of nature at night, and induced to sleep and dream of love before meeting an old woman, a prostitute, who promises to help her, but will ensnare her in a lifestyle the very opposite of her romantic ideal.

In the middle of Debussy's *Pelléas et Mélisande* (1902) the hero leads the frail heroine into a cave in nocturnal darkness. The moon emerges, lighting up the cave mysteriously, revealing three beggars sleeping there, victims of the famine in the land. The fusion of the motifs of night, cave, moonlight and human suffering/tragedy capture in symbolic essence the thrust of the story and its dreamy meandering exploration of the unconscious. The cave is a semi-subterranean place, in touch with the deepest secrets of the earth, and is coupled with the moon. The lunar body, with its changeable half-illumination, represents mutability and passion and the futility of human desire. This is also reflected in Rubinstein's *The Demon* (1871), where the denouement takes place at midnight in the moonlight. The tempter enters Tamara's conventual sanctuary at the hour of the devil's influence, and in ambiguous lunar light implants his deadly kiss. This leads, as in Heliane's story, to heavenly assumption, a rising above the futile passions of this world. In Offenbach's *Les Contes d'Hoffmann* (1881), the protagonist's fateful entanglement with the Venetian courtesan Giulietta reaches its climax of disillusionment in the light of the "silvery kindly moon". Giulietta seeks possession of his "reflection", his very soul, in the midst of the uncertain shadows of moonlight and on the insubstantial medium of the watery lagoon, before eventually floating away in a gondola with her new lover, leaving Hoffmann abandoned and ruined, realizing the truth of his own observation that "Love is damnation".

Poisons and Potions

The role of poisons and magical potions that change the narrative are a frequent method of driving on the drama. The most obvious example is the death/love potion of *Tristan und Isolde.* A *Liebestrank* or a *Zaubertrank* allows Faust to see Helen in every woman. In Gounod's opera Faust is given potions at the beginning and end of the opera. Here the potion is an agency of demonic deception. Richard Strauss in *Die Aegyptische Helena* (1928) has potions of forgetfulness and recollection; Meneleus takes a potion to forget Helen's adultery.

It is the magic formula which keeps Emilia alive so long in *The Makropulos Case.* A modern variant, David Sawer's *Skin Deep* (2009), is about a Dr Needlemeier who is developing an elixir of youth, which requires fragments of human tissue and, importantly, the testicle of a young man (who spends the rest of the opera looking for it). In *Die Gezeichneten* the passions are inflamed with drugs. In the opera *Rappaccini's Daughter* (1991), by Daniel Catán (1949-2011), based on the play by Octavio Paz, the scientist's daughter lives in a garden filled with poisonous plants; she becomes poisonous herself and poisons others. Cilea's Adriana Lecouvreur is poisoned by kissing faded poisoned violets; and in Meyerbeer's *L'Africane* Sélika kills herself by inhaling the vapours of the toxic Manchinel Tree. In a similar fashion Lakmé, the daughter of a Hindu priest, kills herself with jimsonweed. In *La Muette di Portici*,

Masaniello is poisoned and loses his reason. In *L'amore dei tre re* the King poisons the dead Fiora's lips hoping that her adulterer will kiss her, but with tragic irony her own betrayed husband is the one who dies from this plan.

In Wagner's operas we have in addition to the "Vergessenes güt'ger Trank" of Tristan and Isolde;[23] the "betäubenden Trank" that puts Hunding to sleep; and the "tödlichem Tranke" given to Siegfried in *Götterdämmerung* causing his amnesia (found in the Nordic *Völsung Saga*) which is then reversed, leading Siegfried to reveal the truth and the apparent violating of the mendacious blood-brotherhood. As a variant we have the blood of dragon Fafner, the taste of which alters Siegfried's life forever: Mime's potion was strictly to poison![24]

It may be said that every good tragedy involves a drinking song expressed with Dionysian energy, wine as an aphrodisiac yet bringer of violent passions. In Haydn's *Orfeo* the Bacchantes seek to comfort the grieving Orpheus with a potion which is in fact a deadly poison.

> Drink, drink from this cup
> Drink the nectar of love.
> The potion will bring you
> All the greatest joys.

Drugs and alcoholism explain the death of Edgar Allan Poe (1976) in the opera of that title by Dominick Argento (1927-2019), in which Poe hallucinates the voice of his long dead wife Virginia before he dies. Such elixirs can bode ill (in Greek *pharmacon* means both cure and poison), but from another angle wine is the source of Dr Dulcamara's fraudulent 'magic' elixir in Donizetti's *L'Elisir d'Amore* (1832). The association of love with self-inflicted death through potions and poisons gains strength from the examples of Romeo and Juliet (in both Bellini's 1830 and Gounod's 1867 operas), as well as Leonora in *Il Trovatore* (1853) and Ponchielli's *La Gioconda* (1876). Both the latter take poison rather than give themselves to someone other than the beloved: both entail suicide after offering themselves to allow their loved ones to escape (just like Sélika and Lakmé). In Verdi's *Luisa Miller* (1849) the deceived Rodolfo poisons himelf and his beloved Luisa so that they can die together in a kind of *Liebestod*, rather than become pawns in the social engineering of his masterful father and his disreputable henchman.

In the operas on Franceca da Rimini by Rachmaninov (1906) and Zandonai (1914) there is a confluence of several powerful images that are linked in association with fated love and related death. There is a 'fatal kiss' ("Ah! Spring, Take my soul/Take my mouth"), a one-eyed brother-in-law Malatestin consumed with jealousy, and a *Liebestod* finale with a double stabbing. Wine is used as a potion (Francesca gives it to Paolo in a way not unlike Isolde proffering the drink to Tristan), a spell to induce the loss of memory.

The concept of forgetfulness and social oblivion is integral to Schreker's *Die Gezeichneten* (1918). It is very much part of the secret world of wild abandon cultivated by the aristocratic but hunchbacked Salvago in his secret island retreat where he seeks to realize dreams of wild forbidden ecstasy. In Act 3 a celebration of Dionysus takes place, where the heroine Carlotta appears in the middle of the Bacchantes, all her earlier fears and inhibitions overcome, as though she had "drunk wine mixed with bewitching herbs, or a potion brewed from the blood of a love-sick maiden".

Conversely, in Offenbach's *Les Contes d'Hoffmann,* where the celebration of drinking is a feature of the Prologue, the Epilogue and the middle Giulietta Act, wine is stimulus to Hoffmann's telling his tales. Such bacchic celebration is also delusory though, as in Hoffmann's disastrous encounter with the Venetian courtesan. Eventually in the Epilogue he is too drunk to rise from the table, and Lindorf is able to walk away with the poet's muse, Stella.

Eyes and Faces

We introduced the power of eyes in Chapter 1: the look and looking are everywhere. Wagner was very interested in eyes and their potent dramatic effect, emerging first in *Die Feen*. The theme is recurring in the first meeting of between Senta and The Dutchman. In the text of *The Ring* there are 60 direct references to eyes, and anyone who has paid attention to the opening of *Rheingold* should notice two very important things about Wotan as he first appears: he is asleep, and he has a damaged eye, both features that thread a tapestry throughout the cycle.[25] In our interpretation, the importance of the left eye being damaged is fundamental to the whole of *The Ring* narrative, the left eye representing the lost affective part of Wotan's nature, that which was suppressed by the dogma of the rules and runes that ultimately cost him the world.

In poetry the power of the eyes between lovers is a powerful theme in Metaphysical Poetry. We earlier noted John Donne as a good example:

Let us possess one world, each hath one, and is one.
My face in thine eye, thine in mine appears,
And true plain hearts do in the faces rest.

Souls are united by those twisted eye-beams upon one double string.[26] Eyes mesmerized Edgar Allen Poe.

In Boito's *Mefistofele*, during his Classical incarnation, Faust observes that the pupils of Helen's eyes wander like the moon. The mystery of the eyes is a part of the mystery of Mélisande. The power of eyes drives the whole of the plot in *Tristan und Isolde.* As Isolde advances towards Tristan with the fatal cup, they fixedly gaze at each other, their glance remaining throughout the moments as they sing, one after each other: "Isolde! … Tristan!" (Act 1 scene 5). In their earlier history, it was the moment that Tristan's eyes met those of Isolde, after the death of her betrothed Morold when she was about to take revenge and slay him, that her sword dropped and she healed Tristan's own wound with her magic (Act 1, scene 3).[27] Also recall Fasolt's final plea before losing Freia, whose "starry eye" continued to shine on him, visible only through a tiny crack in the pile of gold mounted up around her. He would not give her up while he could still see her "glance and this lovely eye" (*Das Rheingold*, Scene 4). Herman's eyes possess "sinister fire", enough to put fear in the Countess, as they have captivated Lisa in *The Queen of Spades*. Heliane's explains her attraction to the Stranger through the power of sight: "God has touched me through your eyes".

Of the gaze, Roger Scruton perceptively wrote, "a deep contrast between the gaze of the stranger and the gaze of intimacy [is that]… in the gaze of intimacy freedom is not asserted against the other but offered to him".[28]

Eyes also have other connotations. The implications of the action can be further reinforced by recurrent motifs of watching (and overhearing, of recounting and warning), as in the manner of a Greek tragedy, where decisive events happen offstage and are lived vicariously (often in ironical or fateful illusion of the truth) in the actual time of the drama. This is bound up with seeing and deceiving, concealing and revealing, and eventually with confusion and illumination of mind and heart. Secret windows and hidden vantages, blindfolds, veils, and night, are used in hiding and laying bare truth and motivation in a process of mystery and revelation. Examples are in *Les Huguenots* (1836). Nevers is observed through an oriel window in colloquy with a mysterious and striking lady visitor. She turns out to be Raoul's distant beloved, but is thought to be another of Nevers's mistresses (Act 1). Raoul is led blindfolded through the gardens of Chenonceaux into the presence of the Queen; his blindfold is removed, and he is entranced by this beautiful, unknown woman, as she is by him. And let us not forget the Rose-Tinted spectacles of Hoffman's Dr Coppelius.

Premonitions and Re-collections

There is more to the eye than just the sensory visual input. There is the world of difference, as we have emphasized, between perception and apperception. And there is the tricky concept of the *a priori* nature of at least some of our construction of and knowledge about the world. We have re-collection that relates to predetermined premonitions that we earlier discussed, going back to Plato, and embodied in a number of religious and philosophical traditions. It implies a prevoyant seeking of the psyche, linked with pre-embodied memories, or even pre-Adamic resonances and ideas of reincarnation.

Des Grieux, in Puccini's *Manon Lescaut* (1893), increasingly ensnared by Manon's captivating allure, speaks of a premonition of their passionate love ("Donna, non vidi mai"). Even more mystically, the Demon in Rubinstein's opera elevates premonition to a type of epic predestination, declaring that Tamara's image was imprinted on his soul from the beginning of the world, even within the emptiness of eternal time and space.

In the Berlioz version of *Faust*, Marguerite is fixated on him, having seen him in her dream. In The *Queen of Spades*, Herman, while looking at the portrait of Muscovite Venus, refers to the secret force which binds their destinies together. The Governess has a "strange connection" to Miles in *The Turn of the Screw*. The portrait of the Dutchman is central to Senta's infatuation with Vanderdecken, He too can see before his eyes her image, dreamt of for restless ages. Their mystical affinity is marked by Senta's cry when they come face to face, like a dream realized from eternity. Other Wagner operas find similar precognitions: Elsa, has dreamt of her hero, knowing he had come to her following God's counsel.

Such premonitions are crucial to the developing story in *Die Walküre*, as reminder:

Siegmund
Whatever I longed for

I saw in you;
in you I found
whatever I lacked!
I laugh aloud in hallowed delight
as I hold you within my embrace
and feel your beating heart!

Minds and hearts together, and the theme of incest, another well-visited Romantic trope, is interlaced throughout the *Ring* Cycle. From the myth of Oedipus through to the psychological theories of Freud, in Wagner's opera the very idea is unconscionable for Fricka, the guardian of wedlock. Thus she sides with Hunding in his forthcoming fight with Siegmund. As Wotan pleads for Love's enchantment which has united the couple in Spring, Fricka reminds him of the inviolability of the covenants he safeguards. She also mentions the others he has fathered out of wedlock, including the Valkyries and, with "heart-quaking and brain-reeling", she portends the end of the "blessed immortals" (Act 2 scene 1). This is one of the pivotal movements in the Cycle, when Wotan caught in his own plight has become "the saddest of all living things" (Act 2 scene 2).

When Brünnhilde is awoken by Siegfried, she 'knew' it was he: "it is Siegfried who woke me … how I have always loved you … Your own self am I" (she is his half-sister and aunt) (*Siegfried*, Act 3 scene 3). Likewise, Gurnemanz when he first encounters Parsifal, observes, "I think that I do know you [*recht erkannt*]" (Act 1).

The Kiss

If the gaze and intertlocking of the eyes are veritably torques for the plot of several operas, even more significant are kisses and curses. From List 1, six subjects (12 per cent) have a kiss at crucial dramatic moments. When the Demon kisses Tamara the Angel appears; the Stranger kisses Heliane and she enters into a trance-like state and he then kills himself. The kiss between Pelléas and Mélisande is seen by Golaud: shortly afterwards he strikes down Pelléas with his sword and kills him. In contrast, in *A Village Romeo and Juliette,* the kiss between Sali and Vrenchen changes their surroundings to the Paradise Garden, with moonlight and singing angels· In Puccini's *Turandot* it is Calaf's kissing of the Princess that saves his life. Otello dies upon a kiss, united with Desdemona. Moments of fatal decisiveness are marked when Francesca and Paolo kiss (in Rachmaninov and Zandonai), and also Werther and Lotte in Massenet's opera (1892). In Montemezzi's *L'amore dei tre re* (1913) Fiora's secret lover Aviso dies kissing her dead poisoned lips, as unintentionally does her betrayed but doting husband Manfredo.

The Undine/Rusalka operas bring only fatal kisses. The Prince begs Rusalka to kiss him until he dies, “Kiss me a thousand times”. In *Undine*, she enters Huldbrand’s bedchamber and kisses him passionately, knowing that her kiss, following the destiny written in the world of magic, will bring him death. Wagner trumps all with the dramatic scene of the kiss between Parsifal and Kundry. Her kiss is reported to be the longest kiss in musical history and is as significant as the kiss which awakens Brünnhilde. The saddest of all for her, however, is that of Wotan who, with a last tender moment kisses her godhead away (Act 3).

Hair

Hair is another Romantic cipher. Our First Lady Eve had unadorned golden tresses that “hung down to her slender waist, dishevelled, but in wanton ringlets wav’d”.[29] Hair is used in perhaps the first novel ever written, *The Golden Asse of Apuleius*, a wonderful bawdy travelogue akin to a tale from Ovid. Apuleius falls in love with Fotis, whose hair hung about her shoulders, leading to a kiss:

> If you spoile and cute the heare of any woman or deprive her of the colour of her face … [no matter what other attributes of beauty she may have] …she could nowise please, no, no her owne Vulcanus.[30]

Boadicea, the Queen of the British Celtic Iceni, had hair cascading down both sides of her war-like frame, while the beautiful *Venus* of Botticelli has long golden hair blown by the breezes modestly covering the left side of her body down to cover her pudenda. One of the most stunning wooden sculptures of the Renaissance, that of *Saint Mary Magdalen* (The penitent Magdelen, 1454/55) by Donatello (1386-1466), shows her emaciated figure enswathed in rags with lengths of disordered hair sweeping down both sides of her body.

Hair can be erotically charged. Mélisande’s cues Pelléas to use it fetishistically. The wonder of Heliane’s long golden hair competes with that of Mila, which captivates Živný (*Osud*) as it falls around her brow emphasising her shining eyes. Sierva (*Love and other Demons*), like Heliane and Mélisande, has hair which cascades down her body. The Rusalki are usually portrayed with long flowing hair (as seen in many a Pre-Raphaelite image). Kundry uses her glowing hair to dry the feet of

Parsifal. Hair can also portray more sinister events such as the dishevelled appearance that reflects the madness of Elektra.

The above features which permeate the drama of opera do not stand isolated: they are all one way or another entwined with three most important themes of Romantic opera, to be discussed in the next three chapters: namely the Transcendental, Redemption, and the *Liebestod*—all bound up with the nature of love.

Notes

[1] Eliot, T.S. 'Rhapsody on a windy night'. *Collected Poems 1909-1935* (London: Faber and Faber, 1936), 24.
[2] Eliot, T.S. 'Dry Salvages': *Four Quartets* (London, Faber and Faber, 1943), 26.
[3] Poizat, M. Ibid., 1992, 134-35.
[4] Poizat, Ibid., 135.
[5] The physical sound needs to be increased exponentially in order for equal steps to seem linear, a fact that is used in current electronic devices to control volume. The *Helmholtz resonator* identified the various frequencies or pitches of the pure sine wave components of complex sounds containing multiple tones, particularly the overtones.
[6] Fourier analysis is a transformation of wave forms is such a way that general functions can be represented by simpler trigonometric expressions.
[7] Cahan, D. Ibid., 197.
[8] We now know in more detail those aspects of musical appreciation which Helmholtz was working out. Pitch and timbre processing activates cerebral areas for emotion. Both the auditory cortex and subcortical systems are involved in different ways by both "pleasurable" and "non-pleasurable" musical sounds. This appears to relate in particular to harmonicity: the closer the timbral and harmonic character of the music is to the structure of the harmonic series, the more "pleasurable", the further away, the less "pleasurable". Harmonicity appears to be deeply embedded in the functioning of Heschl's gyrus, an important part of the auditory cortex. For further reading: Blood, A. J. and Zatorre, R. J., "Intensely pleasurable responses to music correlate with activity in brain regions implicated in reward and emotion." *Proceedings of the National Academy of Sciences USA,* 98 (2001): 11818–11823; Griffiths, T. D.; Buchel, C.; Frackowiak, R. S.; Patterson, R.D. "Analysis of temporal structure in sound by the human brain." *Nature Neuroscience, 1* (1998):422–427; Hesdorffer, D.; Trimble, M. "Music and the brain: the neuroscience of music and musical appreciation."*British Journal of Psychiatry International*, 1;14, 28-31. eCollection, 2017; Koelsch, S.; Fritz, T.; Schlaug, G. *"Amygdala activity can be modulated by unexpected chord functions during music listening."* Neuroreport, 3;19(18) (2008):1815-9. doi:10.1097/WNR.0b013e32831a8722, 2008.
[9] Steege, B. *Helmholtz and the Modern Listener* (Cambridge: Cambridge University Press, 2012), 228. On the influence of neuroscience on the arts see:

Trimble, M.R. *The Intentional Brain: Motion, Emotion, and the Development of Modern Neuropsychiatry* (Baltimore: Johns Hopkins Press, 2016). Helmholtz visited Bayreuth, and met Wagner on several occasions. He appreciated Wagner's music, and was brought to tears by it on one occasion, "suggesting a Romantic dimension in him that is otherwise only occasionally visible" (Cahan, D. Ibid., 483-34).

[10] On pre-Darwinian influences of evolution, see: Stott, R. *Darwin's Ghosts. The Secret History of Evolution.* (New York: Spiegel & Grau Trade Paperbacks, 2012). Mary Shelly used the term galvanism, after the pioneer Luigi Galvani (1737-1798) who discovered animal electricity, for the "spark of being" which infused life into Frankenstein's creature. She did not use the word electricity. The full title of the book is: *Frankenstein; or The Modern Prometheus*. For much more on these developments see: Trimble, M.R. Ibid., 2016.

[11] See also Robertson, J. Ian S. *Doctors in Opera*, Ibid. 2016.

[12] Shakespeare, *Macbeth* 1:3,156.

[13] Merleau-Ponty, M. Ibid, 2012, 113. For an excellent discussion of the meanings of time: Lakoff, G. and Johnson, K. *Philosophy in The Flesh: The Embodied Mind and Its Challenge to Western Thought* (New York: Basic Books, 1999).

[14] From Tennyson's *In Memoriam*, a requiem for the poet's Cambridge friend Arthur Henry Hallam, who died at the age of 22. Tennyson, W.A. Ibid. 1883, 277.

[15] See Ashbrook, William. *Donizetti* (London: Cassell, 1965), 308-339.

[16] Lorusso, L.; Franchini, A.F.; Porro, A. "Opera and Neuroscience. Progress in Brain Research." 216:389-409; doi: 10.1016/bs.pbr.2014.11.016. Epub 2015 Jan 20. 391.

[17] Based on the play *Juliette, ou La clé des songes [Juliette, or The Key of Dreams*] by the author Georges Neveux (1900-1982).

[18] Temkin, O. *The Falling Sickness: A History of Epilepsy from the Greeks to the Beginnings of Modern Neurology.* 2nd ed. (Baltimore: Johns Hopkins Press, 1971), 8.

[19] Dostoyevsky, F. *Demons*. Trans. Maguire R (London: Penguin Classics 2008), 653. Dostoyevsky described several moments of 'the presence of eternal harmony' in his novels. He was diagnosed as having epilepsy in 1848, although Freud thought his attacks were psychogenic. Medical opinion was in favour of epilepsy. For the relation between neuropsychiatric illness and creativity, see Trimble, M. R., Ibid. 2007.

[20] See Temkin, O. *The Falling Sickness*, Ibid., 15, 92-96; Graves, R. *The Greek Myths* 2 vols (London: Penguin Books, 1960), 1:115. The association of epilepsy and madness with gods and other evil influences was debunked by Hippocrates, but the ideas still linger on. The history is well laid out by Temkin.

[21] Keats, J. *Endymion.* In *The Complete Poems*. Ed. J. Barnard (London: Penguin Classics, 2006), 107, 729. Diana put him into a deep sleep so she could kiss him undiscovered (as she was Goddess of chastity)—Jupiter gave him perpetual youth, and he could pass time on Mount Latmos contemplating beauty.

[22] See "Moon" in Cirlot, J. E. *A Dictionary of Symbols*, ibid. 1962, 204-207. "The symbolism of the moon is wide in scope and very complex. The power of this satellite was noted by Cicero, when he observed that 'Every month the moon

completes the same trajectory executed by the sun in a year … It contributes in large measure to the maturation of shrubs and the growth of animals.'" "This helps to explain the importance of the lunar goddesses such as Ishtar, Hathor, Anaitis, Artemis" (204).

[23] Robert Greenberg alerts one to a motif of 'descending chromatic notes which play with the words "Ich trink sei Dir" as the drink goes down.

[24] A superb discussion of potions in Wagner's operas is "The Potion as Symbol and Prop in Richard Wagner's Musical Dramas" in Müller, U.; Panagl, O. *Bayreuther Festspiel Programme* (1998): 126-138. Ian Robertson and Gunther Weitz in their paper "Siegfried's Amnesia" speculated about the medical possibilities of the extract he had imbibed. A key theme was the use of belladonna (alkaloids of atropine and hyoscine, known as 'beautiful woman' as it led to pupil dilatation making women more attractive). It causes marked autonomic alterations sometimes including dramatic changes of the mental state. Wagner himself said that Siegfried's transformation could only be explained by taking a poison (as opposed to more psychological explanations). Robertson, J. I. S. and Weitz, G. "Siegfried's Amnesia: An Attempt at an Explanation." *Core Image, Scottish Opera,* 2015.

[25] Holman, J. K. *Wagner's Ring: A Listener's Companion and Concordance* (Milwaukee: Amadeus Press, 1996), 264-68. For the importance of Wotan's damaged eye as an important key to understanding the Ring Cycle see: Trimble, M. R.; Hesdorffer, D.; Letellier, R. I. "The Mystery of Wotan's Missing Eye". *The Wagner Journal*, 7 (2013): 26-37. Wagner himself also had a very prominent eye disorder, unnoticed by nearly every critic or commentator, namely a squint (lazy eye), which affected his health and compositional abilities. See Trimble, M.R.; Hesdorffer, D.; Letellier, R.I. "In Wagner's Eyes: Casting Light on a disputed portrait". *The Wagner Journal*, 13 (2019):20-31.

[26] Gardner, H. *The Metaphysical Poets* (London: Penguin Books, 1972): John Donne, *The extasie,* pp. 74-77.

[27] Reverting to psychophysics, it has been shown using measurements of brain waves (with the Electroencephalogram, EEG), that eye contact with direct but not indirect gaze between mothers and babies alters the EEGs of both, synchronising wave forms. See Leong, V.; Byrne, E.; Clackson, K.; et al, "Speaker gaze increases information coupling between infant and adult brains." *PNAS*, 114 (2017): 13290-5.

[28] Scruton, R. *Death-Devoted Heart: Sex and the Sacred in Wagner's 'Tristan und Isolde'* (Oxford: Oxford University Press, 2004), 220. It is interesting to note that in late November 1863, two years before the première of *Tristan und Isolde*, Wagner visited the von Bülows in Berlin. While Hans was at work one day, Wagner and Cosima went out for a drive. Years later, in his autobiography, Wagner described the moment that changed their lives: "Our jesting died away in silence. We gazed speechless into each other's eyes; an intense longing to speak the truth overpowered us and led to a confession, of the boundless unhappiness that weighed upon us. With tears and sobs we sealed our confession to belong to each other alone". See R. Greenberg, *The Music of Richard Wagner*, https://www.goodreads.com/book/show/ 40099072-the-music-of-richard-wagner.

[29] Milton, John. *Paradise Lost* Book IV. In *The Portable Milton*. Ed. D. Bush (London: Penguin 1976), lines 304-306.
[30] Adlington, W. (trans.). *The Golden Asse of Apuleius* (London: Grant Richards Ltd, 1913), 35, 36 (late 2nd century AD); Vulcanus, the God of fire.

CHAPTER 21

LOVE, OTHER WORLDS AND THE NATURE OF REDEMPTION

> It is a question, you see, of pointing out the path to salvation, which has not been recognised by any other philosopher, and especially not by Sch., but which involves a total pacification of the will through love, and not through any abstract human love, but a love engendered on the basis of sexual love, i.e. the attraction between man and woman.
>
> (Richard Wagner, Letter to Mathilde Wesendonck 1 December 1858)[1]

1. Introduction

In opera, the music will always reflect, through beauty, peacefulness or exultation, a transcendental implication.

In his analysis of the musical structure of *The Ring,* the American classical pianist Jeffrey Swann draws attention to what he refers to as the 4-note cell, a potential *Grundtheme* found throughout *The Ring* (but also in many other Wagner operas). It consists of a falling second, followed by a falling fourth or tritone, and continuing with a rising third—found in both major and minor modes. He suggests that it is almost unique, the *Grundtheme* having no fundamental leitmotivic form, and, therefore, no concrete underlying dramatic association. He writes, "If the cell is a musical symbol for love—and to some extent, such an association is inevitable—then it is natural that Siegfried should hear it in the song of the Forest Bird. Siegfried is longing for love, and the Bird has become a reflection of his dreams and desires. And it will indeed be the Forest Bird who will reveal to Siegfried an object for his love, Brünnhilde, and guide him to her at the end of this Act". It appears as a common cadential or sequential tag or extension to other motivic material.

> The Ring is, indeed, a magnificent kaleidoscope, in which the basic elements continuously collide and escape, interacting in a wondrous and intricate dynamic and organic web. The cell is a primary element in this new form, in which every moment is vital with the present, laden with the past, and pregnant with the future.[2]

In earlier chapters we have drawn attention to the importance of the Elemental Being in characters' search for peace and redemption, the contrasts and conflicts between dark and light, night and day, being thematic as redemption through Love becomes epiphanic. The *femme fatale* and the *Ewig-Weibliche* form part of a palimpsest of sacred and profane images through the operas we have been considering, embroidering Love in its various historical and artistic images.

In Puccini's *Turandot* Calaf must answer three riddles set by the Princess, but in the end she must know his name. She names him 'Love'.

2. Varieties of Love

Love is a word with so many nuances, yet coupled with death in the greatest tales of the operatic genre. This covers not only love of country (Auber's *La Muette de Portici* 1828, Rossini's *Guillaume Tell* 1829); that of a mother for children (Bellini's *Norma* 1831); or that of a father for a daughter (Orevesco for Norma, and in several Verdi operas, such as *Rigoletto,* 1851, or *Simon Boccanegra,* 1857). There is the love of the Father in *Hansel and Gretel*, and of a mother for her son—Fidès for John of Leyden in *Le Prophète*. Then there is the lure of the *femme fatale* (Carmen, Thaïs); love versus lust (*Die Gezeichtenen*); erotic love (as in *The Demon*); love between siblings (Siegmund and Sieglinde); and what may be referred to as transcendental love. The desire of the Duke Mantua contrasts with the purity of Gilda's loving devotion in *Rigoletto*. The Faust operas explore the differences between lust and love, in the ambivalent attitude of Faust for Marguerite and her pure sacrificial love that is ultimately redemptive. Transcendental love explores the transcendent nature of the power of love, suffering and sacrifice as themes, and religion so often of considerable significance: God as an active principle of redemption in the world. What is significant is that death and loss always embroider different kinds of love. This will bring us to the definitive use of a *Liebestod* as an operatic device, considered shortly.

3. Love in Opera

Gluck's *Alceste* (1767) is a pre-Romantic redemption opera, with Hercules the saviour figure. There is almost a *Liebestod* as the couple decide to enter Hades together, when Hercules as *deus ex machina* intervenes to save Alcest and Admetus. They are hailed as models of conjugal love, a story that points much towards *Fidelio*.

Wagner questions us as to whether Romantic and sexual love can be combined, following these topoi in operas such as Bellini's *La Sonnambula* or *Norma*, and Donizetti's *Lucia di Lammermoor*. Sexual love was viewed in Victorian times as a powerful but disruptive force, but so were the harmful effects of arranged marriages in the repressive age of domesticity. By the mid-19th century the plays of Henrik Ibsen (1828-1906) and Johan August Strindberg (1849-1912) explored social and sexual dynamics in ways that were shockingly exposed by writers such as Émile Zola (1840-1902), and in many of the operas we have been discussing. Female characters such as Carmen, Thaïs, Manon, and then Salome with full nudity, were portrayed on the operatic stage for the first time.[3]

Wagner's love pairs are not in the usual run of 'love music'. In *Das Liebesverbot* 'free love' is not quite sexual. The affairs involving Senta, Elisabeth, Elsa, and Kundry are unconsummated, while that between Siegmund and Sieglinde is incestuous (erotically associated with Spring). Brünnhilde is Siegfried's aunt and cousin, and Siegfried's enforced link to Gutrune is deceptive, driven by a potion.

The earlier incestuous couple in the oeuvre lies in Rienzi's interest in his sister, but this is non-sexual, and Rienzi's downfall relates amongst other faults to his pride (*hubris*) and the malediction of the Papal delegate. The ideas of pride and cursedness are repeated in *Der fliegende Holländer* and *Parsifal* (Kundry). In *Die Meistersinger von Nürnberg,* Eva and Hans Sachs have a love duet, and Walther does win his bride. With *Tannhäuser* we have an exploration of sexual love, the depth of the sexuality differing with different versions. The later Paris version (1861) has the long Venusberg sequence and other changes which enhance the erotic aspects. Yet Wolfram's song to Elisabeth (Act 3), evoking the Evening-Star Venus ("O Du meine holder Abendstern"), pleads for a non-sexual love, perhaps a fusion between sexuality and spirituality.

Going beyond these considerations, Wagner's operas explore creative aspects of art imbricated with the nature of love and the erotic. The song contests (in *Tannhäuser* and *Die Meistersinger*) are much about tradition, but also imply inspiration related to romantic love. In *Tannhäuser* the characterisation of Venus shifts to become a more active, complex vector. The contrast is not between a bad and a good woman, but about internal tensions, sexual tensions as creative forces, the fission between the sacred and profane, accentuated in that song of Wolfram.

The earlier 'grand opera' style shifts to a more personal view, more intimate, sexual attraction being mysterious: Tristan is unable to explain to King Marke why things happened the way they did; elective affinities—the unalloyed adherence of nature's objects that mutually affect one another—affined, are in place.[4] The politics we find in the operas of Meyerbeer and Verdi are not so important: the emphasis is much on the night, rather than the day; personal identities fuse, compassion is foregrounded.

Wagner's heroes are not the tragic heroes of the past, those whose destinies are plotted by fatal flaws (*hamartomas*), but with tragic lives and repetition as a tragic trap (The Dutchman, Tannhäuser, Lohengrin). Klingsor's solution is castration. But encounters with the sexual other (Tristan, Parsifal, Siegfried) lead to self-knowledge, and for Parsifal compassion for the other knights.

The turning point in *The Ring* is the moment in *Die Walküre* (Act 2) when Brünnhilde understands Siegmund's sacrifice through love for Sieglinde and the child within her. Before that everything was following Wotan's way (nearly), but the dawning of compassion in Brünnhilde leads her and her father to sacrifice so much. Hans Sachs, one of Wagner's greatest creations, is full of melancholic sadness and compassion, yet the opera opens to us how art and creativity flourish through these emotions.

4. Love and Wagner

Barrie Emslie has written on the centrality of Love in Wagner's life and music.[5] When *Tristan und Isolde* was composed, Wagner was in Zürich at the Asyl, a small house in the grounds of the Wesendonck villa provided for him by Otto von Wesendonck. Wagner was preoccupied by the works of Schopenhauer. He wrote to Franz List in 1854

> … of the philosopher's principle idea … the final denial of the will to live, is of terrible seriousness, but it is uniquely redeeming … it is the sincere and heartfelt yearning for death: total unconsciousness, complete annihilation, the end of all dreams,—the only ultimate redemption!

He continued:

> …since I have never in my life enjoyed the true happiness of love, I intend to erect a further monument to this most beautiful of dreams, a monument in which this love will be properly sated from beginning to end … I shall then cover myself over, in order—to die.[6]

There has been much speculation about Wagner's relationship with Mathilde, wife of Otto, the details of which are beyond the scope of this book. To Mathilde von Wessendonck, 7 April 1858, using the *Du* form, Wagner wrote:

> In the morning … I was able to pray to my angel from the very depths of my heart; and this prayer is love, which is the well-spring of my redemption …

And on 3 September 1858 (now in Venice):

> To work then! Tristan the hero, Isolde the heroine! help me! Help me my angel! … From here the world shall learn of the sublime and noble distress of supreme love, the lamentations of the most sorrowful bliss …

Wagner referred to Mathilde as his Muse, his only love. He dreamed about her waking him with a kiss, they wrote romantically to each other, and while her role as an inspiration for the opera has often been minimized, it is hard to believe that she did not believe she was Isolde.[7] Swann contends that the music of *Tristan und Isolde* is some of the most erotic ever composed.[8]

Love is seen in Wagner's oeuvre as "prima facie—in the context of two separate and arguably opposed categories: the spiritual and the sensuous ... although love is not restricted to the privileged unit of the heterosexual pair, the importance of that unit will be repeatedly underlined …."[9] Emslie largely ignores the first three operas, with a major concern on the 'redemption-based' dramas. He devotes an early chapter to *Music and the Eternal Feminine*, a theme we have already introduced. Schopenhauer's ideas took Wagner further in directions he had already expressed in embryo (we noted, for example, in the operas composed before *Tristan und Isolde*), not only with regards to music but also conceptually. Emslie recognizes Wagner's problems with reconciling his views with those of Schopenhauer, and discusses the importance of Hegel's aesthetic and his dialectic.

Wagner was impressed by Hegel's insistence on history as a basis for philosophy (his first acquaintance with Hegel's works being *Vorlesungen über die Philosophie der Weltgeschichte* (*Lectures on the Philosophy of World History*, 1837). Unlike Nietzsche, Hegel would not dismiss God from his world view. As discussed earlier (Chapter 3), in Hegel's dialectic, tensions between opposites (thesis and antithesis) produce something greater than the two (synthesis) which becomes the new thesis. This is referred to as *Aufhebung*, from *aufheben*: to lift, counterpoise, to save up, to preserve, to transcend, related to *becoming*.[10] Much was made out of this idea by the post-Hegelians, especially Marx and Feuerbach, but it was the latter who united religion with philosophical idealism: goodness, the idea of God and love emanating from the human mind.[11] Hegel draws us to the contrasts between the universal and the particular, reason and desire, spirit and flesh, and freedom and necessity. These he writes are not invented, but

> … have from all time and in manifold forms preoccupied and disquieted the human consciousness … [but] … truth only lies in the conciliation and mediation of the two … resolved antithesis … art has the vocation of revealing *the truth* in the form of sensuous artistic shape, of representing the reconciled antithesis … in its representation and revelation.[12]

Wagner elevated the Eternal Feminine as an *Ursprung* for creativity, art and love: that which draws us onwards ("Zieht uns hinan"), which some would translate as 'towards heaven'.[13]

5. The Music

In *Rienzi* Wagner moved from the mellifluous harmonies of the French/Italian schools to Grand Opera (Spontini, Auber, Halévy and Meyerbeer). The latter was for him a revolt against the German art of his time, especially the aesthetic of its music. Hans von Bülow, who regarded seeing *Les Huguenots* as "One of the most wonderful moments of my life", ironically described *Rienzi* as "Meyerbeer's best opera". Wagner considered that he, at last, had become an artist, and conceived "the Spirit of Music as aught but *Love*".[14] Rienzi is attracted to his sister, but this GSA is not developed as the sexual impulse found in *Die Walküre*. Their mutual death in the conflagration of the flaming Capitol echoes through to *Götterdämmerung*, just as the Papal malediction in Act 4 of *Tannhäuser* reminds us of other curses in Wagner's operas.

These imprecations include those of the Dutchman who cursed Christ while he was carrying the Cross, and Kundry's similar trespass. Tristan's curse of the "terrible drink" ("Verflucht sei, furchtbarer Trank!", Act 3 scene 1) flows throughout the work, as does Alberich's curse, which weaves together the whole of the *Ring* to Brünnhilde's "Welches Unholds List liegt hier verhohlen?" ('What demon art lies hidden here') (as the fateful music bursts out on the English horn and strings, *Götterdämmerung,* Act 2, scene 5). In her reduced humanity Brünnhilde is unable to discern the underlying deception at work, and is surprisingly obtuse about the destructive power of the Ring. Her Vow and Vengeance Trio with Gunther and Hagen is also a terrible malediction of murder, and extension of the inescapable malfeasance of Alberich's primordial curse.[15]

It is interesting to examine the type and style of love music in Wagner's oeuvre. In *Rienzi* the love is between Adriano (a *travesti* role) and Irene. They share a duet, but it must be said that Irene's interest always seems more focused on her brother's welfare, his political cause, his wellbeing.

In *Der fliegende Holländer,* the Dutchman and Senta are entirely absorbed in each other, but their motives are lofty, concerned with sacrificial commitment and 'redemption' beyond any physical attraction (which is *pour le monde*). This latter earthly interest is embodied in the wretched huntsman Erik and his passion for Senta, who regards him with sympathy but is entirely immune to the urgency of his love. His mellifluously charming declarations are couched in the language of the traditional love music of early German Romanticism, even of Italian *bel canto,* and are quite different from the soul-searching fixedness of the Dutchman and Senta.

In *Tannhäuser,* the world of erotic love is presented in the Venusberg: voluptuous scenes unfold, the sirens call, but Venus needs to ask Tannhäuser for some sign of his involvement. He dutifully sings the passionate melody of his Hymn of Praise, but his mind is elsewhere, in the free open spaces of the mortal world, and in a freedom from (erotic?) enslavement. Each stanza becomes more detached, and the duet, far from celebrating sensuality, become a series of recriminations and pleadings. Eventually the freedom Tannhäuser is seeking will be found only in the embodiment of virginal love and Immaculate Conception, the Virgin Mary herself.

Enunciation of her name, and that of Elisabeth, are key moments of 'salvation' in this story. The Virgin's earthly surrogate is the saintly

Elisabeth, and her love duet with Tannhäuser in Act 2 is deeply touching in the power of reminiscence and the joy of reunion. But there is no question of any erotic attraction or fulfilment. The conventional tripartite division underscores the Italianate melody and hence the more 'conventional' approach. Elisabeth's love is full of passionate anticipation (in her Greeting to the Hall of Song), full of sorrow when Tannhäuser reveals his erotic passion in his impatience with the pallid love songs of the *Minnesänger.* But it is in Elisabeth's pleading for the erring knight's life with the outraged company (Act 2), and especially in her extended prayer to the Virgin (in Act 3), that her true heavenly alignment is revealed. Her vision is one of celestial beatitude, not the cloying sensuality of the Venusberg.

In *Lohengrin* the hero develops the register of motive and imagery expounded in Elisabeth. He is sent as a saviour figure, and is distinctly aligned with the Holy Spirit. He offers to love Elsa in return for her acceptance of his mysterious origins and name. He is celibate, of an order in heavenly service, but is prepared to love Elsa with the all the commitment of his devotion—the moral state we call chastity. The famous Wedding Chorus and the long duet that follows embody his vocational understanding of chastity and his self-offering. But Elsa is by now actually not interested in any love he proffers, erotic or chaste. Her underlying fears expose her psychological weakness, espied and played on by Ortrud, and result in the mounting loss of control that becomes hysterical as she demands to know the forbidden truth. Romantic love, let alone erotic concourse, are swept aside.

In *The Ring* the issues are of power in all its manifestations, and on the proper order of the world. Search for power or control leads Wotan in primordial times to commit the original sin of cutting a branch from the World Ash Tree. All nature and motivation are involved in this, and it will flow through the ages to damage and progressively ruin the divine order that controls all creation.

Love is needed, but is constantly subverted and subjected to power and control. The passionate love music of Siegmund and Sieglinde is indeed erotic and torrential, but is almost a piece of social engineering by Wotan, anxious to find the truly detached and fearless hero who can restore right order. So the glorious love music is fated, an incestuous violation of nature (GSA), and, more significantly, in the disregard for the state of matrimony sanctioned universally by contract of commitment and loyalty. Fricka points this out with complete righteousness and irresistible logic: in condoning it Wotan has further seriously contradicted the laws of contract and covenant, which he has already done in stealing the stolen gold of the Ring from Alberich.

The real love duet is the Farewell Scene between father and daughter, full of profound sadness in loss, and yet prescient of a possible new world order. The huge extended duet at the end of *Siegfried* sees the hero and heroine brought together and living in compression, the whole gamut of their personal development and emotions associated with the various stages of love. He must overcome his lack of experience; she must let go of her old self. They eventually find the pitch of personal expression in the closing rapture of the duet. But the eroticism is not explored in the music: this is all about their heroic roles in the vocation both are called to. Their mystical "laughing death" is embraced uncomprehendingly and presciently, a fact underscored in the Prologue of *Götterdämmerung,* where the heroine sends the hero off to complete his personal development in great feats of quest and adventure. The music is exultant but not erotic.

The sexual love considered is rather in the conventional tropes of courtly love by the limited Gunther and Gutrune. She in her simpering music (like Erik in *Der fliegende Hollander,* and the *Minnesänger* in *Tannhäuser*) does not have the vaguest inkling of heroic, let alone salvific calling. The celebration of her 'marriage' to Siegfried is realized in similarly conventional style, and is a massive act of deception induced by Hagen's potion. This potion is a central theme of betrayal, the inverted Grail of duplicity leading only to confusion, befuddlement and the further unravelling of the strands of proper order. Hagen becomes Judas at the Last Supper.

Only *Tristan und Isolde* explores the nature and ecstasy of erotic love. But from the very beginning, with the Sailor's folk-song of the Irish Maid, the love is fated. Its force is irresistible and hypnotically beautiful, but at the cost of peace of mind and the probity of social, political and familial order. The eroticism destroys Marke, Morold, Tristan and Isolde, and must find its resolution in death. Is this a release? Is Isolde's final rapture a promise of salvation in another realm, or an intense bittersweet acceptance of the inevitability of loss and futility?

The interactive music for Tristan and Isolde reveals the nature of their involvement, both externally and internally. There is loud, confrontational exuberance in their early Act 1 encounter. Tristan remains aloof, prompted by subconscious desire that he cannot admit to; Isolde is embittered in her processes of recall, as she recounts her early meeting and healing ministrations to the wounded Tristan. Her mind is on medications and potions, pulled between life and death, and in the midst the ambiguity of her feelings is her outrage at Tristan's apparent disregard; his murmured restrain against her impassioned railing. The tumult immediately subsides with the imbibing of the love potions, a hair's breadth away from the intended mortal poison.

The ingestion works an instant magic of truthfulness, with psychological convolutions dissolved in the profound epiphany of love. This sees the music shift into another register, already heard in the prelude: the opening motifs are identified as they utter each others' names in a hushed ecstasy. This continues to grow as they become lost in each other, entirely oblivious to the outside world with its demands and expectations. The tumult of arrival now replaces their own formerly externalized angry repression. They are now dangerously indifferent to their circumstance, and so enter into the situation that leads inevitably to their deaths: love and death are inextricable in the expression of passion. This self-absorption is explored at length in the tryst of Act 2. The magic of night and its rustling mysteries set the scene for their secret encounter, driven by impatience and passion.

This bursts out into a sustained torrent of greeting, a loud, almost frenzied expression of the importunity of physical passion, completely self-centered in the vaunting of selfish feeling, almost arrogant in its tumult. The inner more mystical nature of their complete involvement is unfolded and explored in the extended main body of the duet, the hypnotic celebration of total self-absorption, where physical union becomes a type of mutual spiritual osmosis. The key characteristic is utter indifference to the world around them, where even Brangäne's voice of warning, of social sense and precaution, seems to become part of Isolde's very self in the tonal dissolving of the vocal lines into the mesh of orchestral harmonies. The unison singing soon reaches its apogee in the emergence of the theme of Love-Death, the full implications of which are understood only in Isolde's final soliloquy.[16]

In *Parsifal* there is no love duet, only the birth of compassion in the context of growth and development, themes all too familiar in *Der fliegende Holländer*, *Tannhäuser*, *Lohengrin* and *The Ring*. Parsifal is called to disregard the erotic blandishments of Kundry which have ruined Amfortas and threatened the whole social order in succumbing to unguarded and irresponsible passion (as in all the synonymous prototypes she is likened to). Klingsor has tried to obviate the ongoing cycle of eroticism in his self-castration, but this is no answer to true vocation which is found not in continence, nor in celibacy, but in the profoundest apprehension of chastity as a metaphor for purified intention and a gateway to salvation. This is embodied in the Grail, the symbol of the Eucharist, which in turn is a metonym of the totality of the divine plan of reconciliation, perfection and fulfilment that will mean a New Heaven and New Earth.

6. Love and Death: The Cross and Transcendence

Love and Death are bound together inseparably by the universal symbol of the Cross of Calvary, whereby supreme love is freely given in sacrificial pain and death for the life of the world; "Did e'er such love and sorrow meet, or thorns compose so rich a Crown?"[17] This continues to be rehearsed in the concept of martyrdom, most strikingly in the death of St Sebastian, pierced by arrows of hatred and death (that has for many become the disturbing locus of some sort of erotic icon). This combination of passionate love and ecstatic pain is also famously combined in St Teresa of Avila's mystical union with Christ, with a fiery phallic dart the vector for sublime transfiguration.

The image of St Teresa by Bernini is one of the most famous sculptures in the world, depicting St Teresa's out-of-body experience. No-one had ever before portrayed a saint this way. Beneath the angel in human form, with a golden spear ready to plunge it again into her heart, she is levitating, with half closed eyes, already in a state of ecstasy, with her bodice flowing down in liquid sensuality. The physical intensity of her emotional transcendence awaits the return of the arrow aiming, as Bernini positioned it, not at her heart but lower down. (See Chapter 1, **Fig. 1.20**)

Some of the operas we are considering have purely transcendental epiphanies. In *Die Feen,* Arindal brings Ada to life from the underworld by the power of music, and lives with her thereafter in fairyland. His lovely life there is akin to that of Tannhäuser in the Venusburg. But Arindal's delusions turn out to be manipulated by the fairies. In this opera, the female principal is turned to stone. To be whole, Arindal has to bring her back to life, and together they transcend to immortality. Ada is saved in a redemption scene which might even qualify as a *Liebestod*: it is at least a transfiguration. This is akin to Gluck's *Alceste* in the saluting of the conjugal fidelity, and like Lortzing's *Undine* in the attainment of some kind of supernatural beatitude.

We have noted how Jupiter's thunderbolts burn and consume Semele in Handel's opera (1743).

Ah I feel my life consuming,
I burn, I burn, I faint, for pity I implore,
Oh help, oh help, I can no more!

In Messiaen's *St François*, at the sanctuary of La Verna, the heavy pounding sounds in the percussion and strings suggest an entry of grace and foreshadow Francis' acceptance of the *stigmata* by the Angel. In Act 3 scene 7 he says his farewell and sings the last verse of his *Canticle of the Sun.* His last words are:

Dazzle me always by the overflowing of your Truth....

Light appears where his body had been, increasing to a blinding intensity.

[**Figs. 21.1-5**]

Notes

[1] Richard Wagner, quoted in N. Parly, *Vocal Victories*: *Wagner's Female Characters from Senta to Kundry* (Museum Tusculanum Press) (Chicago: University of Chicago Press, 2011), 167.
[2] Swann, J. *A four-note cell as a basic element in Wagner's 'Ring'.* The Juilliard School (April 1980), 188 and personal communication.
[3] See Williams, S. *Wagner and the Romantic Hero* (Cambridge: Cambridge University Press, 2004) for some of these ideas.
[4] Goethe, J. W. *Elective Affinities.* Trans. R. J. Hollingdale (London: Penguin Classics, 1971), Chapter 4, 46-57.
[5] Emslie, B. *Richard Wagner and the Centrality of Love* (Woodbridge: The Boydell Press, 2010).
[6] Wagner, R. Letter to Franz Liszt (16 December 1854).
[7] Wagner, R. Letters to Mathilde Wesendonck (7 April and 3 September 1858). The story of the relationship between Wagner and Mathilde is well outlined in Cabaud, J. *Mathilde Wesendonck Isolde's Dream* (Milwaukee: Amadeus Press, 2017).
[8] The authors are grateful to Jeffrey Swann for his wonderful insights into the music of Wagner.
[9] Emslie, B. 2010. Ibid., 3-4.
[10] *Aufheben*, a somewhat difficult word to tie down, *Aufhebung* is the noun.
[11] There is no evidence that Wagner ever read Marx. However Hegel's *Philosophe der Geschichte* was in his Dresden library.
[12] Hegel, G.W. F. *Introductory Lectures on Aesthetics* (London: Penguin Classics, 1993), 60-61.

[13] Johann Wolfgang Goethe, *Faust, Der Tragödie, Zweiter Teil, Berschluchten, Chorus Mysticus*, final lines, "Das Ewigweibliche/Zieht uns hinan."
[14] Wagner, R. *A Communication to my Friends* (1851). Trans. William Ashton Ellis (Wokingham: Dodo Press), 31.
[15] Other dramatic curses in opera include those of the Commendatore in *Don Giovanni*, the Countess on Herman in *The Queen of Spades*, the Gypsy on Di Luna in *Il Trovatore*, Gertude in *Bánk Bán*, The Water Goblin on the Prince (*Rusalka*), Rigoletto, Othello, Elektra, Simon Boccanegra, Valentine against Marguerite (in *Faust*) and Billy Budd.
[16] For a different approach to the subject, see Charles Arnould Tournemire (1870 – 1939), a French composer and organist, a pupil of Messiaen. He set two of the themes pertinent to this study: Tristan and Isolde and St Francis as *La Légende de Tristan* and *Il poverello di Assisi* (1939). He was obsessed with Tristan, naming one of his houses Iseut, and the other Tristan. His second wife was nicknamed Iseut. His *Tristan* was not based on the tale Wagner used (Gottfried of Strasbourg), but on that of Chrétien de Troyes, although with significant variations. Tristan marries Iseut, but in this opera she is a mere cipher. He concentrates on Tristan. Brangien sleeps with King Marc so that Iseut can remain a virgin.
[17] The hymn by Isaac Watts (1674-1748):

When I survey the wondrous cross
On which the Prince of Glory died
My richest gain I count but loss
And pour contempt on all my pride

See from His head, His hands, His feet
So much sorrow and love flow mingled down
Did e'er such love and sorrow meet
Or thorns compose so rich a crown.

See Watts, I. *Hymns and Spiritual Songs: The Three Books of Great Christian Bible Hymns— Complete* (CreateSpace Independent Publishing Platform 2018).

CHAPTER 22

REDEMPTIVE LOVE: WAGNER, AND THE ISSUES OF GENDER AND SEXUALITY

Such sweet compulsion doth in music lie.

(John Milton, *Arcades*)[1]

Gender and Sexuality

In the last chapter we discussed various aspects of the way Love is contextualised in opera, one emergent theme being the contrast of restrained sexuality versus eroticism. This is prominent throughout much of Wagner's output, from *Das Liebesverbot* onwards, and most notably in *Tannhäuser*, *Die Walküre* and *Tristan und Isolde*. In each opera, the self-abandonment to love brings the lovers into mortal combat with the surrounding social order. In *Das Liebesverbot*, because it is a comedy, the outcome is a happy one: unrestrained sexuality. Isabella's chaste soul is pitted against the lustful Friedrich (who will pardon her brother only if she succumbs to his erotic attentions).

Six Women

Nina Parly, in her encomium *Vocal Victories,* refers back to Clément's book, the latter espousing the misogyny she considers inherent in the treatment of women in opera: "at the end … she has usually been raped, murdered, or has committed suicide". Calling on Carolyn Abbate's rebuttal of Clément's perspective in *Unsung Voices*, Parly draws attention to "the locus of women's operatic triumph: the female operatic voice's overwhelming sound and the musical gestures which enclose this voice in the totality".[2] She praises the Wagnerian soprano, and devotes her text to seven Wagnerian heroines: Senta, Elisabeth, Elsa, Isolde, Brünnhilde, Eva and Kundry. We have discussed the operas related to these dramatis personae, some in more detail than others, and now explore further some aspects of the characterisation of those that have taken up most of our attention. We move from Love and transcendence to gender and sexuality.

Senta

Wagner, via *Rienzi,* moved his musical style and thematic material away from the influence of the French and Italian composers towards something entirely new in *Der fliegende Holländer*. His ideological concerns with the evocation of nature and the sea, influenced by Undine and her sisters, are by this time inextricably bound into so much of the music. The sea is both the all-pervasive natural presence and the locus of dark enchantment, at once beneficent and peaceful (the *Südwind* of the Steersman's Song), and also the tempestuous frenzy of the supernatural storm summoned up the unseen crew of the Dutchman's ship in Act 3. The Dutchman, longing for calm amidst the storms of life, is confounded by the limits of nature, unable to sail around the Cape of Good Hope, and in part his sin was hubristic. The Dutchman is seeking a release from an eternal striving; he is lost among the living and the dead.[3]The concept of premonitions and precognitions is much developed in this opera. Senta dreams of the Pale Stranger, and his image is embedded on the portrait which so entrances her. Erik in his dream also has the image of her and the Dutchman out at sea. Senta has trance-like states, and when she emerges from one of them she knows she must perish with him. The Dutchman is an embodiment of her dream, there is a fatalistic attraction, locked through the gaze. In their duet, Senta and the Dutchman sing of a common death: for Senta this affords a rescue: "I am the one who will redeem you with her loyalty". Caught between three men, she only has one choice, putting an end to all dreams; resolution via transcendence.

The duet is central to the opera, just as the duet for Valentine and Marcel in Act 3 of *Les Huguenots* is at the heart of the dramaturgical and spiritual truths explored in that work: prejudice and partisanship give way to a heart opened in new understanding and transformed by love. One mode of existence and understanding is broken down and another elevated vision of life and its purpose entered into. It is not surprising that the soprano-bass cadenza, expressing the outpouring of the deepest emotion, with its sublime postlude, is replicated in transformation by Wagner for Senta and the Dutchman.

The harmony and orchestration of the ending of the overture and of the opera itself emphasise the transcendence reflected in *Tristan und Isolde,* and *Götterdämmerung.*

Some critics construe the Dutchman as a sexual predator, caught within the economic forces of society and thus turning Senta into a possession, to be sold by her greedy father. But Vanderdecken would seem to be truly in love, even if bidding for salvation, battling with the sea for eternity. Wagner considered him to be a mythical folk-figure with a "primal trait of human nature speaking out with heart-enthralling voice. This, in its most universal meaning, is the longing after rest from the storms of life".[4] The ideas that Wagner developed in *Die Feen* and *Das Liebesverbot* were adopted here for Senta.

Der fliegende Holländer was written and composed long before Wagner had read Schopenhauer, but in the libretto one finds the concept of the unending misery of the striving Will, as the Dutchman wants eternal annihilation (*Ew'ge Vernichtung*). Conceptually there were changes made in a revision of the opera in 1860, just after completing *Tristan und Isolde,* with ever deeper reflection on Schopenhauer.

But Senta is no Heliane, nor Odysseus' Penelope. A dreamer perhaps, but Wagner refers to her as "a *Woman* who, of very love, shall sacrifice herself for him, but is the quintessence of womankind; and yet still unmanifested, the longed-for, the dreamt-of, the infinitely womanly Woman—*the Woman of the Future*".[5] Rejecting domesticity, and like one of several of Ibsen's characters (Hedda Gabler or Nora in *A Doll's House*, 1879), we find Senta's attributes developing all the way towards Brünnhilde. For Parly she "is the only one able to unite the two worlds", Love has become a redemptive force...[6]

Elisabeth

In *Tannhäuser* we are involved in at least three different worlds: that of the *Minnesänger* (perfect and unattainable); that of the tormented Orpheus (Tannhäuser as the singer); and that of St Elisabeth of Thuringa (1207-1231), famous for her compassion and self-offering. There are the visits to an 'underworld', back to Orpheus, but symbolically in *Tannhäuser* (under the Venusberg). This should be compared with the dungeon (*Fidelio*); the haunted forest (*Der Freischütz*); the graveyard in a ruined monastery (*Robert le Diable*). When asked by the Landgrave where he has been for such a long time, he replies "I have journeyed in far-distant realms—there where I never found repose nor rest. Do not ask!" (Act 1, scene 2). At the mention of Mary (the new Eve, the First Disciple, the Mother of us all, the Eternal Feminine), the Venusburg vanishes and Tannhäuser is in another world. Later he is jolted out of his state of mind by the name 'Elisabeth' uttered by Wolfram. We meet such significance in the pronouncement of names for dramatic effect and thematic implication in *Lohengrin* and *Parsifal*, unravelling a whole stream of connotations. The holy world of Elisabeth is epitomised in her extended prayer to the Virgin, in Wolfram's paean to the Evening Star, and eventually in the ecstatic chorus of the returning pilgrims bearing the news of the blossoming of the Pope's staff, the final symbol of redemption reached in this opera.

Elsa

The worlds depicted in *Tannhäuser* and in *Lohengrin* seem to differ substantially. In *Lohengrin* Elsa explains how, lonely in troubled days, she prayed to the Lord her heartfelt grief: her groans poured out in prayer. But the latter became a 'plaintive' sound that grew into a roar as it echoed through the skies. She listened as it receded into the distance, closed her eyes and fell into a deep sleep. Her prayers and lamentations were heard across a space by one from another world.

The contrasts between the Christian and Pagan, and light and dark (the polarities of Elsa and Ortrud; A flat major versus the relative F minor), are found in *Lohengrin* and in *Tannhäuser*. But in the latter these conflicts are within the protagonist himself. In the music the safe diatonic tonal world and closed, recognized musical forms (aria, love duet, oration, march, strophic song, extended ensemble, formal chorus) are used to illustrate the safe, ordinary and orthodox world of Medieval Christianity. This is contrasted with the fluid forms, shifting chromaticism, glittering orchestration and passionate oratory of the Venusberg and the colloquy between Venus and Tannhäuser. The impassioned fluidity marks all Tannhäuser's references to Venus, and eventually characterises the extended 'formlessness' of his Rome Narration and final exchange with Wolfram.

It was with *Lohengrin* that Wagner further developed his wish for opera to return to legends, and here he provided a complete portrait of the Middle Ages. Wagner tells of the origins of some of his thought thus:

> One primal, manifold-repeated trait runs through the Sagas of those peoples who dwelt beside the sea or sea-embouching rivers: upon the blue mirror of the waters there draws nigh an Unknown-being, of utmost grace

> and purest virtue, who moves and wins all hearts by charm resistless; he is the embodied wish of the yearner who dreams of happiness in that far-off land he cannot sense. This Unknown-being vanishes across the ocean's waves, so soon as ever questioned on his nature. Thus—so goes the story—there once came in a swan-drawn skiff, over the sea to the banks of the Scheldt, an unknown hero: there he rescued downtrod innocence, and wedded a sweet maiden; but since she asked him who he was and whence he came, he needs must seek the sea once more and leave his All behind.—Why this Saga, when I learnt it in its simplest outlines, so irresistibly attracted me that, at the very time when I had but just completed *Tannhäuser*, I could concern myself with naught but it, was to be made clearer to my feeling by the immediately succeeding incidents of my life.[7]

He was also exploring the meaning of love.

> Lohengrin sought the woman who should *trust* in him; who should not ask how he was hight [*sic*] or whence he came, but love him as he was, and because he was whate'er she deemed him. He sought the woman who would not call for explanations or defence, but who should *love* him with an unconditioned love. Therefore must he cloak his higher nature, for only in the non-revealing of this higher [*höheren*]—or more correctly, heightened [*erhöhten*]—essence, could there lie the surety that he was not adored because of it alone, or humbly worshipped as a Being past all understanding—whereas his longing was *not* for worship nor for adoration, but for the only thing sufficient to redeem him from his loneliness, to still his deep desire,—for *Love*, for *being loved*, for *being understood through Love*. With the highest powers of his senses, with his fullest fill of consciousness, he would fain become and be none other than a warmly-feeling, warmth-inspiring Man; in a word, a *Man* and not a God—i.e. no 'absolute' artist. Thus yearned he for Woman,—for the human Heart. And thus did he step down from out his loneliness of sterile bliss, when he heard this woman's cry for succour, this heart-cry from humanity below.[8]

Wagner saw in Elsa the 'desired antithesis' to Lohengrin and he introduced us to the Swan, a poetic motif, with much symbolic power. From the mythology of *Leda and the Swan* to Strindberg's play *Svanehvit* (*The Swan Maiden*, 1902), then Jung's use of the swan as a symbol of rebirth, we find the bird also in *Parsifal.* Yeats tells us the swan, sacred to Venus, is a symbol of purity and love (as previously discussed in our discourse on Undine).

Lohengrin's farewell to the Swan ("Leb' Wohl meine Liebe Swann"), echoes through to Wotan's Farewell to Brünnhilde in *Die Walküre.*

At the end of *Lohengrin,* Godfrey will be protected by the ring, a horn and a sword bequeathed to him. Here we straddle events between the death of Elsa and the objects that are so pertinent in the life of Siegfried. *Lohengrin* brings us the mystery of the Grail, and the sublime music of the prelude as a musical narrative of the Grail's descent to earth and then its return. The key of A flat major is open and most suited to the extensive string writing in as many as eight sections that imparts to the 'vision' a sense of luminosity and mystical intensity. This music is in fact meant to represent the creative, transforming breath of the Holy Spirit.

It is generally considered that *Lohengrin* in one of the most pessimistic of Wagner's operas. The love that the Knight offers Elsa is chaste and very special, spiritual rather than based on sexual needs. But she is not able to accept this, and transgresses, breaking her covenant with her saviour. Her love demands to know his name. In *Tannhäuser* it is the faithful Wolfram who sings of spiritual love, but in both this opera and *Der fliegende Holländer,* the protagonists are redeemed by female sacrifice. In *Lohengrin* the situation is much more ambivalent.[9]

Wagner wrote as follows:

> The character and situation of this *Lohengrin* I now recognise, with clearest sureness, as the *type of the only absolute tragedy*, in fine, of the *tragic element of modern life*; and that of just as great significance for the *Present*, as was the 'Antigone'—though in another relation—for the life of the Hellenic State. From out of the sternest tragic moment of the Present one path alone can lead: the full reunion of sense and soul, the only genuinely *gladsome* element of the Future's Life and Art, each in its utmost consummation.[10]

With *Lohengrin*, Wagner had pursued his ideas about producing a new style of opera, a drama with historical themes from the sagas and myths, but with characters revealing human motives and emotions: he was also developing his ideas of reminiscence motives.

Kundry

Parsifal is Lohengrin's father. In *Parsifal*, Amfortas was originally the main protagonist (an intensification of Tristan), with a wound and wishing to die, but with *Parsifal* the religious intent is taken to its highest point. Continuing with Schopenhauer's ideas of life as continual suffering, it is the mysterious and complex Kundry who dies at the end, a *femme fatale,* Eve, Mary Magdalene and all the ciphers of the *Ewig-Weibliche*, bound together. She comes from afar, fades into sleep under bushes (Act 1), reappears in sleep then awakens and utters a cry, disappears, reappears (Act 2), emerges lifeless from bushes and is found by Gurnemanz next to flowing waters, and at the end sinks lifeless to the ground (Act 3).Wagner struggled with the characterization of the "world-demonic Kundry", a dual-figured messenger of the Grail, guilty of Amfortas's wound, compared by Wagner to the Wandering Jew, under a curse for laughing at the condemned Christ. She has trances, and calls out for "sleep, deep sleep", and death.[11] The kiss of memory (Amfortas) is the wound which burns in Parsifal's heart ("Die Wunde! Die Wunde! Sie brennt in meinem Herzen"). The next time they meet is Good Friday, the Grail Temple has been depleted of divine sustenance (cf. the deprivation of Freia's Golden Apples in *Das Rheingold*). Kundry dries Parsifal's feet with her glowing Mary Magdalene hair; he closes the wound of Amfortas; and a white dove descends. Death releases Kundry from the neverending cycles of reincarnations (as with Dutchman), and she dies with her eyes fixed on Parsifal. Cosima Wagner noted (1875):

> R. saw God the father, masculine; the Holy Ghost, feminine; the Redeemer as the world stemming from them; will, idea, and world, the world emerging from the division of the sexes.

In Parsifal, the sexes are united by the male spear and female cup, as sometimes made explicit in stage productions.[12]

Kundry's character is an amalgam from many different mythological and religious congeries, Kunde, 'customer' but she also must serve ("dienen, dienen") and for Klingsor she is the *Urteufelin* (Prehistoric female devil). Eve, Mary Magdalene and the Virgin Mary: Parly christens her "a musical saint".

The mystery of the Eucharist infuses the whole opera, which reaches truly transcendental numinosity, filled with Christian symbolism. Wagner's intention, as written about in *Religion and Art*, was to save the essence of religion through art, revealing hidden truths such as compassion and Love as compensation for human suffering.[13]

The Eternal Feminine

Brünnhilde and Isolde

The two other paragons of Goethe's *Ewig-Weiblicheit*, the essence of the feminine, who have been referred to several times in our book, are Isolde and Brünnhilde.

In the *The Ring*, where the transcendent topic is most fully expounded, all eventually devolves on Siegfried and Brünnhilde. Siegfried in his innocence is unwittingly betrayed and Brünnhilde, having lost her quasi-divine status, is likewise victim to the maleficence of the Curse. Hence her wisdom and knowledge of runes do not help her to understand that Siegfried has been envenomed by the potion. Like Siegfried, she too suffers betrayal. Siegfried is then murdered, a sacrificial death analogous to the Crucifixion. He is now a saviour figure assumed by Brünnhilde in the music of her final self-immolation. She is then a participant in the redeeming process. Her riding into the fire means that (like the Assumption of the Blessed Virgin), she completes/complements the Divine Godhead in releasing salving love, removing the curse of the original sin initiated by Wotan and compounded by Alberich, enabling nature to be restored and renewed.

The development of the enveloping leitmotivic continuum represents the complex of motives and sinfulness of the characters, and their eventual recapitulation in the purified and transcendent simplicity of Brünnhilde and Siegfried in the orchestral postlude. This is also true of the musical situation at the end of *Tristan und Isolde* with Isolde's rapt dying.[14]

With Isolde and Brünnhilde Wagner's women have sung their final song. Physically, spiritually and philosophically they die, perhaps like St Teresa, in ecstatic bliss.

In Parly's estimation, Wagner's female characters:

> … are, from the outset, active subjects whose singing and music has an immense influence on the musical structure and development of the work. … far more than simple metaphors for the music; the majority of them are also artistic creators of the music … Wagner's heroines develop from being redeemers in the early operas to a position of exacting redemption through the man in later operas.[15]

In the next chapter we return to consider Redemption and associations with transcendence.

Notes

[1] Milton, John. *The poems of John Milton*. Ed. Helen Darbishire (London: Oxford University Press, 1958), 442.
[2] Parly, N, *Vocal Victories: Wagner's Female Characters from Senta to Kundry*, Ibid. 2011, 16, 18: Abbate, C. *Analysing Opera, Verdi and Wagner*. Abbate, C.; Parker, R. (eds) (Berkeley: University of California Press, 1989), Preface IX-X.
[3] In the story of the memories of Herr Schnabelewopski, the hero was the captain who swore by all the devils that he would sail around a cape (not identified) in spite of a most violent storm which was continually raging, even if he had to keep sailing until the day of judgement.
[4] Wagner, R. Ibid., 1851, 32.
[5] Wagner R. Ibid., 1851, 33.
[6] Parly, N. Ibid., 2011, 45/50. In Harry Kupfer's Bayreuth production in 1978, the image of a neurotic disturbed women à la Ibsen seems to have the intended interpretation.
[7] Wagner, R. Ibid., 1851, 56.
[8] Wagner, R. Ibid., 1851, 60 (Italics in original).
[9] Wagner was aware of criticism that the ending seemed too harsh on Elsa, and considered changing it, to one in which Lohengrin and Elsa stay together in one way or another.
[10] Wagner, R. Ibid., 1851, 61 (Italics in original)
[11] Letter, Wagner to Mathilde von Wesendonck 1858. In Garten, H.F. *Wagner the Dramatist*. (London: Overture Publishing, 1977), 137. "*Parzifal* has occupied me a lot: in particular a curious creation. A wonderfully world-demonic woman (the Grail's messenger) ... Have I told you that the fabulous wild Grail's messenger and the seductive woman of the second act are one and the same being."
[12] *Cosima Wagner Diaries*, Vol 1. Ibid., 6 Jan. 1875, 1:817.
[13] Diary or letter to Mathilde von Wesendonck, 1 Oct.1858. Garten, H.F. Ibid., 149.
[14] Ross, A. *Wagnerism*. Ibid., 2020, 69.
[15] Parly, Ibid. 2011, 381. Parly's book is a superb analysis of the musical embroidery accompanying the operas. She also includes for discussion other significant females, such as Ortrud and Venus.

CHAPTER 23

REDEMPTION AND THE NATURE OF LOVE

Non, mais l'âme
De paroles vacante et ce corps alourdi
Tard succombent au fier silence de midi:
Sans plus il faut dormir en l'oubli du blasphème,
Sur le sable altéré gisant et comme j'aime
Ouvrir ma bouche à l'astre efficace des vins!

No, but the soul
Emptied of words and this body grown heavy
Succumb late to the proud silence of the noon:
Enough, let us sleep, forgetful of the blasphemy,
Lying on the thirsty sand and as I love,
Open my mouth to the effective star of wine!

(Stéphane Mallarmé)[1]

Having explored in some detail the nature of Love and Death in operas, and the implications of the transcendental reflections of many of them clearly revealed through the librettos and the music, we now want to continue our explorations of *Unexpected Death* with Redemption, a theme we introduced previously in several chapters. [**Figs. 23.1-6**]

1. Myths and Names

At the top of List 1 is the opera *Undine* in which we encounter two *Unexpected Deaths*, namely Undine herself and Huldbrand. The concept of elemental power, starting with *Undine* and then continued in other *Rusalka* operas, emphasises the dangers which arise when the human and spirit worlds come into contact. Wagner was very influenced by the works of E.T.A. Hoffmann and all his operas involve at least two, sometimes more, levels of experience with the inevitable tangle and aporias that arise when different worlds collide. This is especially so when Platonic intentions intermix with mortal frames embedded in social entanglements, contingent or fatefully outlined. The symbolism of water as cleansing but consuming, the fatal nature of the kiss, contrasts between light and dark, day and night, good and evil, and magic moments, as in Weber's *Der Freischütz* (1821), where the young hunter gains access to enchanted bullets obtained from the Dark Huntsman during a lunar eclipse at midnight, emphasize this legacy.

The struggle to explore the division between our inner and outer worlds, the subjective and objective, has been a feature of philosophy from early Greek speculations. Since Kant's Copernican revolution, the quest has accelerated and is still a major concern to the present time. The French psychoanalyst Jacques Lacan (1901-1981) coined the term 'extimacy', an English translation of the French neologism (*extimité*). Instead of the poorly defined dualities, extimacy posits an essential identity between the exterior and the most interior of the psyche, the outer world and the inner world of the subject.

There are the trials, contesting the powers of different realms. The Wanderer and Mime spar with each other over three questions (3 x 2) about differing orders of existence (*Siegfried,* Act 2). Actions are forbidden, stemming back to the legend of Orpheus. Arindal is not allowed to ask Ada's origins in *Die Feen*. Names are forbidden. Lohengrin's identity must remain unknown to Elsa, as is that of the Stranger in *Das Wunder der Heliane*, a theme which later recurs in Hugo von Hofmannsthal's (1874-1929) *Die Frau ohne Schatten* (the libretto for Richard Strauss's opera, 1919). "You may not ask my name or where I came from". This refers to the revelation of the ineffable, unutterable Holy Name to Moses at the Burning Bush (Exodus 3:7-14): YHWH (*hayeh asher hayeh*, "I Am who I Am"). Like Moses and later Manoah, one is commissioned but may not ask the Divine Name. Later the Angel of the Lord says to Samson's father: "Why do you ask my name, since it is wonderful?" (Judges 13:18). The name *Je-sus* ("God saves") may be spoken because in the Incarnation we see the Face of the Unseen Father revealed.

Myth asks not if a story is true, but why is this story being told. Parsifal kills a swan, but does not consider suffering. Does man, does mankind, need *redemption* ("Erlösung dem Erlöser")? *Die Wunde* is also *das Wunder*: why do we still need the Grail, not just as Christian symbol (it was 'Christianized' around 1200 AD) but as an interfaith symbol, embedded in stories about pathways to enlightenment and salvation?[2]

Wagner believed use of myth and symbol was a function of art (also in music) to save the truths that underlay religious observation:

> One might say that when religion becomes artificial, it is reserved for art to save the spirit of religion by recognising the figurative value of the mythic symbols which the former would have us believe in their literal sense, and revealing the deep hidden truth through an ideal presentation.[3]

As Scruton put it "Myths do not speak of what was, but of what is eternally".[4] These ideas become central for the development of Wagner's operatic craft and are all found in one way or another from the first three operas through to *Parsifal*, with redemption and love overriding all other themes, building blocks for the development of the poems and music of his later works.

Scruton describes this in more detail:

> For Wagner, as for the Greeks, a myth was not a decorative fairy tale but the elaboration of a secret, a way to both hiding and revealing mysteries that can be understood only in religious terms, through ideas of sanctity, holiness and redemption.[5]

2. Redemption and Transcendence

Of 50 operas with *Unexpected Deaths* in List 1, 41 have realisation through Transcendence (28 of 31 females and 13 of 19 males), 28 have Redemption, and 27 have both Redemption and Transcendentalism. Further analysis reveals that there are 16 deaths that have *Liebestode* (9 female and 7 male). These are shown in List 2 (page 478): Of these, 15 (94 per cent) have transcendental implications, a highly significant association that cannot be ignored. Further, 4 of these (27 per cent) are found in the operas of Wagner. Likewise, three of four of his operas quoted convey Redemption, found in 12 of the total of 16. Thus, with regards to all composers we have looked at, Wagner tops the list for Redemption, but also for Transcendence and *Liebestode*. Transcendentalism and Redemption are closely entwined in those tales of *Unexpected Death.*

For Wagner, the agency of Redemption is via the unification of the masculine and the feminine principles. This reaches an apotheosis in *Tristan und Isolde*, but is interwoven in the other operas, obvious in the climax of *Parsifal,* for example, with the unified symbolism of the Spear, which pierced Christ's side, and the Chalice of the Last Supper. The body, which is the receptor from our senses, internal and external, moves beyond the material to the transformative self of the Hegelian synthesis, and the experience is relayed to us by Wagner above all in the music.

Emslie points out the difficulties Wagner had in fusing the two aspects of the Eternal Feminine into the plots and staging of, for example, Venus and Elisabeth, which he considerers was solved by the character of Kundry:

> Exactly this synthesis, produced when both female polarities are selectively suffused into each other and raised to a higher level, enables Wagner to wilfully colonise religion, turning it into the site of an immaculate sexuality … it is a largely unblemished example of the Hegelian notion of Aufhebung…[6]

Aufhebung relates to enhancement, not negation. Art for Hegel was an emblem of the rise of human self-consciousness, being a part of world history with the flow of 'absolute spirit' and the urge of the mind to realise the inner nature of things: "To call forth an echo of the mind from the depths of consciousness".[7]

Joachim Köhler, in his study of the Romantic tradition in *The Ring*, examined Hegel's influences on the plot of the cycle which "resembles the Hegelian theory of world history, which proceeds through antagonism of dialectical opposites … it ends only when all the contradictions, appearing as thesis and antithesis, are finally brought to a synthesis." Nietzsche referred to Wagner as 'Hegel's heir': in *The Ring* he heard Hegel's philosophy set to music.[8]

For Hegel, the inorganic as well as the living displayed spirit, as did history and art, but the latter transcends nature, revealing latent 'truths'. Pablo Picasso (1881-1973) famously observed: "Art is a lie that makes us realize truth, at least the truth that is given us to understand".[9]

There are several who would emphasize a Hegelian interpretation of the great love relationships in *The Ring*: Sieglinde and Siegmund; Wotan and Brünnhilde; and Siegfried and Brünnhilde, the latter unity being celebrated as 'laughing death'.[10]

Love and redemption elevate the meaning of life, but alone the one-dimensional man in the dramas cannot be fulfilled without the feminine, the eternally feminine. Recall that Wagner conceived that in opera words were the masculine, and music the feminine. *Natur* (nature) is a feminine noun in German: the Eternal Feminine had the upper hand in interpretation, Wagner wrote, in his letter to Rökel (24 January 1854):

> Love in its most perfect reality is only possible between the sexes; it is only as man and woman that human beings can truly love … that only in the union of man and woman does the true human being exist, and only through love, therefore, do man and woman become human … Siegfried alone is not then a complete human being, he is merely the half; it is only along with Brünnhilde that he becomes the redeemer. To the isolated being not all things are possible: there is need of more than one, and it is woman, suffering and willing to sacrifice herself, who becomes at last the real conscious redeemer; for what is love itself but the 'eternal feminine'?

Wagner emphasises the eternal nature of Love, when 'I' passes to 'thou':

> It is only by love that man and woman attain the full measure of humanity … it is only in the union of man and woman by love (sensuous and super-sensuous) that the human being exists … the transcendent act of his life is this consummation of his humanity through love.[11]

Emslie explores Wagner's representations of love, as we have discussed, much adrift from that traditionally displayed on the stage, transcending conventional heterosexual pairing: much more than lustful desire. Free from Schopenhauer's misogyny, and conceptually conciliant with Goethe's *Ewige-Weiblichkeit*, the dramas portray "a concept of heterosexual love that turns out to be the royal road to a complex of virtues": included are self- knowledge ('Know thyself'), physical bliss, suffering, redemption and finally renunciation of the world.[12]

Sexual union has not only the association with orgasm, itself affiliated with death (notably sudden, *Unexpected Death*), but it has a metaphysical symbolism, long used by poets, but taken by Wagner into the aesthetics of Romanticism at its most profound. Donne composed several poems on the theme of love and dying with the orgasm as a mode of transition/ transcendence.

> Call us what you will, we are made such by love;
> Call her one, me another fly,
> We're tapers too, and at our own cost die,
> And we in us find the eagle and the dove.
> The phœnix riddle hath more wit
> By us; we two being one, are it.
> So, to one neutral thing both sexes fit.
> We die and rise the same, and prove
> Mysterious by this love.[13]

For do not space and time become one during the orgasm?

Parsifal and Amfortas struggle with desire, desire which affects Klingsor so much that he castrates himself. In an early version of the libretto, the wound of Amfortas was in the genitals, but in the final version it is euphemistically moved to his side. He is striken with guilt: there is no way to the Grail for the unchaste. Kundry is also guilty but cleansed in baptism. The wound of Amfortas, symbolic of his bleeding soul, is redeemed through suffering (Christ-like)—yet the redeemer himself needs to be redeemed. Redemption is necessary for not only Parsifal but for all humanity. As in the older pagan myth of the Fisher King (the Wounded King), whose waste land is sick like the King himself, Parsifal redeems not just Amfortas, Kundry, and himself, but metonymically all of mankind, fostering redemption through love.[14]

In Chapter 2 we wrote "Contrasts and conflicts between dark and light, night and day, reason and inspiration, are thematic as Redemption through love becomes epiphanic". In List 1 there were 19 females and 9 males whom we considered had themes of Redemption related to the *Unexpected Deaths*—well over half the total. One of the most loved of the Renaissance images is Masaccio's (1401-1428) fresco of Adam and Eve in *The Expulsion from the Garden of Eden* (1425), found in the Brancacci Chapel, Florence. [**Fig.** Masaccio]. The utter grief expressed in the figures has hardly been bettered. Simply look at the agony of Eve and the hidden tears of Adam, the utter shame portrayed in both, unable to face the world. Examine the posture and anatomical details of Adam's chest and rib-cage, showing the details of ANS activity leading to inspiration, the breathing pattern that immediately precedes emotional crying, as he presses his thumbs deeply into his eye sockets. But there is much more. The Angel is escorting them out of Paradise, and the portal of Paradise is showering dark rays, and light floods down from the right. They are emerging into the light, having been caught in the act, the focus on their genitals emphasizing their new sexual awareness and sense of culpability. Tristan and Isolde experience the same exposure, to their bitter cost. Importantly the chapel is dedicated to St Peter, and the other frescos around the altar (not all by Masaccio, others are by Masolino, 1383-1440) reflect the overriding themes of salvation and redemption.

3. Covenant and Grace

Redemption is always an act of grace, but usually with a chosen figure who obeys the commission unquestioningly. Throughout the Bible, Redemption is associated with a divine plan of salvation, incrementally infolded in a series of covenants. There is always a chosen recipient called to function in an office of salvific efficacy: Noah, Abraham, David, the Prophets, until fulfilled in Jesus who is the covenant embodied. The Cross (whether the caduceus, the bronze serpent borne aloft on the pole in the Wilderness by Moses [Numbers 21:9], or the Cross of Calvary raising Jesus on high [John 3:14]) becomes the central image—of poison and healing, of death and life, of lament and triumph.

When Gurnemanz sings of time becoming space (“Zum Raum wird hier die Zeit”), the philosophy is played out in the grandiose processional music and the sudden growing ringing and clashing of bells, hanging in timeless space between heaven and earth. *Great mysteries, including* death, are beyond our feeble Newtonian understanding of the universe and universals. The apophthegm asserts a breakdown of Kantian Categories, and looks towards the Schopenhaurian *Will*. There can be no space or time in a realm we cannot visualize.

4. The Concept of *Gesamtkunstwerk*

Daud, examining Tristan’s shadow, draws attention to the erotic within the *Gesamtkunstwerk*. The academic position behind this concept was that at one time the various arts were not split from each other, as became the case by the 19th century; their ‘organic’ unity had been lost.[15] The concept of the complete artwork existed long before Wagner’s attempts to reinvigorate it. This was the aspiration of the Camerata, the aesthetic of the *tragédie lyrique* at Versailles, the dramaturgy of French *grand opéra*. Indeed, Wagner and others remind us that it was integral to Greek drama.

Here is Jean-Jacques Rousseau (1717-1778):

> Through poetry, one speaks to the brain, through music, to the ear, through painting to the eyes, and all of these ought to be united in order to move the heart and bring to it simultaneously the same impression through different organs.[16]

Hilda Meldrum Brown notes that the term *gesamt* has less to do with the plurality of the arts than the completeness of the process of integration or fusion of two or more major forms.[17] Johann Gottfried Herder had discussed the potential union of the dramatic arts, with poetry, music and dance, and the emotive power of sounds, vowels in contrast to consonants, alliteration intensifying word tone synthesis. Wagner, picking up on this linguistic concern, considered vowels to be the ‘*roots of speech*’ but came to dislike iambic verse; his aim was to create 'verse-melody' (*Versemelodie*) to produce an impression of feeling.

Wagner closely explored the relationship between the word and tone, *Tone speech* becoming "the organ of expression proper … turning from understanding to feeling … deep roused human feeling", enhanced with *Stabreim* and alliteration. *Urmelodie* was the mother-element, a "condensation of this womanly into a manly":

> … this charm is the influence of the eternal womanly … the feeling is the *purely-human,* that which makes out the essence of the human species as such … nurtured by both the Manly and the Womanly, which only *by their union through Love become first the Human Being* ... Music the glorious loving woman … melody the love-greeting of the woman to the man, and the open-armed 'Eternal Womanly'.[18]

The ideas of unity and universality appealed greatly to those of the Romantic persuasion, as presented by Schelling or as in Schiller's *The Homage of the Arts,* which is all about the integration of the arts. Lessing in his (unfinished) *Laocoön oder über die Grenzen der Malerei und Poesie* (1766) critiqued the relationship between poetry and the visual arts, based on an analysis of the Greek statue of Laocoön [**Fig**. Laocoön]. He compared this image of the struggle and death of the Trojan priest and his sons with the poetry of the incident from Virgil's *Aeneid*. Circling around the idea of the 'Moment', unique for the visual arts, whilst in poetry and drama there are no such spots of time (*Augenblicke*) only sequences of actions, the art forms gave different emotional effects, a lacuna he wished to explore. So did Wagner, although he considered Lessing's exposition flawed. Not explicitly calling up the *Leitmotif* (a term Wagner disliked), *melodische Momente* (Melodic Moments) can extend time and space, accentuated by dynamic rhythms, another *Urphenomenon,* bringing dramatic climaxes and unification of verse and music, an *Aufhebung*.

In opera, Gluck, Grétry, Hoffmann, Weber and Meyerbeer led the way, and although *Gesamtkunstwerk* is discussed in several of Wagner's texts, including *Das Kunstwerk der Zukunft* (*The Artwork of the Future*) (1849) and *Oper und Drama* (*Opera and Drama,* 1852), he only used the word twice. His ideas for a music of the future required distancing music from the looser Italianate operas with their disconnected vocal elements; it was with verse-melody that thoughts and feelings needed to be synthesized. Drama moved outwards from within, rather than inwards from without, the character of romance, the former gave a much richer development of individuality.

Gesamtkunstwerk was not understood as a static art form but rather an aesthetic (almost physically experienced) action: production and re-production at the same time. The sensory experience, enhanced by the drama, gave the audience unconscious sensations, not unlike the involuntary memories which are invoked by moments of sensory accentuation for Marcel in Proust's novels.[19]

Love as the unification of subject and object: Platonic with mighty Eros, erotic expression fully embellished, along with other glorifications of 19th century Romanticism—the *Gesamtkunstwerk* is an idea of the anthropomorphic re-presentation of the ecstatic experience, rapture and transformation through the enactment of Love. As A. Daub phrased this: "Unlike Schopenhauer, love for Wagner can push beyond the representation of individuality. It is an instinct towards transcendence … Transcendental intuition of the absolute did not require schooling, but lurked inside any authentic human being."[20]

List 2: Operas with a *Liebestod* from List 1 (n=50)

transc = transcendental A= autonomic redemp = redemption

Undine (1816) transc redemp
La Muette de Portici (1830)
I Capuleti e I Montecchi (1830) transc A redemp
Lucia di Lammermoor (1835) transc A redemp
Faust (Gounod) (1859) transc redemp
Queen of Sheba (1883) transc Male
Thäis (1894) transc A redemp
The Wreckers (1906 transc redemp
A Village Romeo and Juliet (1907) transc
Der ferne Klang (1912) transc redemp
Das Wunder der Heliane (1927) transc redemp
Daphne (1938) transc redemp
Wagner:
Lohengrin (1850)? transc A redemp
Tannhäuser (1845) transc A redemp
Tristan and Isolde (1865) transc
Götterdämmerung (1876) transc A redemp
[of interest of all Unexpected Deaths (50), there are 16 with Liebestode, 15 with transcendental implications and 4/16 are from Richard Wagner.]

Notes

[1] Mallarmé, Stéphane. *The Poems in Verse.* Trans. Manson P. (Glasgow: The Hunterian University, 2012), 91.
[2] Wagner understood that the Grail story led back to Arabic texts of Spanish Moors, of pagan origin.
[3] Wagner, R. *Religion and Art.* Trans. William Ashton Ellis [1897] (Lincoln: University of Nebraska Press 1994), 213); Wagner, R. *Opera and Drama,* Trans. William Ashton Ellis [1900] (Lincoln: University of Nebraska Press,1995),156
[4] Scruton, R. *Death-Devoted Heart: Sex and the Sacred in Wagner's 'Tristan und Isolde'* (Oxford: Oxford University Press, 2004), 5.
[5] Scruton, R. Ibid., 2004, 7.
[6] Emslie, B. Ibid., 234-235.
[7] Hegel, G.W. F. Ibid., 44.
[8] Köhler, J. "The Ring and the Romantic Tradition". *The Wagner Journal* 2 (2008): 29-30, 33. Köhler makes an additional point that Schelling also had a considerable influence on the *Ring.* Nietzsche quotes from Köhler, 33. Mark Berry, in his paper "The Positive Influence of Wagner upon Nietzsche" looks at Siegfried and Wotan as possible examples of what Hegel refers to as "world-historical individuals … created not by the existing, sacrosanct order, but from another source, whose content is concealed and does not flourish in contemporary existence, by the Inner Spirit, still subterranean, which bursts through the shell of the external world, for it is a kernel different from that belonging to the shell". Berry, M. *The Wagner Journal* 2 (2008): 11-28; 23. It is worth quoting Thomas Carlyle here: "Forms which *grow*, if we rightly understand that, will correspond to the real nature and purport of it, will be true good; forms which are consciously *put* round a substance, bad. I invite you to reflect on this. It distinguishes true from false in Ceremonial Form, earnest solemnity from empty pageant, in all human beings." Carlyle, T. *On Heroes, Hero-Worship and the Heroic in History* (London: Chapman and Hall Ltd, 1907), 205.
[9] Tracing the origin of this is difficult, but it was quoted in "Picasso Speaks", *The Arts* (New York, May 1923).
[10] The relationship between Wagner and philosophy is well discussed in Bowie, A. *Music and Philosophy* (Cambridge: Cambridge University Press, 2007), 210-260. Magee B, Ibid. is also highly recommended on these matters. Mark Berry explores associations to *Parsifal* in: Berry, M. "Is it Here that Time becomes Space? Hegel, Schopenhauer, History and Grace in *Parsifal*". *The Wagner Journal*, 3:3 (2009):29-59. Richard Bell looks at Hegel's influence in the *Ring* Cycle: Bell, R. H. "Teleology, Providence and the Death of God: a New Perspective on the *Ring* Cycle's Debt to G.W.F. Hegel". *The Wagner Journal*, 11:1 (2017): 30-45.
[11] Wagner, R. *Richard Wagner's Letters to August Röckel* (Forgotten Books, 2012), 84, 85, 90, 99.
[12] Emslie. Ibid., 291.
[13] Donne, J. "The Good-Morrow". In: *John Donne: The Penguin Poets* (London: Penguin Books, 1950), 23.

[14] Stefan Herheim's *Parsifal* in Bayreuth (2008-2012) made much use of the large mirror which descends from the roof and at the end reveals the audience to themselves. Mark Berry gives an interesting analysis of this production: Berry, M. *After Wagner: Histories of Modernist Music: Drama from Parsifal to Nono* (Wooodbridge: The Boydell Press, 2014), 210-233.
[15] Daub, A. *Tristan's Shadow: Sexuality and the Total Work of Art after Wagner* (Chicago: University of Chicago Press, 2014).
[16] Rousseau, J.-J. *Dictionnaire* (1768). Quoted by Laudon, R.T. *Sources of Wagnerian Synthesis: A Study of the Franco-German Tradition in 19th Century Opera* (München–Salzburg: Musikverlag, Emil Katzbicher,1979),154.
[17] Brown, H.M. *The Quest for the Gesamtkunstwerk and Richard Wagner*. (Oxford: Oxford University Press, 2016).
[18] Wagner, R. *Opera and Drama,* Trans. W. Ashton Ellis (London: University of Nebraska Press, 1995), 230-236, 285, italics in the original. As Michael Dyson points out "The tricky relationship between music (female) and musician (male) deserves further analysis". Dyson, M. "Sea, Mirror, Woman, Love: Some Recurrent Imagery in Opera and Drama" *The Wagner Journal,* 5:3 (2011):16-33.
[19] The relationship between Proust's *In Search of Lost Time* and Wagner's operas is elegantly explored in Swann, J. "Wagner and Proust". *The Wagner Journal*, 12 (2018): 34-55. He draws attention not only to frequent references to Wagner in the books, but also to the thematic nature of Vinteuil's musical motifs in his Violin Sonata and Septet. The tapestry of various artworks threading throughout the narrative is a veritable *Gesamtkunstwerk*. For other interesting references to the *Gesamtkunstwerk*: Millington, B. "All in it together: *The Gesamtkunstwerk* revisited". *The Wagner Journal*, 11 (2017): 46-61; Millington, B. "Manifestations of the *Gesamtkunstwerk* in Fin-de-Siècle Vienna". *The Wagner Journal*, 12 (2018): 4-25. Stein, J. M. *Richard Wagner and the Synthesis of the Arts.* (Detroit: Wayne University Press, 1960).
[20] Daub, A. Ibid., 7,159. *Gesamtkunstwerk* was not in keeping with the ideas of Schopenhauer. This theme is explored in Richard Strauss's *Capriccio*, 1942), *Prima la musica e poi le parole* (First the Music and Then the Words) (Salieri, 1786). We do not wish to minimize the importance of Schopenhauer for Wagner, the philosopher gave the meaning of music a transcendental lift that embraced the Platonic, ideas as we have noted which were latent and being explored by Wagner before he was introduced to Schopenhauer. He seemed to justify for Wagner what he had been moving towards aesthetically.

CHAPTER 24

REFLECTIONS

> Music, such music, is a sufficient gift. Why ask for happiness; why hope not to grieve? It is enough, it is to be blessed enough, to live from day to day and to hear such music—not too much, or the soul could not sustain it—from time to time.
>
> (Vikram Seth, *An Equal Music*)[1]

1. Introduction

In List 3 (page 497) we have drawn up those operas from List 1 in which we have not identified a *Liebestod.* Difficult decisions were made with regards to categorising the various forms of Love-Death as *Liebestode*: for example, some of the Faust operas (where Marguerite sometimes dies before assumption) and the endings of *Salammbô* (where both lovers choose to end their own lives), *Le Villi* (the duo-finale suggests a Love-Death), and *St François d'Assise* (where the saint's death is associated with an intensity of love that draws him heavenward).

List 1 emphasised the excess of operatic *Unexpected Death* in females, especially with autonomic dysfunction, and the overall subsumption of Redemption and transcendentalism, somewhat more so in females. List 2 exposed an even closer association of *Liebestode* to both these themes. In List 3 are listed 19 females and 15 males in operas from List 1 with no *Liebestod.* Twenty-five (73%) have transcendental events integrated in the librettos compared with 94% in those with *Liebestode* (List 2). Comparing the two groups for Redemption and transcendence together, the figures are 11 of 16 with *Liebestode*—(69%) and 16 out of 34 without *Liebestode*—(47%), a significant association.[2]

As we have previously suggested, and borne out by the operas in Lists 1-3, the inclusion of transcendental deaths and then Redemption in opera became particularly prevalent in the19th century. There is a remarkably close association of these themes presaged in the operas we have reviewed, starting from our own observations of *Unexpected Deaths* on the stage. This leads us towards our explanations of the events.

2. Redemption and Transcendence

These happenings had especial significance for Wagner. They are to be found in one form or other in his earlier works and can be followed through to *Parsifal.* It cannot have been his intention to create 'happy endings': indeed his well-known pessimism always casts doubt in the minds of commentators and audiences on how, for example, the end of *Götterdämmerung* should be viewed.[3]

'Redemption' is a complex concept, overwhelmingly associated in general understanding with strong theological associations. It is hence regarded with a certain caution, if not reticence, and can induce a sense awkwardness in an age of demythologized mystery. But it has many usages, and has its origins in a material level, a transaction involving the regaining of possession of something in exchange for a payment, or the clearing of a debt. The concept was expounded in Biblical times in social terms. The laws governing the Year of Jubilee (every 49 years, 7 x 7), carefully controlled the possibilities of retaining or acquiring personal property and personal freedom from loss or servitude (Leviticus 25, especially verses 47-53). The process is one of *retrieval*, *recovery*, *reclamation*, *repossession and return.* If one is personally unable to fulfil the requirements of *recoupment*, then a redeeming agent (*go'el*) would be required to fulfil the role of 'redeemer'.

There is just a step into a more metaphysical dimension, with the notion of liberation, the action of buying one's freedom, or even more importantly, the action of being set free. Again the concept was formed in Antiquity and first finds significant expression in relation to the enslavement of the Ancient Israelites in Egypt, and God's promise to Moses to set them free: "I will bring you out from under the burden of the Egyptians, and I will deliver you from their bondage, and I will redeem you with an outstretched arm and with great acts of judgment" (Exodus 6:6). The physical act of liberation now takes on a 'spiritual' implication, an action of saving or being saved from sin, error, or evil, a harmful ideology. The 'redeemer' now becomes a person, action or thing that saves someone from error or evil. The '*saving*' now entails a *vindication*, or an *absolution* from guilt, and (as with the Ancient Israelites) implies a transference to a different sphere, a place of safety, or even transformation beyond the confines of normal experience. The momentous Crossing of the Red Sea, the Years in the Wilderness, and the Crossing the Jordan into 'a Land Flowing with Milk and Honey' become portentous metaphors for freedom, transference and elevation into a special location. Here transformation of purpose, place and personality can provide transcendence into another realm of existence.

The concept of redemption/liberation thus comes to have an intimate association with the notion of a transference out of one lowly, difficult, or oppressive situation to a transformed circumstance of untrammelled existence. This transforming elevation is 'transcendence', which literally can mean surpassing others, becoming preeminent or supreme, and, more metaphysically, entering or lying beyond the ordinary range of experience or perception.

'Transcendence' since the late 18th century has become associated with philosophy. The term alone, with 'transcendental', may sound allusive and illusive, but implies rising above or transcending the Aristotelian categories (as in Kant's theory of knowledge). Transcendence implies being beyond the limits of sensory experience and hence entering the realms of the unknowable. For the philosopher Emmanuel Levinas (1906-1995) the transcendental is that which "cannot be encompassed". *Jouissance* finds first place in his 'phenomenology of the other', the metaphysical other; this is to do with intersubjectivity. In the encounter with another, "face to face is the irreducible and ultimate relation … metaphysics is the desire of a person". His philosophy was to overthrow the dominance of 'egology' and to insist that metaphysics comes before ontology: an ethical dimension takes precedence embedded within face-to-face encounters, I-thou relationships, the *Augenstern* and eye to eye trans-ascendence.[4]

Taking a line perhaps from Wagner, Levinas writes:

> The I-thou in which Buber sees the category of interhuman relationship is the relation not with the interlocutor but with feminine alterity … this exercises its function of interiorisation only on the ground of the full human personality, which, however, in the woman, can be reserved so as to open up the dimension of interiority.[5]

For Wagner it was not only protagonists who were redeemed but, as he implied in *Das Kunstwerk der Zukunft* (*Artwork of the Future*), German culture itself needed salvation. Redemption for all, including Wotan—who renounced his powers and wills his own ending—all the gods (Waltraute's "Erlöst wär Gott und die Welt" the gods and the world redeemed) and of course Brünnhilde and Siegfried. The Ring itself finally is returned to the Rhine daughters and purged of the Curse. The enigmatic ending of *Parsifal*—"Erlösung dem Erlöser" (Redemption for the Redeemer)—likewise is endlessly debated.[6] The Grail is freed from the nexus of human folly represented by Amfortas, through the saving ministry of the guiless Fool made wise. All now benefit from the untrammelled, encompassing radiance of divine grace.

In the operatic experience, the occurrence of sudden death when *Unexpected*, is in part *Explained* by the autonomic reactions to situations of extreme emotional and physical stress, and intertwined with the rapture of love or the anguished consequences of loss and disappointment: love with death with an almost unique consistency. The idea of Love-Death is inevitable, aspirational, associated with freedom of liberation from 'the vale of tears' with its limitations, brokenness and mortal mutability. The process of dying is invariably in the context of the expression of great love, or self-sacrifice, or sublime yearning. These are transcendental aspirations, in many cases associated with the aspirations of faith for transference to a higher reality (be it Platonic or religious) in a process we can only call 'redemptive'.

When describing Redemption in opera we can therefore infer spiritual realities. However, these are only rarely couched in overtly theological terms, and shade over in some operas simply into the manifestation of a love so strong it assumes a sacrificial and hence transcendental character. This is often the case where the heroine dies sometimes of illness, or the hero takes his own life, but where the context of love is so palpable as to be transformative. Within these narratives, *Unexpected Death,* as we have framed it, becomes *Explained.*

How do we associate Redemption with transcendental moments? The latter can be understood in various differing scenarios as a consequence of, a cause of, a metaphor for, the manifestation and consequences of love and death, where the power of melody and harmony generate musical contexts associated with the powerful stimulus and the uplifting of the emotions we sometime call rapture. Here the musical language becomes the expression and vector of the transcendent experience, with or without necessarily the association of liberation we call redemption—relationships we explored through our analysis of Lists 1-3.

Both the religious and secular uses of the terms transcendence and Redemption will have special meanings for some composers, like Wagner, and differing implications for potential audiences. The value of these terms for those who have deep religious feelings will be different from those with a materialistic interpretation of what they hear (material transcendentalism). Yet perhaps both may (will?) respond from a phenomenological perspective in similar ways: all humans have or aspire to have similar needs for assurance, hopefulness, aspirational happiness and love.

> As human beings we know how we are limited, confined within the constraints of a biological body that is mortal and finite, inhabiting a tiny temporary speck of the vast cosmos. Scientific understanding can track the history of this tiny biological creature, investigate its workings, map its brain activity, trace out its behaviour from conception to death. And yet none of this scientific mapping unlocks the domain of subjective conscious-awareness, that fragile and fleeting private space that each of us inhabits when we close our eyes and hear the pulse of our thoughts and feelings and sensations—fragile, fleeting, transient, wholly finite, yet ranging out to the infinite world of which, while it endures, it remains, mysteriously, the subject and the centre. There is a kind of transcendence here, one that is rooted in something none of us can deny, our own conscious presence in the world as enduring centres of awareness, and graspers of meaning and value.[7]

In opera this emotional charge is generated by the magic of the music, exploring, and expressing scenarios that touch the deepest well-springs of human experience and yearning. Such is the way the ANS affects us. Schiller observed on the sublime: "It is a composition of melancholy which at its utmost is manifest in a shudder, and of joyousness which can mount to rapture ... this combination of two contradictory perceptions in a single feeling demonstrates our moral independence in an irrefutable manner."[8] Poizat introduced us to the word *jouissance,* used also by Levinas. The term is associated with a feeling "known to opera lovers, the thrill or shiver that courses through the body in those supreme moments of musical ecstasy … tears of joy".[9] These tears are common to opera goers, and we earlier quoted 'Wagner Moments' in well-known personalities. But these affect us all, and of course, are not confined only to Wagner's operas.

3. Tragedy and Compassion

In *Why Humans Like to Cry,* a distinction was made between Tragedy and tragedy. Tragedy is much associated with the art form identified with the Classic Greek era, Tragedy as drama. But tragedy is a fundamental aspect of the human condition. The 'tragic feeling', if the experience of witnessing Tragedy might be so termed, is one which in that book was referred to as 'tragic joy', Nietzsche used 'tragic pathos'. Perhaps it is closely related to Poziat's *Jouissance,* but it is not a 'catharsis' in the way that most people have interpreted Aristotle's explanation for the feelings. It is associated with tears. What is not much discussed is that *Homo sapiens* is the only living species that displays emotional tears: emotional crying is exclusively human.

There are many classifications which try to look at different aspects of our emotions (e.g. anger, happiness, sadness, shame) yet none seem to include crying or even love. The tragic feeling is simply not a combination of known feelings, but is a different emotion infused with the vibrant remnants of our evolutionary past. There is no need to invoke 'fear and pity', the two characteristics Aristotle cited as underlying the 'purgative' effects of Tragedy, to explain *tragic joy*. Aristotle does not explain the term, but Critchley asks if it is "something closer to purification, with religious and ritualistic connotations …or … a transformation of passion?" Perhaps it just is the way we feel at the end of seeing a play or opera. Or has it and does it embody moral sentiments, as it did for the Greeks and can still do today?[10]

Art does not fill in gaps left by nature, but "the work of art has its true being in the fact that it becomes an experience that changes the person who experiences it".[11]

Many commentators are searching after the emotions aroused in the spectator, some suggesting that Tragedy induces a sense of the sublime, or of aesthetic pain, with a combined emotion of debasement and at the same time awe, elevation or grandeur. Or is it a religious transformation? Roger Scruton conceives of tragic theatre recreating the experience of the liturgy, a "recreation of this sacramental moment", a communion, but one which evokes "primeval feelings of guilt, threat, and collective vengeance, and also the transition from sacrificial victim to sacred presence which is the gift of so many religions". The unity between tragic and religious feelings, "the experience of sacred (being) a human universal [is] bound up with our very existence as self-conscious, rationally choosing subjects".[12]

4. '*Durch Mitleid wissend, der reine Tor…*'

"Knowing through compassion, the poor fool" is Gurnemanz's description of Parsifal (Act 1). In preliterate societies, the development of feelings for and with another and the ability to appreciate that they have minds like our own (referred to as Theory of Mind), would seem to have been crucial to higher hominid and human evolution. Through the medium of early religious ceremonies and rights, fear became shared, and fear of death became mollified with the potential of the other world and an afterlife; mourning was part of this process. As a new type of consciousness emerged with the development of human language, 'I', 'now' and 'here' reified the individual's location in time and space. Artefacts representative of our cognitive projections were created and this new form of creativity blossomed into art, initially of a religious nature, enabling a binding of the individual to the past and the future. Sympathy, empathy and compassion have become a part of our language, implying some sort of emotional contagion between individuals. The greater interest has come from philosophy (phenomenology) and via psychoanalysis exploring variants of psychological therapies. But the terms have caused confusion of the etymological boundaries, especially in social psychology. They do not simply refer to pity. Sympathy (*Mitgefühl* or 'fellow feeling') is not the same as compassion, which is better allied with the German *Mitleid*—'suffering with', a way to knowledge, unconscious yet invoking powerful prosocial feelings, explored by Levinas.[13]

Emotions are triggered by external and internal bodily events. What the neuroscientist Antonio Damasio calls "emotionally competent stimuli" act on cerebral circuits, parts of brain development echoing back millions of years. These change our bodily state concordant with patterns generated in the brain, his Somatic Marker hypothesis. "Our brains receive signals from deep in the living flesh and thus provide local as well as global maps of the intimate anatomy and intimate functional state of the living flesh". The repertoire of actions developed come from unlearned and learned brain programmes of automatic actions and cognitive strategies related through a lifetime's learning and memories, serving to bias responses consciously or unconsciously to outside influences. A reflex part of such responses is lachrymation.

Construction of meaning involves inner-world metaphorical transformations of the outer world, and since much metaphor is shaped by our bodily interactions with the world around us, our lived experience is embodied. Viewing an emotion in another, in a daily encounter or on stage, thus activates the neuronal core of our own repertoire of emotions: in opera, with music, we have the capacity to re-sound with feelings. Compassion comes through intuition, learning and experience, hence the tears of tragedy and Tragedy.[14]

5. Wagner's Compassionate Characters

The concept of compassion is powerful in Wagner's work from its earliest stages. The relationship between Rienzi and his sister Irene, despite undertones of GSA, is essentially based on Irene's deep sympathy for her heroic brother's visionary plans and their foiled nature. Their duet in Act 5 is a manifestation of their affinity. Irene completely identifies with Rienzi in his hour of total abandonment after the Papal malediction and excommunication. The situation takes on an almost sacrificial character in her eschewal of flight and love with Adriano, and her preparedness to die in the flames of the Capitol, participating completely in the fate of her brother. In *Der fliegende Hollander* it is Senta's almost supernatural prescience and sense of destiny that transforms the recitation of an old ballad into a heroic and explosive vision of compassionate self-identification with the legendary victim's fate and suffering. The uncomprehending nature of her milieu only intensifies the power of her sense of predestined calling, that like Irene in *Rienzi*. She must cast aside the conventional demands of romantic love for the work of self-offering and redemption. The situation is ratified and deepened with the entry of the Dutchman and the rapt spirituality of their duet together. Her compassion takes on the power of grace in her self-sacrificing death at the end and the transfiguration of the lovers.

Tannhäuser presents what are perhaps the most moving of Wagner's portraits of *Mitleid*, the figures of Elisabeth and Wolfram. Both are linked in their common devotion to the erring and complex Tannhäuser, with the added strand of Wolfram's unacknowledged and unfulfilled love for the saintly heroine. The articulation of a nexus of compassion and sympathy is bound up with the articulation of Elisabeth's name, which has the power to still the ire of mortals and cast an aura of peace and hope.

Both are aware of Tannhäuser's failing, perhaps even tacitly understanding the nature of his long disappearance. Both are depicted directing and saving Tannhäuser from himself and from the outrage of society. Wolfram detains Tannhauser from proceeding on his travels by recounting the mortal sadness of Elisabeth at his long absence (although this is at the expense of his own hopes of winning her love). Elisabeth throws herself between Tannhäuser and the outrage of the Court and society at the revelation of his true absence in the erotic blandishments of the Venusberg. Elisabeth's plea for his life ("Ich fleh' für ihm, ich fleh' für sein Leben") is among the most heartbreakingly beautiful music Wagner ever wrote and the very tonal embodiment of *Mitleid*. This will continue in her desolation at his absence and failure to return from the Roman pilgrimage. Wolfram's grief for both Elisabeth and Tannhäuser finds its perfect expression in his vigil and famous Song to the Evening Star which distils his sorrow, his love and renunciation in the transcendence of prayer and the unattainable remote beauty of the planet Venus. His compassion continues as he listens to Tannhäuser's sad Rome Narration—despite the sinner's anathematization by the Pope. As before, he holds Tannhäuser back from returning to the Venusberg, by invoking Elisabeth's holy name, which the exhausted Tannhäuser repeats. The rest of the opera sustains the concept of compassion in the chorus of mourners accompanying Elisabeth's bier and in the ecstasy of the chorus of the retuning pilgrims with the flowering papal staff which becomes the sublime paean of the concluding chorus of grace.

The whole of *Die Walküre* is framed in the concept of *Mitleid.* The opening scene in Hunding's hall resembles the situation in *Der fliegende Holländer* in the intuitive magnetism between Siegmund and Sieglinde. The scene is steeped in the growing sense of attraction and sympathy between the fugitive and the lonely wife overwhelmed by their compassionate telepathy. Even when the implications of GSA are manifested, they are seemingly subsumed into the overwhelming compassion that characterizes their relationship. This is manifested again in the flight from Hunding, in Sieglinde's frenzy of fear, and Siegmund's resolution to save her, even to the extent of foregoing his entrance into Valhalla. It is the power of this self-sacrificing love that stimulates Brünnhilde's compassion and her determination to save Sieglinde, even at the risk of brooking Wotan's anger. This same *Mitleid* dominates the finale in the great colloquy between father and daughter. Brünnhilde's love and compassion for both the Volsungs and for the foiled Wotan eventually overcomes all his righteous anger and sense of betrayal and turns the judgment into a great outpouring of the supreme pain in parting from a loved one. Retribution now becomes the hope for the future in Brünnhilde's saving of Sieglinde and her child, and in the compassion she has generated in Wotan's heart. Their farewell, couched in Magic Fire and Enchanted Sleep, is the opening to a new future born of sympathy and love.

Elements of Wolfram's renunciation and Wotan's compassionate understanding underpin the exploration of *Mitleid* in *Die Meistersinger von Nürnberg* in the figure of Hans Sachs. Only he has the ability (like Senta and Elisabeth) to see beyond the confines and expectations of his own society governed by the dictates of time and place. While respecting the value and place of tradition and venerability, he is able to discern creativity in the young, as shown in his invitation to innovation in the fresh vision of the impulsive Walther von Stolzing and the pure but naïve Eva. Like Wolfram, he is able to act on his spiritual understanding even beyond his own human desire, sacrificing as he does any possibility of love with Eva. Sachs's ability to see into the heart of the matter, the nature of the person, the essence of artistic wholeness and inspiration, means that his compassion can encompass the past and the present, and provide a vision into the future (as does Brünnhilde in her key actions of compassion in *Die Walküre*). Both the characters Siegfried and Parsifal have a parallel development that traces the whole of their emotional maturation. This development can be measured by their engagement with compassion.

Siegfried is shown growing from an unformed and impetuous youth through his encounters with Mime, Fafner, the Woodbird, the Wanderer, and eventually Brünnhilde. The scene on the Loge's Rock compresses the whole of his development to manhood in the various stages of the duet (continued into the Prologue of *Götterdämmerung*). Eventually we see him as a fully-fledged hero, setting out to find adventure and fulfilment in the wider world. This process is cruelly interrupted and diverted in the tragedy of his betrayal by Hagan giving him the fatal potion that deprives him of his memory and true identity. Only when the antidote is applied can he be himself again. The narration of his youth captures the feckless joy and bliss of his life-changing love with Brünnhilde.

The first and only time we see Siegfried able to be fully himself and speak in the voice of his complete psychic entity is after his stabbing, in the moments of his death. He sees and seems as never before: he speaks in depth, with the voice of true heroic insight, of his love and compassion in greeting his bride again in the midst of his failing strength and the rising waters of death. This is the saddest music in the whole *Ring* Cycle, fraught with perception, empathy, and loss—in the fullness of the tragic experience.

For Parsifal it is different. We also meet him as an unformed, naïve, callow adolescent (lacking Siegfried's deep intuitions, his oneness with nature, his innate creativity). Unlike Siegfried he has violated nature in hunting the swan, without an inkling of the wholeness and unity of all creation. But as with Siegfried, he begins a process of maturation through his meetings with Gurnemanz, with the wild Kundry (who awakens his memories of his lost mother), through his visit to the Grail and uncomprehending experience of Amfortas and his suffering. The *Lebenslauf* continues with the entry into Klingsor's Magic Garden, with its carnal temptations (to which he remains impervious). Only with Kundry, unveiled as a youthful seductress, do things change. During her narration of the sorrows of Herzeleide, and specifically at the moment of her kiss, his emotions become palpable to him and to the audience. This moment is intended as the moment of irresistible erotic blandishment, but is transmuted into a spiritual current of illumination, a gleaming dart of epiphany, as the nature of life and sorrow are shatteringly revealed to him, and he finds himself living the agony of Armfortas. This is the lighting flash of compassion. It is a moment of consecration, a baptism of desire, a pneumatological ordination that empowers him psychically, emotionally and spiritually. He is now a hero, graciously enabled to stop the Spear in its flight and end Klingsor's reign of enchantment.

While the rest of Parsifal's development is hidden from us (his wanderings described musically in the prelude to Act 3), when he next appears years later as the Black Knight, he is fully-fledged in his heroic calling. He is prophetic in fulfilling his destined vocation of compassion on Good Friday, and priestly in his effortless assumption of ministration in baptizing Kundry. He is now at one with the sentient universe as the whole of nature blooms radiantly around with him at the hour of the supreme Sacrifice of Calvary. He is now able to assume his kingly role, to enter the Grail Temple, administer the Sacred Vessel, consecrate the dead Titurel, bring healing to Amfortas, assuage the desperate yearning of the Brotherhood, and end Kundry's endless cycle of repetitious suffering.

6. Construction, Deconstruction and Reconstruction

In his analysis of Greek Tragedy, the philosopher Simon Critchley summarises as follows:

> Tragedy presents a conflictually constituted world defined by ambiguity, duplicity, uncertainty and unknowability, a world that cannot be rendered rationally fully intelligible through some metaphysical first principle or set of principles, axioms, tables of categories, or whatever. Tragedy is the experience of transcendental *opacity*.[15]

He continues: "As such, the experience of tragedy poses a most serious objection to that invention we call philosophy". Music too seems such an anathema for philosophy, opera even worse. The professor of music Julian Johnson, in his book on the 'language' of music (based much on his analysis of the music of Debussy), rails against the trend of musicologists to obsess about technical analyses of sound, music as the product of vibrations in the air. He opines that philosophy needs a musical turn as much as music needed a philosophical reevaluation. Attempts to downgrade the operatic experience or even ridicule it by *Regie* directors is a process whereby the application of the Meyerhold/Brechtian *Verfremdungseffekt*, the dehumanizing and deconstruction of texts (and librettos) deprives the music of its proper purpose, so that it becomes the Kantian *Zweckmässigkeit ohne Zweck* (purposefulness without purpose). Productions achieve dramatic notoriety by their infidelity to the music. "The alliance of arts which make up the form is in danger of collapse".[16] This is directly in opposition to the careful and considered understanding of the operatic form, not only from a historical point of view, but particularly from the Romantic perspective. This is not simply a sentimental experience, but an Aristotelean process of melancholic and emotional purification.

We seem to have got to the *Regie* via structuralism, post-structuralism, modernism, post-modernism and beyond, undermining concepts of the transcendental and the self, in part a reflection on the limited interest that philosophers have in music.[17] After all, language is the only bulwark known to philosophers. Johnson wants us to "think *through* music—not thinking about music", aesthetic experience arising from "the experience of the sensible body, pre-figured and pre-thought in music". The embodied mind barely gets into philosophic discourse, and few philosophers have embodied music into their theories.[18] This coincides with our insistence throughout this book on a phenomenological approach to aesthetic experiences that are important to our understanding of opera and *Unexpected Death.*

It is futile to consider music as an object with perceived properties independent from emotions and a world saturated with meanings. Self-consciousness along with music resist objectivation, yet who would deny the reality of both for humanity? The contentions of the Vienna Circle that metaphysical statements that cannot be verified are meaningless, eschews the Romantic aspirations of understanding music and that "longing plays a central epistemological and ontological role in Romantic philosophy".[19] The sonic landscape is internal as well as external, the ANS (even in someone who is perfectly still when listening to music) re-sounding within us as "an avatar of the human body—a virtual being which enacts new forms of motion ...": a metaphysical sense of the physical that remains, sometimes referred to as the sixth sense.

Husserl's phenomenological reduction and the Cartesian *cogito*, suspended in an independent realm of experience, devoid of the body, avoid sound vibrating within. Andrew Bowie uses the expression "pre-conceptual engagements with things" to refer to the embodiment of our acting and feeling in the world. For Johnson, the "incarnate mind" of the "enactive subject" has, with the world, especially music, a mutually transformative dialogue, to the attentive, "allowing things to come to light". Damasio's Somatic Marker hypothesis underpins these events, revealing a change of state in the body proper and the brain structures that map the body, unfolding perhaps Proustian involuntary memories but giving us a way to knowledge of the world beyond a purgative catharsis.[20] "Music is not the representation of an idea, but the idea itself".[21]

Conceptual art seems to start with an idea (*Konzept,* the concept) which becomes the entire basis for the creation, an *idée fixe* more relevant than the work itself. The intentions of the artist are irrelevant, and as deconstruction became part of fashionable philosophical discourse (a movement that began with language, but then was applied to quite diverse fields), the deconstructed could be reconstructed. Although much of Western culture (especially politics) functions with binary oppositions (such as we have emphasized in many operatic themes), we are now encouraged to consider that everything is unstable, things fall apart, the centre cannot hold. In opera there is the libretto, the music, the conductor, the director, and the audience. The music and the libretto must form the kernel of the work, carefully crafted by the creators. The conductor develops the musical themes within the confines of the composer's intentions but must remain true to the score (there are exceptions of course). The unfettered ability to re-construct seems the prerogative of the director. Post-modern themes, with biased political conceptions, will deny that any piece of literature has a single meaning, meta-narratives are not allowed, we have only 'points of view'. These are so often presented in a complicated and obscure fashion, wrapped in verbal and visual rhetoric. The Frankfurt School's intention to resuscitate Marx, along with the attacks of Positivism, tried hard to guard against metaphysics and the transcendental. The audience experience of music became irrelevant: Schönberg famously was disdainful:

> As for consideration for the listener, I have as little for this as he has for me. All I know is he exists, and so far as he is not 'indispensable' for acoustic reasons (since music does not sound well in an empty hall), he's only a nuisance.[22]

Yet in advanced capitalist countries, philosophers such as Jürgen Habermas (1929-) and Herbert Marcuse (1898-1979) realized that a Marxism that could only foresee social emancipation in economic terms and that

Proletarian art must be revolutionary, had passed its sell-by date. Marcuse, influenced by Freud, wanted to bring back *Eros* alongside *Logos*. He recognized that the Enlightenment was based on Judeo-Christian values, but there was a rent that needed to be repaired: "Among modern societies, only those that are able to introduce into the secular domain the essential contents of their religious traditions which point beyond the merely human realm will also be able to rescue the substance of the human".[23] The influence of advances in technology and the monolithic culture industry creating the one-dimensional man in a one-dimensional society have limited art, literature and personal freedom, and what it is to be human.[24]

The foundations of Western culture are to be found in Greek and Roman civilizations, and our values are predominantly those of Christianity, yet there is a pressure to limit the scope of understanding about the Classics, disabling the humanities, and demanding originality: post-structuralism's deposit on humanism. The contradictions in the universe are manipulated as the impoverished world of academic correctness invalidates myth, even if ghosts continue to haunt, and Eros and Thanatos cannot be sublimated. "The utopia of great art is never the simple negation of the reality principle but its transcending preservation (*Aufhebung*) in which past and present cast their shadow on fulfillment … grounded in recollection".[25]

Personeregie, which literally implies putting emphasis on "the direction and interaction of characters"[26] seems to have emerged as *Regietheater* in the 1960s, sometimes attributed to Harry Kupfer (1935-2019). Famously the composer-conductor Pierre Boulez (1925-2016) wanted to blow up the opera houses in Germany. But earlier aberrations would include *Victory over the Sun* (1913), an opera created by the anti-rational Russian Futurists who called for a dissolution of language, liberating the unconscious, providing a libretto without meaning—Apollo extinguished.[27] Brecht's collaboration with Kurt Weill (1900-1950) in *The Rise and Fall of the City of Mahagonny* (1930) was referred to by Theodor W. Adorno (1903-1969) as the first surrealist opera, showing the Bourgeois world as absurd, the music "hammered and glued together with a fetid mucilage of a soggy potpourri of operas".[28] Stephen Hastings in his pleading for greater humanity in opera settings regrets that "enlightened humility is … largely forgotten to the more prominent directors today … who often prefer to flaunt their own restricted understanding of the work …" He imputes directors who are technically efficient but musically nescient and seem disconnected from the musical dramaturgy.[29]

The directions that these trends have taken opera may not have been anticipated. Poor Orpheus has come in for disturbing metamorphoses. Wikipedia lists 71 versions, stemming back to Peri's *Euridice* of 1600. Productions were common in the 17th and 18th centuries, with many fewer in recent times. At the time of writing we have seen only four productions in this century. An example is Emma Rice's *Orpheus in the Underworld* (2019) at the English National Opera (ENO). Her first work in opera, she decided to rewrite the story and change the libretto. Daniel Kramer's production of Birtwhistle's *The Mask of Orpheus* (ENO, 2019), an allegorical non-linear confusion of several versions of the Orpheus myth, made Orpheus a drunken pop-star who hanged himself, and confused the audience with multiple avatars and four hours of sensory overload.

Controversies abound with the works of Wagner. The great *Ring* cycle is not allowed to be like Dante's *Divine Comedy,* a journey of epochs from the world's beginnings revealing the rise of human consciousness, with characters enfolding together over time, present and past rejoined as *Leitmotifs* and symbols united in a holistic story of life, death and love. More recent settings reveal how far from Wagner's intentions it is possible to go. *The Ring* of Patrice Chéreau collaborating with Boulez (1976), and that of Harry Kupfer (1976, 1991) were controversial enough, but the latest Bayreuth contribution from Frank Castorf, variously set in a motel, Baku, Berlin's Alexanderplatz and Mount Rushmore, with Kalashnikovs and pink alligators, seemed to many incomprehensible.[30] The *Ring* of Richard Jones (Covent Garden, 1994) came over almost as a comic parody, Wotan as a traffic warden. In our paper "The Mystery of Wotan's Eye" we emphasized the great significance of Wotan's damaged *left* eye as a key to understanding the entire *Ring,* yet Wotans still appear with the right eye patched, and when researching for that piece we asked singers about the significance of the left eye, most had no idea.[31]

Mariusz Treliński's *Tristan und Isolde* (Metropolitan Opera, 2016) had Isolde slitting her wrists. We have seen other productions in which she has been forcibly dragged off the stage to King Mark's castle. In Stef Lernous's production, after the 'love-death', Tristan gets up from his death bed and the couple walk off together (Flemish Opera, Antwerp, 2013). We have had *Lohengrin* productions with no swan or Elsa refusing to die, just wandering off the stage at the end; *Der fliegender Holländer* with no ship or portrait vie with discombobulating *Parsifals* (the Grail a mirror turned on the audience, or set in a hospital ward). There is now an inevitable need to use video projections to embellish or confuse the action, and clutter the stage to distract from the music, notoriously in Christoph Schlingensief's

Parsifal at Bayreuth (2004). In Katharina Wagner's *Meistersinger* at Bayreuth (2007, 2010) everything is turned upside down, including paintings of Nurnberg. Sachs is the outsider, Beckmesser the hero, and the Meistersingers, along with cultural icons of German art (Goethe, Schiller, Bach, and Wagner), are paraded across the stage wearing grotesque masks. Sachs also burns doublets of both a conductor and stage director. *Entartete Kunst* had something to do with it, but there was no love between Eva and Walther.

Our book is concerned with the *Unexpected Deaths* in opera. Nila Parly in her analysis of death in *Tristan und Isolde* suggests that since Clément's feminist study of the way women are treated in productions, directors have let heroines live, irrespective of the libretto and composition.[32]

Roger Scruton considers Wagner's attempt to create a new musical public that would see the point of idealizing the human condition:

> … kitsch culture was already eclipsing the romantic icon of the artist as priest. Since then, Wagner's enterprise has acquired its own tragic pathos, as modern producers, embarrassed by the dramas that make a mockery of their way of life, decide in their turn to make a mockery of the dramas … The producer strives to distract the audience from Wagner's message and to mock every heroic gesture.[33]

Jonathan Miller (1934-2019) observes:

> One is constantly fighting a battle between two forms of idiocy … mindless traditionalism … [and] relevance, one of the most ghastly words which ever comes up in the arts. All great works are bound to be relevant because they deal with fundamental and inextinguishable human situations which are the same in any period, but nevertheless you have got to refer to and honour the tone of the historical period in which it was created.[34]

List 3: Operas from List 1with no *Liebestod* (n=34)

[None] quasi-Liebestod

Opera	Character	Composer	Year
Females			
Anna Bolena	Anna	Donizetti	1830 transc
L'Ange de Nisida/La Favorite	Countess Sylvia de Linares /Lenore de Gusman	Donizetti	1839/1840 transc/redemp
La Nonne Sanglante	Agnes	Gounod	1854 transc/redemp
Rusalka	Natasha/Rusalka	Dargomyzhsky/Dvořák	1856/1901 redemp
Hamlet	Ophelia	Thomas	1868 transc
Manon Lescaut	Manon	Auber/Massenet/Puccini	1856/1884/1893 transc/redemp
The Demon	Tamara	Rubinstein	1871 transc/redemp
Les Contes D'Hoffman	Antonia	Offenbach	1881 transc/redemp
The Snow Maiden	The Snow Maiden	Rimsky-Korsakoff	1880/81 transc/redemp
Parsifal	Kundry	Wagner	1882 transc/redemp
Le Villi	Anna	Puccini	1884 transc
The Queen of Spades	Countess	Tchaikovsky	1890 [None]
Salammbô	Salammbô	Reyer/Mussorgsky	1890/1892 transc/redemp
Pelléas et Mélisande	Mélisande	Debussy	1902 redemp?
Elektra	Elektra	Richard Strauss	1909 transc/redemp?
La Vida Breva	Salud	De Falla	1913 transc
Die Gezeichneten ("The stigmatized")	Carlotta	Schreker	1918 [None]

The Makropulos Case	Emilia Marty	Janáček	1926 transc/redemp
Das Wunder der Heliane	Heliane	Korngold	1927 transc/redemp
Males			
Undine	Hulbrand	Hoffman/Lorzing	1816/1845 transc/redemp
Rusalka	The Prince	Dargomyzhsky/Dvořák	1856/1901 transc/redemp
Bánk Bán	Bánk Bán	Erkel	1861 [None]
Die Walküre	Hunding	Wagner	1870 [None]
Boris Godunov	Boris	Mussorgsky	1869/1874 transc
Le Villi	Roberto	Puccini	1884 transc
Pelléas et Mélisande	Prince Golaud	Debussy	1902 [None]
The Wreckers	Mark	Smythe	1906 transc/redemp
A Village Romeo and Juliet	Sali	Delius	1907 transc
Osud	Ẑivny	Janáček	1907 transc
Das Wunder der Heliane	The Stranger	Korngold	1927 transc/redemp
Lulu	Dr. Goll	Berg	1937 [None]
The Turn of the Screw	Miles	Britten	1954 redemp
Death in Venice	Aschenbach	Britten	1973 transc
St Francis	Francis	Messiaen	1983 transc/redemp

Notes

[1] Vikram, Seth. *An Equal Music* (London: Phoenix London, 1999), 484.
[2] For those interested in statistics, for the last calculation, the chi-squared statistic with Yates correction is 5.3, the p-value is < .03.
[3] See Bell, Richard H. "Are Wagner's views of 'Redemption' relevant for the Twenty-first Century?" in *Das Kunstwerk der Zukunft: Perspektive der Wagnerrezeption im 21. Jahrhudert*. Ed. Sven Friedrich (Würzburg: Königshausen & Neumann, 2014), 71-86. Bell considers Wagner's approach to redemption under five interrelated aspects: 1. social/political/economic; 2. psychological; 3. existential; 4. cosmological; 5. religious. "Wagner's myths, like all myths, only work in a mythical reception. And one key aspect of his myths is the presentation on stage of a 'ritual' (e.g. the deaths of Isolde or Brünnhilde). But the sacrifice that interested Wagner more than any other was the death of Jesus of Nazareth. We find this in letters, essays, diaries, the sketches for *Jesus von Nazareth*, and above all in *Parsifal*. He viewed this death as one in which the human person participates; indeed it is one in which the whole creation participates. It is not that creation is forced into a state of suffering by Jesus. Rather there is a mutual identification: Jesus identifies with human misery and the human beings related to him identify with his suffering in a pattern of what Hübner would call "identical repetition". This mutual identification of the suffering Jesus and nature is also reflected in a remarkable comment Wagner made in the rehearsals for *Parsifal*, and noted down by Heinrich Porges, at the words of Parsifal in Act 2 "O torment of love" ("Oh! Qual der Liebe"): "Now all at once Parsifal sees how the whole world is a sacrificial slaughter". Wagner appears to have understood the redeemer's sacrifice not as one which excludes us ("ausschließende Stellvertretung") but rather as one which includes us ("einschließende Stellvertretung") (84-85).
[4] Levinas, E. *Totality and Infinity: An Essay on Exteriority*. Trans. Alphonso Lingis (Pittsburg: Duquesne University Press, 1969), 52, 293,295,299. Exteriority —the other, Interiority—the self. As with much philosophy the task is to unite the two.
[5] Levinas, E. Ibid 155.
[6] For a most illumined and informed analysis of this, see Scruton, R., Ibid., 2020, 18.
[7] Cottingham J. *In Search of the Soul: A Philosophical Essay*. (Princeton: Princeton University Press, 2020). 132.
[8] Schiller, F. von. *On the Sublime*, in *Two Essays*. Trans. J. A. Elias (New York, 1966), 198.
[9] Poizat, M, 1992, xiii.
[10] Critchley, S. *Tragedy, The Greeks and Us* (London: Profile Books, 2019), 188.
[11] *Aristotle. Translated by Theodore Buckley* (New York: Prometheus Books, 1992); Gadamer, H.-G. *Truth and Method*. 2nd Revised Edition (London: Continuum, 2004),125. Aristotle seems to have been very unclear as to what he meant by *Katharseos,* and there have been many discussions of the nature of this feeling. James Joyce referred to it as "the tragic emotion", for Hans-Georg Gadamer (1900-2002) it was tragic pensiveness. Joyce described the emotion as Janus-faced,

looking two ways, towards terror and towards pity; but it was for him a static emotion, not one like desire or loathing that demanded kinesis. The philosopher and critic Walter Kaufmann (1921-1980) was unhappy about the word 'pity', and when expressing the tragic emotion as one in which we are seized and shaken by the whole misery of humanity and he settles on the word 'ruth'.

[12] Scruton, R. Ibid., 2004: 164, 169, 180.

[13] Scruton, R. Ibid., 2004: 105, See also Coplan, A. and Goldie, P. *Empathy: Philosophical and Psychological Perspectives* (Oxford: Oxford University Press, 2011). *Einfühlung* (empathy) imagines the narrative of another person's feelings, but this can be through an understanding but not a feeling, i.e. cognitive or emotional. Empathy has recently become a subject explored in neuroscience, especially with brain imaging, beyond the scope of this book. See Trimble, M. R. Ibid., 2012.

[14] Damasio, A. *Looking for Spinoza: Joy, Sorrow and the Feeling Brain*. (London: Heinemann, 2003), 128. See also for the underlying reasons why only humans like to cry, Trimble, M.R., Ibid., 2012.

[15] Critchley, S., Ibid. 2019, 137. Italics in the original.

[16] Conrad, P. *A Song of Love and Death*, Ibid. 1987, 278. "Thanks to a new army of avant-gardists, opera has never been theatrically livelier; nor has it ever been so disputatious and so gratuitously bizarre." Conrad gives an excellent balanced discussion of the changes that have occurred in opera productions, for better or for worse, in more recent times.

[17] Johnson, J. *After Debussy, Music, Language, and the Margins of Philosophy* (Oxford: Oxford University Press, 2020). There are exceptions: he cites Wittgenstein, Bergson, Roland Barthes, George Steiner (1929-2020), and Nancy. From earlier generations are included Schleiermacher, Hegel, Schelling, Schopenhauer, and Nietzsche. Kant compared the effect of music on listeners to that of perfumed handkerchiefs. For further reading see: Sorgner, S. L. and Fürbeth, O. *Music in German Philosophy: An introduction* (Chicago: University of Chicago Press, 2010), 38.

[18] Johnson, J. Ibid., 2020: 21, 27.

[19] Bowie, A. *Music, Philosophy, and Modernity* (Cambridge: Cambridge University Press, 2007), 92.

[20] Bowie, A. Ibid., 2007: 378. Johnson, J. Ibid., 2020: 232.

[21] Cosima Wagner, *Diaries,* 3 April 1870, Ibid. Vol 1, 206.

[22] Goeher, L. *Dissonant Works and the Listening Public.* In *Cambridge Companion to Adorno*. Ed. T. Huhn (Cambridge: Cambridge University Press, 2004), 226.

[23] See Jefferies, S. *Grand Hotel Abyss: The Lives of the Frankfurt School.* (London: Verso, 2016), 379, for a brilliant analysis of the Frankfurt School and its membership. This included some of the most influential philosophers of the mid-20th century: Walter Benjamin (1892-1940), Adorno, Max Horkheimer (1895-1973), Habermas and Marcuse. A number left Germany and fled to America but some returned after the Second World War. Their influence waned and they did not seem to believe in the revolution anymore. The corruption of human consciousness with 'the freedom to choose what was always the same' (Ibid., 10) reflects the concern that many of us now have, surrounded by social media, 'fake

news' and misinformation. Their influence on music comes through, for example, Berg and Schoenberg, and their influence on many opera performances has to this day remained controversial and brought a legacy of bewilderment.

[24] The term comes from the book *One Dimensional Man* (1964) by Marcuse.

[25] Marcuse, H. *The Aesthetic Dimension: Toward a Critique of Marxist Aesthetics*. Trans. H. Marcuse and J. Sherover (Boston: Beacon, 1978), 73.

[26] Millington, B. "Performing Wagner's Music Dramas in the Modern Age", *Wagner News*, 14 (2020): 235, 9-20.

[27] *Victory over the Sun* is a Russian Futurist opera. The libretto was written in 'zaum language' in part by Aleksei Kruchonykh (1886-1968), the music composed by Mikhail Matyushin (1861-1934), with a prologue by Velimir Khlebnikov (1885-1922). The stage designer was Kazimir Malevich (1879-1935). The latter's famous alogical Black Square painting was included in the stage design.

[28] Jefferies, S. Ibid., 2016: 132.

[29] Hastings, S. "Shallow Brilliance", *Opera* (May 2017): 585-590.

[30] Chéreau himself used terms such as 'ganz schlecht', 'furchtbar' and 'entsetzlich' about various of the scenes of his production (very bad, terrible, horrible): Interview in *Die Zeit: Wagneropera.net, 2013*.

[31] Trimble, M. R.; Letellier, R. I.; and Hesdorffer, D. Ibid., 2013. The only person who clearly understood was John Tomlinson, see quote page 30,31.

[32] Parly, N. "Doing the Diva Dying: Performative Studies of Death in *Tristan und Isolde*". *The Wagner Journal*, 10 (2016): 45-55. Her analysis of Wagner's heroines in *Vocal Victories* refers to them as 'artistic creators' with their strong soprano voice offering considerable strength of character and determination, being more commanding than some may conclude. She suggests that Wagner was ahead of his time in such matters as equality between the sexes (revealed in his operas and writings) even if, echoing Clément, asking "Why must women suffer in order to redeem the world?" Parly, N. *Vocal Victories: Wagner's Female Characters from Senta to Kundry* (Museum Tusculanum Press) (Chicago: University of Chicago Press 2011), 35. Kate Hopkins in a recent analysis of Wagner's heroines has nicely examined Nila Parly's text of *Vocal Victories* and opened up a further exploration herself of the librettos and music of Elisabeth, Elsa, Brünnhilde, and Kundry. "They demonstrate empathy's noble potential … Wagner—whatever his views on women in life—has created some of the most morally courageous, humane characters in opera" (46). Hopkins, K. "Agents of Loving Empathy: Wagner's Heroines Reassessed". *The Wagner Journal,* 14:3 (2020): 31-46.

[33] Scruton, R. "Desecrating Wagner," *Wagner News* (2005):11-17.

[34] Tomlinson, J.; Fairman, R.; Miller, J. *Opera* (Feb 2020):173.

CHAPTER 25

CODA

> Through singing, opera must make you weep, shudder, or die.
>
> (Vincenzo Bellini)[1]

> In ancient times a story could only end in two ways: having passed all the tests, the hero and the heroine are married, or else they died. The ultimate meaning to which all stories refer has two faces: the continuity of life, the inevitability of death.
>
> (Italo Calvino, *If on a Winter's Night a Traveller*)[2]

The author Julian Barnes asks: "What is it about the present that makes it so eager to judge the past? There is always a neuroticism to the present, which believes itself superior to the past but can't quite get over a nagging anxiety that it might be … the further the past recedes, the more attractive it becomes to simplify it …'[3]

We simply love to predict the past.

As we noted in our Prologue, something happened in the arts, especially from the perspective of Western music, around 250 years ago: great works by ingenious composers raised the sonic horizon to new levels. Although we do not suggest that historical eras, mostly identified retrospectively, have clear boundaries, or do not overlap, it is surely of some significance that many pieces from those times are still in what might be termed 'the canon' and remain in demand worldwide. It must be of further interest that with regards to opera, its popularity has travelled far and wide, in contrast to imported compositions from any other continents. The reception of Western 'classical music' in countries such as Japan, China, Malaysia, Turkey and Egypt has been remarkable, in contrast to the adoption by Europeans and their diaspora of other cultures' variations.

> What we do know is that Western classical music has somehow managed to be foundational to the entire world, regardless of national or native cultures, gender, or race.[4]

The more radical art of the last century has had little to do with classical refinements of beauty or romantic philosophical intentions. As Critchley put it, the former is "an art of desublimation that attempts to adumbrate the monstrous, the uncontainable, the unreconciled, that which is unbearable in our experience of reality". The concept of the divine has been diminished or lost completely, a view that can be seen beyond the operas we have examined.[5] Certainly opera (inspired by Greek Tragedy) has many tragic endings, but the meanings always far outshine the stories. The symbolism and metaphor embedded in the librettos and dramas must be alive, not stale, and adumbrated by and with the music, the ensemble embodied, re-sounding within.

In earlier chapters we have discussed the development of opera from the pre-classical era through to the present, but devoted our exploration of what can be called the Romantic Era with the huge shifts in space and time that occurred mainly in the 19th century. In the 18th century, the ancient association of Tragedy to catharsis became rather untenable, even the meaning or need for catharsis was doubted. As composers moved music towards the 20th century, neglect of the inward as opposed to the outward signification of opera denied the tensions between *being* and *becoming* and cauterised the ability to explore transcendental interpretations and associations with Redemption. Ideological, political, and social preoccupations, modernity as such, struggled with a metaphysical realism, and instead of throwing the gauntlet down, threw the baby out with the (metaphorical) bath water. Friedrich Schlegel aptly said: "In truth you would be distressed if the whole world, as you demand, were for once to become completely comprehensible".[6]

Wagner's tragedies were not those of the Greek playwrights, his transformation of Redemption through Love is not to be found there. The musicologist Burnet James informs us that Dido's Lament (in Purcell's *Dido and Aeneas*), probably never heard by Wagner, almost exactly parallels musically Isolde's *Liebestod,* demonstrating "the totality of music across the ages and of many different kinds, that determines these coincidences and correspondences".[7] The major triad was the foundation stone of the Western harmonic system, triadic harmony having dominated our music for over 500 years. However, according to Philip Ball, the scales of ancient Greek music were composed of notes similar to those in use today. If this is so, we listen to music based on very old tonal appreciation—that is, music with a tune, with a tonal system anchored to tonal centres possessing tonic gravity.[8] Roger Scruton notes:

> We cannot hear musical movement without seeking for points of stability and closure—points towards which the movement is tending or from which it is diverging, and to which it might at some point 'come home' … there are a priori constraints on musical syntax.[9]

We have emphasised the importance of a phenomenological approach to understanding what it is that holds composers and audiences enthralled over centuries. We have dwelt much on why *Unexpected Death* is so frequent in opera. The answer lies in the original myth embodied in the origins of opera: Orpheus. He, the archetypal lyric poet, suffers love, loss, sorrow, and attempts to recover what is lost. His quest, associated as it is with the power of music, provides the essential golden thread of the plot.

The quest is related to a metaphysical or spiritual concept, reality based as it is, in the human realities of love, loss and suffering, but with the desire and capacity, the bravery, to address these and change circumstances by courageous action. This implies belief in a spiritual or metaphysical reality beyond the realm of human frailty: the experience of the sacred is a human universal. Scruton intimated that Wagner "believed in the idea of the sacred as an independent force in human affairs", and "in unity of love and death … we glimpse the meaning of human life"—through art we explore the mystery of sacred things.[10] The Orpheus myth tells us that the most powerful and ennobling human emotion is love. When this is lost through death, treachery, betrayal etc. the mind/spirit refuses to accept this as final and is determined to find and recover what was lost. This is also a mystery of self-offering or even sacrifice, because to act beyond the paralyzing confines of human failing and loss requires great spiritual resolution and physical courage that will confront even death itself.

The Quest and its motivating force are related to transcendence. Scruton reminds us of other transcendental experiences we take for granted, such as those beyond the edge of empirical knowledge that we have on meeting another, the embedded humanity in us, the cocoon of feelings that come together as we feel the presence and feelings of the other. Our emotions transferred between each other have a transcendental quality—a "kind of hunger for the transcendent—a reaching beyond".[11]

This is to embrace the possibility of a spiritual reality beyond the corporeal, and in doing so to enter the metaphysical sphere. This is induced, conjured, sustained, celebrated in the power of music where the cooperation of melody, harmony and tonality often induce a feeling of rapture, even ecstasy, that promises life beyond the corporeal realities. It can carry a transforming meaning since it can infuse an emotional power, even with physical effects of tingling, shivering, tremors—even heart-attack.

Self-offering has always been seen as important, crucial, and sacred. All humanity struggles (because of consciousness) to understand, explain, escape from the sorrows of natural limitation (frailty, illness, misfortunes, disaster, death) to find permanence—immunity from aging and deterioration; to find a place where the power and richness of human consciousness and imagination will find its proper healing fulfilment, immune from mutability, imperfection, evil and loss (cf. Romans 8:18-25).

Gender often plays a determining role in this process. One must look to the Orpheus myth again, where the male hero bears the sorrow of loss, and must find a way to retrieve the stricken beauty and innocence embodied in Eurydice. Jesus at the centre of time brings about the intersection of the mundane and the spiritual (or eternal) in Incarnation, and in freely assumed suffering to the point of death. This is the supreme sacrifice, and opens the human condition to transformation and the immortal. The role of Eve (the fallen woman) is redeemed in Mary (the New Eve) who cooperates through challenge and loss to bring about the Incarnation. She is the First Disciple, present at Pentecost, and is assumed into heaven to complete the process of salvation.

In opera we have conjoined autonomic dysfunction, *Unexpected Death*, and Love-Death conceptually with abnormal mental states, including madness born of an intensity of suffering, of an undue confrontation with the limitations of human experience. In the Romantic period, madness was regarded as a strong motif, a pathway to an alternative reality: Orpheus in his sorrow was viewed as mad with grief. Hoffmann saw madness not as an aberration or illness, but as a unique entrée into an alternative (higher?) reality casting further light on the human condition, and a special preliminary to the immortal realities. It can even serve as a metaphor for salvation/redemption in that it dissociates itself from the predictable and sorrowful and begins to live in a new reality opening the transforming power of love, and therefore doubly vulnerable to its loss, a metonym for the glory but crippling imperfection of human life.

The emphasis in much music on the final resolution in the tonic key perhaps implies some teleological homecoming. Such philosophical meta-narratives are not a feature of 20th century horizons. Nietzsche led the way, yet the importance of *becoming*, a yearning impulse and finding an ending, is paradigmatic of the Romantic Quest. The phenomenology of anticipation, tension and resolution in music, what David Huron refers to as *Sweet Anticipation*, is linked to tonality and 'coming home'.[12] Music has an intentionality, an 'aboutness' of its very own. The special feeling associated with Tragedy is a physiologically aroused aesthetic state but not a catharsis, and not, as was argued by some philosophers, circling on the ideas of Kant, one of passive disinterested contemplation.[13]

It is beyond that scope of our book to go further into these matters from a neuroscience perspective, but we have emphasized the significance of the ANS as a basis for our feelings (shivers down the spine and much more) and for the emergence of *Unexpected Death* on the stage. These autonomic parakinesia can serve as entrées to a world detached from the mundane and can have a transforming effect.[14] Poizat in Chapter 13 described the feeling "known to opera lovers, the thrill or shiver that courses through the body in those supreme moments of musical ecstasy … tears of joy". Others have described such enchantments occasioned by art: Gerard Manley Hopkins' (1844-1889) 'instress', the manifestation of 'inscape'; Rainer Maria Rilke's (1875-1926) 'in-seeing' (*Einsehen*), moving from the surface of things to their heart, linked with 'feeling into' 'in-being' (*Einfühlung*), an active process but which allowed the latter to secularize the power of religious symbolism. For Poe it was 'the effect'. Recall Koestenbaum's *interiority*, "the inside of a body and the inside of a self … those operatic moments when suddenly interiority upstages exteriority, when an inner and oblique vision supplants external verity". Perception as 'apprehension', a power we direct towards the world.[15]

Music, the essential Orphic medium (related to the transcendental bridges of Dionysia, wine, song, dance and prayer) is the breath of life related to the Breath of God (*ruah*, *Spiritus Sanctus*), the vector of inspiration and life itself, the key transmitter of metaphysical reality or ultimate truth. Pre-eminently, being a universal medium able to express itself and move the heart beyond words, music is a unique drive of emotional change that looks to the metaphysical world of new realities untrammelled by corporeality. Music is a source of inspiration and life itself, intimately linked to a search for permanence. The fact that many musical sounds are produced by the human breath poured into (rather than through) wooden and brass media, and fingers plucking strings (invoking Pan and Orpheus) suggests symbols of the Spirit and nature working though us. This is particularly related to the mystery of the power of song and the human voice using air through physicality to produce sounds that entrance our search for constancy and beauty amidst inevitable change and mutability.

In our quest to answer Clemént's problem concerning the unexplained *Unexpected Deaths* of women in opera, we have shown differences between males and females, the main era when deaths occurred (Romantic, especially the 19th century for females), the position in the opera (usually at the finale for females), the associations with transcendence and the close affinity of the latter with Redemption (for both males and females) and the Love-Death, with its cipher the *Liebestod*. Our own observations of many operas and studies of the librettos yielded close enfolding of autonomic instability and affairs of the heart (socio-politically, romantically, or medically)—*Unexpected Deaths* no longer unexplained.

If the music which continues to enchant us has ancient Orphic ancestry, then, as we have discussed in this book, so does the *Liebestod*—the Love Death, the yoke of elective infinities, the I–Thou gaze of embodiment of the self and the other, *Mitleid*.

> The redemption offered by the Liebestod is no illusion, it offers the very thing that redemption is, namely a transcendence of the world of appetite into the realm of values. In the face of this transcendence, death can do no harm.[16]

Humans are distinguished from other animals by several fundamental behaviours, especially the six L's of life: Language, Lying, Laughter, Lacrymation, Lyric and Love. For Wagner, and the rest of us, death is always the present-future—without death, there is no life; without life there is no love; without love there is no art; without art there is no music, and a world without music would be inhuman.

Das Ende

Wagner began having chest pains in December 1881. He and Cosima left Bayreuth on 14 September 1882 to go Venice and took up residence at the Palazzo Vendramin-Calergi. He continued to have several episodes of 'heart spasms', which by 12 January 1883 were lasting up to two hours. He was planning to write an article to be called either 'The Feminine in Human Life', or 'The Eternal in the Feminine'. [**Fig. 25.1-2**]

On 11 February he told Cosima of his dream in which "All my women are now passing before my eyes". She asked him if he still cared for her, to which he replied, "My world contains only Isolde, how could it be without Isolde?" On the 13 February after a stormy row with Cosima (over an invitation to the soprano Carrie Pringle to come to stay), he was writing his text on the feminine in the human, which he said would be his last work, and died suddenly in the arms of Cosima.[17]

Dr Keppler had prescribed massage therapy, Dr Kapp diagnosed stomach neuralgia. Wagner took some valerian, and somehow these ministrations comforted him: the diagnosis of heart disease seems to have been kept from the composer. Dr Keppler described the terminal event as "sudden death". The last words of famous people are often noted, yet several different versions are quite common. As Wagner collapsed his watch fell from his pocket: he was reported to have exclaimed, "My watch" (*Meine Uhr*—which could also be interpreted as 'my last hour') or "Call my wife and the doctor". Unlike hearsay, what people write, however, cannot be disputed. As his pen slipped from his hand and the writing glided from one page onto the next, his last words were plainly visible— *Love-Tragedy*.[18]

Epilogue

> We forgive the man who gave the world *Tristan und Isolde*.
>
> (Hans von Bülow)

Notes

[1] Vincenzo Bellini in Levine, R. *Weep, Shudder, Die: A Guide to Loving Opera* (New York: Harper Collins, 2011).
[2] Calvino, Italo. *If on a Winter's Night a Traveller* (London: Harncourt, 1981), 259.
[3] Barnes, J. *The Man in the Red Coat* (London: Jonathan Cape, 2019), 168.
[4] Mauceri, J. *For the Love of Music: A Conductor's Guide to the Art of Listening.* (London: Weidenfeld & Nicholson, 2019), 185. He contrasts music with what is served in restaurants. In one of our studies, we assessed the emotional responses to music, especially crying, in several countries. When asking about 'classical music' the responses were very similar in the European countries. But it was not possible to get any responses in Japan. One of our colleagues told us that people did not listen to Japanese 'classical' music very much nowadays. The respondents however were able to note their emotions to Western music. My colleague Kousuke Kanemoto wrote to me the following comments: "You ask, 'Could you cry when you listen to Japanese classic music?' The question is very difficult to

answer because there is no such thing in Japanese traditional music. If we are forcibly urged to answer the question, "Gagaku" (雅楽) is a special music played during the ceremonies in the Shinto Shrines may be a candidate for something resembling to the Western classic music, over which no one could cry." Mauceri notes that Japan has 19 official symphony orchestras.

[5] Critchley, S. Ibid., 2019, 219.

[6] Bowie, A. Ibid., 2007,404.

[7] James, B. *Wagner and the Romantic Disaster* (New York: Midas Books, 1983), 125.

[8] There are many good discussions of the development of music for the interested: Ball, P. *The Music Instinct* (London: Bodley Head, 2010); Storr, A. *Music and the Mind* (London: Harper Collins, 1997); Ross, A. *The Rest is Noise* (London: Fourth Estate, 2007); Spitzer M. *The Musical Human: A History of Life on Earth* (London: Bloomsbury, 2021).

[9] Scruton, R. *Understanding Music* (London: Continuum International Publishing, 2009), 13.

[10] Scruton R. *Wagner and German Idealism*. Based on Wagner's Philosophers, BBC Radio 3 (20.05.2013).

[11] Scruton R, Ibid. 2018, 80.

[12] Huron, D. *Sweet Anticipation: Music and the Psychology of Expectation* (Boston, MS: Massachusetts Institute of Technology, 2007).

[13] Kant seems to have considered disinterested pleasure as the aesthetic response to a work of art. Schopenhauer seems also to have considered that contemplation of works of art allowed escape from the burden of the relentless manifestations of the Will that drives the world and everything in it, the aesthetic being a disinterested experience. Dismissing Dionysus, Nietzsche did not agree with such notions.

[14] For the interested, and touched on earlier in the book: key anatomical activities related to the tragic emotion (tragic joy), which distinguish it from fear, are decreased activity of the amygdala nuclei, and the activity of the parasympathetic ANS (parasympathetic rebound), via the vagus nerve modulating emotional activity. These do not, from a physiological perspective, lead to a 'catharsis', but to a sensation of intimacy which is a special combination of arousal and calm.

[15] Corbet R. *You Must Change Your Life: The Story of Rainer Maria Rilke and Auguste Rodin* (New York: W. W. Norton and company, 2016).

[16] Scruton, R. Ibid. 2004, 192.

[17] Cosima Wagner, Ibid. 1980: Vol. 2, 813-1010. Cosima records many of these 'spasms' in her diaries between1882, until his death. They occurred sometimes before or after eating, sometimes at night, some in bed. They were in the earlier phases short-lived. He told Cosima that he preferred to go out alone as he was ashamed of the attacks and needed to walk slowly. They were almost certainly episodes of angina pectoris, eventually leading to a heart attack. During the build-up to the fatal one, Cosima noted many difficulties that Wagner had, including dealing with business relationships, worries about the illnesses of their children, problems with her father (Liszt), and the death of his close friend Arthur de Gobineau (1816-1862). A Viennese physician (Dr Kurz /Dr Kapp?) diagnosed his

malady as stomach neuralgia. Dr Keppler recommended massages, which he reported helped him.

[18] Last words and writings are noted in many biographies. Newman, E. *The Life of Richard Wagner* (Cambridge: Cambridge University Press, 1946 [Vol. 4 1866-1883]), 4:712; Callow, S. *Being Wagner: The Triumph of the Will* (London: William Collins, 2017), 191; Dawson-Bowling, P. Ibid. 2013, Vol. 1, 19; Millington, B. *Richard Wagner: The Sorcerer of Bayreuth* (London: Thames and Hudson, 2012), 131, 243, 245. Dawson-Bowling called the Carrie Pringle tale a canard, it has been questioned by others, including Millington. Valerian is a mild sedative.

BIBLIOGRAPHY

Adlington, W. (trans.). *The Golden Asse of Apuleius* (London: Grant Richards Ltd, 1913), 35, 36.

Alexander, E. *Proof of Heaven: A Neurosurgeons Journey into the Afterlife* (New York: Simon & Schuster, 2012).

Ameriks, K. "On reconciling the Transcendental Turn with Kant's Idealism". In Gerdner, S. and Grist M. *The Transcendental Turn* (Oxford: Oxford University Press, 2015), 45.

Andréossy, V. *L'Esprit du chant* (Plan de la Tour: Éditions d' Aujourd 'hui, 1979). 122.

Apollonius of Rhodes, *The Voyage of the Argo: The Argonautica. Translated with and Introduction by E. V. Rieu* (London: Penguin Books, 1959), 1:1207-1239).

Aristotle. *Aristotle. Translated by Theodore Buckley* (New York: Prometheus Books, 1992).

Ashbrook, William. *Donizetti* (London: Cassell, 1965).

Au, Susan. "The Shadow of Herself: Some Sources of Jules Perrot's *Ondine*". *Dance Chronicle* (Taylor & Francis, Ltd.) 2:3 (1978): 160.

Baker, Theodore. *Baker's Biographical Dictionary of Musicians.*8th ed. (London; Macmillan Publishing Company, 1995).

Baggini, Julian. *How the World Thinks: A global History of Philosophy* (London: Granta, 2018).

Ball, Philip. *The Music Instinct* (London: Bodley Head, 2010).

Barbier, Patrick, *Farinelli: Le castrat des Lumières* (B. Grasset, 1994).

Barbier, Partick. *The World of the Castrati: The History of an Extraordinary Operatic Phenomenon* (Souvenir Press, 1998).

Bar-el, Y.; Durst, R.; Katz, G.; Zislin, J., Strauss Z, and Knobler, H. Y. "Jerusalem Syndrome". *British Journal of Psychiatry*, 176 (2000): 86-90.

Barnes, J. *The Man in the Red Coat* (London: Jonathan Cape, 2019), 168.

Barrett Browning, E. *Aurora Leigh and Other Poems.* Ed. John Bolton and Julia Holloway (London: Penguin Classics,1995).

Barth, M., Voss, H. and Dietrich, E. *Wagner, a Documentary* (Oxford: Oxford. University Press, 1975).

Brasher, Thomas L. Judith Tick, Paul E. Beaudoin (ed.). "Walt Whitman's Conversion To Opera". *Music in the USA: A Documentary Companion* (Oxford: Oxford University Press, 2008), 207.

Baudelaire, Charles. *The Flowers of Evil. Poison.* Trans. James McGowan (Oxford: Oxford University Press, 2008).

Bell, Richard H. "Are Wagner's views of 'Redemption' relevant for the Twenty-first Century?" in *Das Kunstwerk der Zukunft: Perspektive der Wagnerrezeption im 21. Jahrhudert*. Ed. Sven Friedrich (Würzburg: Königshausen & Neumann, 2014), 71-86.

—. "Teleology, Providence and the 'Death of God': a New Perspective on the *Ring* Cycle's Debt to G.W.F. Hegel". *The Wagner Journal*, 11:1 (2017): 30-45.

Berlin, Isaiah. *The Crooked Timber of Humanity: Chapters in the History of Ideas* (London: John Murray, 1990).

Berry, Mark. *After Wagner: Histories of Modernist Music: Drama from Parsifal to Nono* (Wooodbridge: The Boydell Press, 2014).

—. "Is it Here that Time becomes Space? Hegel, Schopenhauer, History and Grace in *Parsifal". The Wagner Journal* 3:3 (2009):29-59.

—. M. "The Positive Influence of Wagner upon Nietzsche" *The Wagner Journal* 2 (2008): 11-28; 23.

Bhaktivedanta Swami Prabhupāda, His Divine Grace A.C. *Bhagavad-Gītā as it is* (Borehamwood: The Bhaktivedanta Book Trust, 1986), 4:9, 230; 7:7, 373; 4:39, 267.

Bible, The. *The Holy Bible*. Revised Standard Version (1894). Catholic Edition (London: Thomas Nelson & Sons, 1966).

Blake, William (1988). *Auguries of Innocence* (1789). Ed. David V. Erdman. *The Complete Poetry and Prose*. Newly revised ed. (Anchor Books, 1988). See also *Selected Poems*. Ed. P. Butter (London: Everyman, 1993), 121.

Blackmore, S. J. *Seeing Myself: The New Science of Out-of-Body Experiences* (London: Robinson, 2013).

Blin, Gilbert. *Orfeo*. Italian Libretto by Alessandro Striggio, with English Translation by Gilbert Blin, mostly copied from the 1615 edition of Monteverdi's score (second printed edition). *Boston Early Music Festival* (Chamber Opera Series, Boston, 2012).

Blood, A. J. and Zatorre, R. J., "Intensely pleasurable responses to music correlate with activity in brain regions implicated in reward and emotion." *Proceedings of the National Academy of Sciences USA,* 98 (2001): 11818–11823.

Bowie, A. *Music, Philosophy, and Modernity* (Cambridge: Cambridge University Press, 2007).

Brontë, E. *Wuthering Heights* [1847]. Ed. David Daiches (London: Penguin Books, 1965), Ch.15:194-196.

Brown, H.M. *The Quest for the Gesamtkunstwerk and Richard Wagner* (Oxford: Oxford University Press, 2016).

Brown, P. *The World of Late Antiquity* (London: Thames and Hudson, 1971).

Cabaud, J. *Mathilde Wesendonck Isolde's Dream* (Milwaukee: Amadeus Press, 2017).

Cahan, David. *Helmholtz: A Life in Science* (Chicago: University of Chicago Press, 2018)

Callaso Roberto. *La folie Baudelaire*. Trans. Alastair McEwan (London: Allen Lane, 2012).

Callow, Simon. *Being Wagner: The Triumph of the Will* (London: William Collins, 2017), 191.

Calvino, Italo. *If on a Winter's Night a Traveller* (London: Harncourt, 1981), 259.

Carlyle, Thomas. *On Heroes, Hero-Worship and the Heroic in History* (London: Chapman and Hall Ltd, 1907), 205.

Carnegy, P. *Wagner and the Art of the Theatre* (London: Yale University Press 2006), 76-81.

Cather, W. S. 'A Wagner Matinée'. *Everybody's Magazine*, 10 (1904): 325-328.

Celletti, Rodolfo. *A History of Bel Canto*. New edition (Clarendon Paperbacks) (Oxford University Press, 1996).

Charlton, David. *Grétry and the Rise of Opéra-Comique* (Cambridge: Cambridge University Press, 1986), 226-250.

Chéreau, Patrice. Interview in *Die Zeit: Wagneropera.net, 2013*.

Cirlot, J. E. *A Dictionary of Symbols* (New York: Philosophical Library, 1962), 199-200.

Clément, Catherine. *Opera: The Undoing of Women* (London: Virago Press, 1989).

Clément, Félix & Pierre Larousse, *Dictionnaire des operas*, 2 vols. (Paris: Librairie Larousse, 1905), 1:695.

Coleridge, Samuel Taylor. *Biographia Literaria* (London: Dent, 1965), 91 (*Esemplastic*).

Conrad, Peter. *A Song of Love and Death* (London: Chatto & Windus,1987).

Cooke, Brett. "Constraining the Other in Kvapil and Dvořák's *Rusalka,*" in Brett Cooke, George E. Slusser, and Jaume Marti-Olivella (eds), *The Fantastic Other: An Interface of Perspectives* (Amsterdam: Editions Rodopi, 1998), 121-42.

Coplan, A. and Goldie, P. *Empathy: Philosophical and Psychological Perspectives* (Oxford: Oxford University Press, 2011).

Corbet R. *You Must Change Your Life: The Story of Rainer Maria Rilke and Auguste Rodin.* (New York: W. W. Norton and Company, 2016).

Critchley, S. *Tragedy, The Greeks and Us* (London: Profile Books. 2019), 188.

Cruz, Gabriela. *Grand Illusion. Phantasmagoria in Nineteenth-Century Opera* (Oxford: Oxford University Press, 2020).

Damasio, A. *Looking for Spinoza: Joy Sorrow and the Feeling Brain* (London: Heinemann, 2003), 128.

Dante Alighieri. *Canzoniere.* Trans. E. H. Plumptre (London: Sir Isaac Pitman and Sons Ltd), Sonnet 31 1-3, 6.

Dante. *Paradiso*. *The Divine Comedy*. Trans. R. Kirkpatrick (London: Penguin Classics, 2012), Canto 33: 143-145.

Dante. *Purgatorio*. *The Divine Comedy*. Trans. R. Kirkpatrick (London: Penguin Classics, 2012), Canto 18: 19-21.

Dante. *La Vita Nuova.* Trans. B. Reynolds (Cambridge: Cambridge University Press 1969), 29.

Daub, A. *Tristan's Shadow: Sexuality and the Total Work of Art after Wagner* (Chicago: University of Chicago Press, 2014).

Dawson-Bowling, Paul. *The Wagner Experience: and its meaning to us*. 2 vols. (London: Old Street Publishing, 2014).

Dent, Edward J. *The Rise of Romantic Opera* (Cambridge: Cambridge University Press, 1976).

Derrida, J. *Writings and Difference* (Oxford: Routeledge, 2001), 31

Dilthey, Wilhelm. *Poetry and Experience: Selected Works, Volume 5.* Ed. R.A. Makkreel and F. Rodi (New Jersey: Princeton University Press, 1985), 253.

Donne, John. *An Anatomy of the World, First Anniversary* (London: W. Stansby 1625), Lines 206-209, 213.

—. "The extasie" in Gardner H. *The Metaphysical Poets* (London: Penguin Books, 1972): John Donne, *The extasie,* pp. 74-77.

—. "The Good-Morrow". In: *John Donne: The Penguin Poets* (London: Penguin Books, 1950), 23.

—. *The New Oxford Book of Seventeenth Century Verse* (Oxford: Oxford University Press 1991), Sonnet XIV, 118.

Dyson, M. "Sea, Mirror, Woman, Love: Some Recurrent Imagery in 'Opera and Drama'. *The Wagner Journal* 5:3 (2011):16-33.

Eckermann, J. P. *Gespräche mit Goethe in den den letzen Jahren seines Lebens*, 3rd ed. (Berlin: Aufbau-Verlag, 1987), 286, 350.

Eliot, Thomas Sterne. *Four Quartets*: *Little Gidding* (London, Faber & Faber, 1944), 41-42.

—. *Burnt Norton. Little Gidding. Four Quartets* (London: Faber & Faber, 1940), 7, 42, 44; *Ash-Wednesday* (London: Faber & Faber,1930), 13, 20, 25.

Emerson, Caryl. "Capek, Janáček, that Makropulos Thing, and a Word about Sacrificed Women in 20th-Century Slavic Opera" in Craig Cravens; M. Fidler; S. C. Crespin (eds.), *Between Texts, Languages, and Cultures. A Festschrift for Michael Henry Heim* (Bloomington, Indiana, 2008).

—."To What End Rusalka? Pushkin's Folk Tragedy and Dargomyzhskii's Opera". *The Slavonic and East European Review*, 97:1, *1917 and Beyond: Continuity, Rupture and Memory in Russian Music* (January 2019):169-200.

Emerson, Ralph Waldo, philosopher. *Nature*.

Emslie, B. *Richard Wagner and the Centrality of Love* (Woodbridge: The Boydell Press, 2010).

Everett, Andrew. *Josephine's Composer: The Life, Times and Works od Gaspare Pacifico Luigi Spontini (1774-1851)* (Bloomington IN: Author House, 213).

Eliot, Thomas Sterne. *Four Quartets*: *Little Gidding* (London, Faber & Faber, 1944), 41-42.

—. *Burnt Norton. Little Gidding. Four Quartets* (London: Faber & Faber, 1940), 7, 42, 44; *Ash-Wednesday* (London: Faber & Faber,1930), 13, 20, 25.

Emerson, Caryl. "Capek, Janáček, that Makropulos Thing, and a Word about Sacrificed Women in 20th-Century Slavic Opera" in Craig Cravens; M. Fidler; S. C. Crespin (eds.), *Between Texts, Languages, and Cultures. A Festschrift for Michael Henry Heim* (Bloomington, Indiana, 2008).

Euripides, *Hecuba* in *Euripides, The Trojan Women and other plays*. Trans. J. Morwood (Oxford: Oxford World Classics, 2001), 86, 87, 438. Ibid., *The Trojan Women*, 1242-1244.

Fichte, J. G. "Die bestimmung des Menschen". In *Fichtes Werke*. Ed. I. H. Fichte. Trans. R. Richards (Berlin: Walter de Gruyter), 1971) 2,73, 229.

Fontana, David. *The Secret Language of Symbols: A Visual Key to Symbols and Their Meanings* (London: Pavilion Books Ltd., 1993), 87.

Fouqué. Friedrich de la Motte-Fouqué, *Undine* [1811] (Stuttgart: Philipp Reclam Jun., 1969), 90-91.

—. *Undine by De La Motte-Fouqué. Adapted from the German by W. L. Courtney and Illustrated by Arthur Rackham* (London: William Heinemann, 1909), 130-132.

Fragos, Emily. *In Music's Spell* (New York: Alfred Knopf, 2009), 43.

Gadamer, H.-G. *Truth and Method.* 2nd Revised Edition (London: Continuum, 2004), 125.

Gardner, H. *The Metaphysical Poets* (London: Penguin Books 1972). See John Donne, *The Extasie*, 74-77.

Garten, H. F. *Wagner the Dramatist* (London: Alma Classics, 1977).

Godfrey Smith, Peter. *Other Minds: The Octopus and the Evolution of Intelligent Life* (London: William Collins, 2018), 200.

Goeher, L. *Dissonant Works and the Listening Public.* In *Cambridge Companion to Adorno*. Ed. T. Huhn (Cambridge: Cambridge University Press, 2004), 226.

Goethe, Johann Wolfgang von. *Die Braut von Korinth* in *Goethe. Selected Verse, with an Introduction and Prose Translation by David Luke* (The Penguin Poets) (London: Penguin, 1964), 159-168.

—. *Elective Affinities.* Trans. R. J. Hollingdale (London: Penguin Classics, 1971), Chapter 4, 46-57.

—. *Westöstlicher Divan Suleika Nameh—Buch Suleika Gingko Biloba. Reich-Ranicki M. 1000 Deutsche Gedichte* (Frankfurt am Main: Insel Verlag, 1994), 367.

Gordon, J. '*Plato's Erotic World From Cosmic Origins to Human Death*' (Cambridge: Cambridge University Press, 2012), 1-3.

Graves, Robert. *The Greek Myths* (London: Penguin Books, 1960), Vol 1, 115.

Green, M. *Nietzsche and the Transcendental Tradition* (Chicago: University of Illinois Press, 2002).

Griffiths.; T. D.; Buchel, C.; Frackowiak, R. S.; Patterson, R. D. "Analysis of temporal structure in sound by the human brain." *Nature Neuroscience,* 1 (1998):422–427.

Hackworth, R. '*Plato's Phaedrus*' (Cambridge: Cambridge University Press, 1952), 10.

Harrison N. *The Wordsmith's Guide to English Song: Poetry, Music and Imagination*. Volume 2 (Oxford: Compton Publishing, 2016), 12.

Hastings, S. "Shallow Brilliance", *Opera* (May 2017): 585-590.

Hegel, Georg Wilhelm Friedrich. *Introductory Lectures on Aesthetics* (London: Penguin Classics, 1993), 60-61.

—. *Philosophe der Geschichte* (Stuttgart: Frommann-Holzboog, 1988).

Hegel's logic; see https://plato.stanford.edu/entries/hegal-dialectics.

Heine, Heinrich. *Heine. Introduced, Edited and Translated by Peter Branscombe* (The Penguin Poets) (London: Penguin Books, 1967), 40-41.

—. *De l'Allemagne* (1833-34), 453.

—. "Die Lorelei" in 'Die Heimkehr', *Buch der Lieder* (1822).

Herman, Arthur. *The Cave and the Light: Plato versus Aristotle, and the Struggle for the Soul of Western Civilisation* (New York: Random House, 2013).

—. *The Cave of Light: Plato Versus Aristotle, and the Struggle for the Soul of Western Civilisations* (New York: Random House, 2013);

Hesdorffer, D. and Trimble, M. "Music and the brain: the neuroscience of music and musical appreciation". *British Journal of Psychiatry International*, 1;14, 28-31. eCollection, 2017.

Hesse, Hermann. *The Seasons of the Soul*. Trans. Fischer, L. M. (Berkeley: North Atlantic Books, 2011), 95.

Holland, Tom. *Dominion: The Making of the Western Mind* (London: Little Brown, 2019).

Holman, J. K. *Wagner Moments: A Celebration of Favourite Wagner Experiences* (New York, Amadeus Press, 2007), 54.

—. *Wagner's Ring: A listener's companion and concordance*, (Milwaukee: Amadeus Press, 1996), 264-68.

Husserl, E. *Ideas: General Introduction to Pure Phenomenology*. Trans. William Ralph Boyce Gibson (1869-1935) (London: Routledge Classics, 2012).

Hume, D. *A Treatise of Human Nature (1739). Reprinted from the Original Edition in three volumes and edited, with an analytical index, by Lewis Amhurst Selby-Bigge, M.A.* (Oxford: Clarendon Press, 1896), 217.

Huron, D. *Sweet Anticipation: Music and the Psychology of Expectation* (Boston, MS: Massachusetts Institute of Technology, 2007).

Huysmans, Joris-Karl. *Against Nature* (*À Rebours*). Trans. Robert Baldick, (London: Penguin Books, 2003): 53.

James, B. *Wagner and the Romantic Disaster* (New York: Midas Books, 1983),125.

James, William. *The Varieties of Religious Experience* [1902] (London: Penguin Books, 1982), 380-382, 416, 427.

Jelliffe, S. E. and White, W. A. *Diseases of the Nervous System: A Text-Book of Neurology and Psychiatry.* 2nd Ed. (London: H K Lewis, 1929), vi vii.

Jefferies, S. *Grand Hotel Abyss: The Lives of the Frankfurt School* (London: Verso, 2016), 379.

Johnson, G. R. "Swedenborg's positive influence on the development of Kant's mature moral philiosphy". In: *Philosophy, Literature, Mysticism: an anthology of essays on the thought and influence of Emanuel Swedenborg.* Ed. Stephen McNeilly (London: The Swedenborg Society 2013), 25-44.

Johnson, J. *After Debussy, Music, Language, and the Margins of Philosophy* (Oxford: Oxford University Press, 2020).

Kakuzō, Okakura. *Book of Tea* (1906) in J. Baggini, *How the World Thinks: A Global History of Philosophy* (London: Granta, 2018). 50, 299

Kampen, Claire van. *Farinelli and the King* (London, 2015; New York, 2017).

Keats, John. *Endymion.* In *The Complete Poems*. Ed. Barnard J, (London: Penguin Classics, 2006), 107,729.

Keller, Gottfried. *A Village Romeo and Juliet* (1856). Trans. Paul Bernard Thomas. From volume XIV *of German Classics of the Nineteenth and Twentieth Centuries* (New York: The German Publication Society, 1856), 56.

Kellner, L. *Alexander von Humboldt* (London: Oxford University Press, 1963).

Kennedy, Michael. *Richard Strauss: Man, Music, Enigma* (Cambridge: Cambridge University Press, 1999), 26,303.

Kerényi, Carl. *The Gods of the Greeks* [1951] (London: Penguin: 1958).

Koelsch, S.; Fritz, T.; Schlaug, G. "Amygdala activity can be modulated by unexpected chord functions during music listening." *Neuroreport*, 3;19(18) (2008):1815-9. doi:10.1097/WNR.0b013e32831a8722, 2008.

Koestenbaum, W. *The Queen's Throat: Opera, Homosexuality and the Mystery of Desire* (Boston: Da Capo Press, 2001).

Köhler, J. "The Ring and the Romantic Tradition". *The Wagner Journal* 2 (2008): 29-30, 33.

Koski, Barry. *Agrippina* (London: Royal Opera House, 2019),13-14.

Kuehn, M. *Kant a Biography* (Cambridge: Cambridge University Press, 2001).

Kupfer, Harry. Bayreuth Programme (1978).

Laan, J. M. van der (ed.). *The Faustian Century: German Literature and Culture in the Age of Luther and Faustus*. Ed. J. M. van der Laan and Andrew Weeks (London: Camden House, 2013).

Lakoff, G. and Johnson, K. *Philosophy in The Flesh: The Embodied Mind and Its Challenge to Western Thought* (New York: Basic Books, 1999).

Langham Smith: "Pelléas et Mélisande", *Grove Music Online* ed. L. Macy.

Larue, Steven C (ed.), *International Dictionary of Opera.*2 vols. (Detroit MI: St James Press, 1993), with extensive bibliographies.

Lattimore, Richmond Alexander. *Greek Lyrics*. Trans. from the Greek. Edited Richmond Lattimore (Chicago: University of Chicago Press) 1960.

Laudon, R. T. *Sources of Wagnerian Synthesis: A Study of the Franco-German Tradition in 19th Century Opera* (München–Salzburg: Musikverlag, Emil Katzbicher, 1979),154.

Leong, V.; Byrne, E.; Clackson, K.; et al, "Speaker gaze increases information coupling between infant and adult brains". *PNAS* 114 (2017): 13290-5.

Letellier, Robert Ignatius. *The Bible in Music* (Cambridge: Cambridge Scholars Publishing, 2017), 359.

—. *Daniel-François-Esprit Auber: The Man and His Music* (Newcastle: Cambridge Scholars Publishing, 2010).

—. *The Operas of Giacomo Meyerbeer* (Madison: Associated University Presses, 2006).

Levine, R. *Weep, Shudder, Die: A Guide to Loving Opera* (New York: Harper Collins, 2011).

Liébert, G. '*Nietzsche and Music'* (Chicago: University of Chicago Press, 2004), 150.

Levinas, E. *Totality and Infinity: An Essay on Exteriority*. Trans. Alphonso Lingis (Pittsburg: Duquesne University Press, 1969), 52, 293,295,299.

Longfellow Henry Wadsworth. *Poems and Other Writings* (New York, Literary Classics, 2000), 700. The poem was translated into English by Longfellow.

Lorusso, L.; Franchini, A.F.; Porro, A. "Opera and Neuroscience. *Progress in Brain Research*." 216:389-409; doi: 10.1016/bs.pbr.2014.11.016. Epub 2015 Jan 20. 391.

McCumber, J. *Time and Philosophy A history of Continental Thought* (Durham: Acumen, 2011).

Magee, Brian.*Wagner and Philosophy* (London: Allen Press, 2000).

Malherbe, Charles. *Auber* (Paris: Librairie Renouard, 1911), 54.

Marcuse, Herbert. *The Aesthetic Dimension: Toward a Critique of Marxist Aesthetics*. Trans. H. Marcuse and J. Sherover (Boston: Beacon, 1978), 73.

—. *One Dimensional Man* (1964) (London: Routledge, 1991)

Mason, E. C. *Goethe's Faust: Its genesis and Purport* (Berkeley: University of California Press, 1967), 209.

Mauceri, John. *For the Love of Music: A Conductors Guide to the Art of Listening* (London: Weidenfeld & Nicholson, 2019), 185.

Merleau-Ponty M. *Phenomenology of Perception* (2005). Trans. Don Landes (New York: Routledge, 2012).

Michelangelo [Buonarotti]. *The Complete Poems and selected letters of Michelangelo*. Trans. Gilbert C. (Princeton: Princeton University Press 1980).

Millington, Barry. "All in it together: *The Gesamtkunstwerk* revisited". *The Wagner Journal*, 11 (2017): 46-61.

—. "Manifestations of the *Gesamtkunstwerk* in Fin-de-Siècle Vienna". *The Wagner Journal*. 12 (2018): 4-25.

—. "Performing Wagner's Music Dramas in the Modern Age", *Wagner News*, 14 (2020): 235, 9-20.

—. *Richard Wagner: The Sorcerer of Bayreuth* (London: Thames and Hudson, 2012), 131, 243, 245.

Milton, John. *Paradise Lost.* Book IV 73-75. In *The Portable Milton*. Ed. D. Bush (London: Penguin 1976), lines 304-306.

Montaigne, M. de. *Essays*. Trans. J. M. Cohen (London: Penguin Books, 1958), 39, 45.

Moore, G. '*Nietzsche, Biology and Metaphor'* (Cambridge: Cambridge University Press 2002), 85.

Masako U. Fidler, & Susan C. Kresin (eds), *Between Texts, Languages And Culture. A Festschrift For Michael Henry Heim* (Bloomington, Indiana, 2008).

Mula, Marco. "Epilepsy in Dante's Poetry", *Epilepsy and Behavior*, 57 (2016): 251-254.

Müller, U. et al (eds.). "The Potion as Symbol and Prop in Richard Wagner's Musical Dramas". In Müller, U. and Panagl, O. *Bayreuther Festspiel Programme* (1998): 126-138.

Nancy, Jean-Luc. *The Muses*. Trans. Peggy Kamuf (Stanford: Stanford University Press, 1996).

Neveux, Georges. *Juliette, ou La clé des songes* [*Juliette, or The Key of Dreams*] *France Illustration*, Suppl. Issue 109 (*theatral et litteraire*), 1952.

Newark, C. *Opera in the Novel from Balzac to Proust* (Cambridge: Cambridge University Press, 2011), 43.

Newman, Ernest. *The Life of Richard Wagner Vol. 2 1848-1860* (Cambridge: Cambridge Library Collection, 1933/2014), 2:333-339.

—. *The Life of Richard Wagner Vol. 4 1866-1883* (Cambridge: Cambridge University Press, 1946), 4:712.

—. *The Wagner Operas* [*Wagner Nights*] (London: Putnam), 1961), 586.

Nietzsche, Friedrich. *Also Sprache Zarathustra.Das Lexicon der Nietzsche Zitate*. Ed J. Prossliner, V. and Ross, W. (Munich: Deutsche Taschenbuch Verlag, 1892),196: *Die Vergangenen zu erlösen und alles 'Es war' umzuschaffen in ein 'So wolte ich es!' – das heiße mir erst Erlösung!*
—. *Beyond Good and Evil*. Trans. M. Fabe (Oxford: Oxford University Press,1998), 63.
—. *The Birth of Tragedy: Out of the Spirit of Music.* Trans. S. Whiteside (London: Penguin, 1993).
—. *The Gay Science* (London: Random House, 1972), 209.
—. *Twilight of the Idols*. Trans. W. Kaufmann. In *The Portable Nietzsche.* Ed. W. Kaufmann (Cambridge Companion 54) (New York: Viking, 1954), 561.
—. *Zarathustra's Roundelay* [*Zarathustras Rundgesang*}. *Also Sprach Zarathustra*, I-IV. Trans. W Kaufmann (Munich: Dünndruck-Ausgabe/De Gruyter, 1988), 404.
Novalis [Friedrich von Hardenburg]. *Heinrich von Ofterdingen* (1802).
—. *Hymnen an die Nacht.* Ed. P. Kluckhohn and R. Samuel. Ibid, Third Hymn, 1:135. Trans. R. J. Richards, Ibid., 33.

Operas

Auber, Daniel-François-Esprit. *Le Maçon* (Orfeo).
—. *Manon Lescaut*, (EMI) (Chant du Monde).
—. *La Muette de Portici* (EMI).
Bellini, Vincenzo.*I Capuleti e I Montecchi* (EMI).
Berg, Alban. *Lulu* (DG).
Berlioz, Hector. *La Damnation de Faust* (EMI).
Bizet, Georges. *Les Pêcheurs de perles* (Prima Voce).
Boito, Arrigo. *Mefistofele* (EMI).
Britten, Benjamin. *Billy Budd* (Decca).
—. *Death in Venice* (Decca).
Britten. *The Turn of the Screw* (Decca).
Busoni, Ferruccio. *Doktor Faust* (DG).
Dargomyzhsky, Alexander. *Rusalka* (Relief).
Debussy, Claude. *Pelléas et Mélisande* (Eurodisc).
De Falla, Manuel. *La vida breve* (EMI).
Delius, Frederick. *A Village Romeo and Juliet* (EMI).
Donizetti, Gaetano. *L'Ange di Nisida* (Opera Rara).
—. *Anna Bolena* (Decca) (EMI) (Decca).

—. *La Favorita* (in Italian) (Decca); *La Favorite* (in French) (RCA Sony Classical Opera).
—. *Lucia di Lammermoor* (EMI) (Decca) (RCA)
Dvořák, Antonin. *Rusalka* (Supraphon) (Decca).
Erkel, Ferenc. *Bánk-Bán* (Hungaraton).
Floyd, Carlisle. *Wuthering Heights*.
Gluck, Christoph Willibald. *Alceste*.
—. *Iphigénie en Tauride* (Philips).
—. *Orfeo ed Euridice* (RCA).
Goldmark, Karl. *Die Königin von Saba* (Hungaroton).
Gounod, Charles. *Faust* (EMI).
—. *La Nonne sanglante* (CPO).
Hermann, Bernard.*Wuthering Heights.*
Hoffmann, E.T.A. *Undine* (Memories) (Bayer Records).
Janáček, Leoš, *The Makropoulos Affair* (Decca).
—. *Osud* (Chandos).
Korngold. *Das Wunder der Heliane* (Decca).
Lortzing, Albert. *Undine* (EMI) (Capriccio).
Massenet, Jules. *Manon* (EMI).
—. *Thaïs* (RCA) (EMI).
Messiaen, Olivier. *Saint François d'Assise* (Orfeo).
Meyerbeer, Giacomo. *Les Huguenots* (Decca)
—. *Robert le Diable* (Brilliant)
Monteverdi, Claudio. *Orfeo, favola in musica*
Mozart, Wolfgang Amadeus. *Don Giovanni* (EMI)
—. *Die Zauberflöte* (EMI)
Mussorgsky, Modest. *Boris Godounov* (EMI).
Offenbach, Jacques. *Les Contes d'Hoffmann*; the Guiraud version (EMI); the Oeser version (Decca).
—. *Orphée aux enfers* (EMI)
Puccini, Giacomo. *Manon Lescaut* (EMI (EMI).
—. *Le Villi* (CBS Masterworks).
Reyer, Ernest. *Salammbô* (MRS).
Rossini, Gioacchino. *Guillaume Tell* (EMI).
Rubinstein, Anton. *The Demon* (Marco Polo) (Melodia).
Schreker, Franz. *Der ferne Klang* (Capriccio).
—. *Die Gezeichneten* (Marco Polo).
Smyth, Ethel. *The Wreckers* (Conifer Classics).
Strauss, Richard. *Daphne* (DG).
—. *Elektra* (Decca).

Tchaikovsky, Pyotr Illyich. *The Queen of Spades* (EMI-Melodya) (Sony Classical)
Thomas, Ambroise. *Hamlet* (Decca) (EMI).
Verdi, Giuseppe. *Don Carlos* (Decca) (Prima Voce).
—. *La Forza del destino* (Decca) (Prima Voce)
—. *Otello* (RCA) (Prima Voce).
—. *Il Trovatore* (EMI)
—. *Les Vêpres siciliennes* (in French, Opera Rara); (in Italian, *I vespri Siciliani*, RCA).
Victory over the Sun (Russian Futurist opera in zaum language)
Wagner, Richard. *Die Feen* (BBC in Wagner: Complete Operas DG).
—. *Der fliegende Holländer* (EMI).
—. *Das Liebesverbot* (BBC in Wagner: Complete Operas DG)
—. *Lohengrin* (EMI).
—. *Die Meistersinger von Nürnberg* (EMI).
—. *Parsifal* (Decca).
—. *Rienzi* (EMI; also in Wagner: Complete Operas DG).
—. *Der Ring des Nibelungen* (Decca) (EMI)
—. *Tannhäuser* (EMI).
—. *Tristan und Isolde* (DG).

Ovid. *Metamorphosis The Death of Orpheus,* Book IX, 1-66.
—. *Metamorphoses*. Book 4: The Pathos of Love.
Paglia, Claudia. *Sexual Personae: Art and Decadence from Nerfertiti to Emily Dickinson* (New Haven: Yale University Press,1990), 96.
Parly, N. "Doing the Diva Dying: Performative Studies of Death in *Tristan und Isolde*", *The Wagner Journal* 10 (2016): 45-55.
Parly, N. *Vocal Victories*: *Wagner's Female Characters from Senta to Kundry* (Museum Tusculanum Press) (Chicago: University of Chicago Press 2011).
Paterson, D. *Reading Shakespeare's Sonnets* (London: Faber & Faber, 2010).
Picasso, Pablo. "Picasso Speaks". *The Arts* (New York, May 1923).
Pittock, Murray (ed.). *The Reception of Sir Walter Scott in Europe* (The Athclone Critical Tradition Series: The Reception of British and Irish Authors in Europe) (London: Bloomsbury, 2006).
Pleasants, Henry. *The Great Singers: From the Dawn of Opera to Our Own Time* [1966] (London: Macmillan, 1981).
—. *The Great Tenor Tragedy: The Last Days of Adolphe Nourrit as Told (Mostly) by Himself* (Portland OR: Amadeus Press, 1995).

Poe, Edgar Allen. *Sonnet to Science. The Complete Poetry of Edgar Allan Poe* (New York: Signet Classics, 1996), 109.

Poizat, M. *The Angel's Cry: Beyond the Pleasure Principle in Opera.* Trans. Arthur Denner (Ithaca: Cornell University Press, 1992).

Pollan, M. *How to Change Your Mind* (London: Allen Lane, 2018).

Porges, S. W. *The Polyvagal Theory* (New York: W.W. Norton, 2011).

Prideaux, S. *I am Dynamite! A Life of Friedrich Nietzsche* (London: Faber & Faber, 2018), 191, 230.

Proust, Marcel. *The Fugitive*. Trans. Bersani. In *Marcel Proust: The Fictions of Life and Art.* 2nd ed. (Oxford: Oxford University Press, 2013), 91.

—. *The Complete Short Stories of Proust.* Trans: Joachim Neugroschel (New York: Coper Square Press, 2001).

Proust, Marcel. *The Way by Swann's*. Trans. L. Davis (London: Penguin Books, 2002), 352-353.

Pushkin, Alexander. *Selected Works in Two Volumes.Volume One: Poetry* (Moscow: Progress Publishers, 1974), "The Water Nymph". Trans. Avril Pymon, 61-64.

Ranos, I. "Sacred Transformations". *British Museum Publications*, Spring/Summer 2020, 24-27.

Raphael (1483-1520) *The School of Athens*, Vatican City (painted between 1509 and 1511)

Richards, R. J. *The Romantic Conception of Life: Science and Philosophy in the Age of Goethe* (Chicago: University of Chicago Press, 2002).

Richardson, J. *Nietzsche's New Darwinism* (Oxford: Oxford University Press, 2004), 229.

Ringer, Mark. *Opera's First Master: The Musical Dramas of Claudio Monteverdi* (Newark N.J.: Amadeus Press, 2006), 217.

Robertson, J. Ian S. *Doctors in Opera.* (*Core Image, Scottish Opera*, 2016).

—. and Weitz, G. "Siegfried's Amnesia: An Attempt at an Explanation." *Core Image, Scottish Opera,* 2015.

Robinson, Michael F. *Opera before Mozart* (London: Hutchinson University Library, 1966).

Ross, Alex. *Wagnerism: Art and Politics in the Shadow of Music* (London: 4th Estate, 2020), 322-354.

—. *The Rest is Noise* (London: Fourth Estate, 2007).

Rossetti, Dante Gabriel. *Collected Works* (London: Ellis & Scrutton, 1886),1:29.

Rousseau, Jean-Jacque. *Dictionnaire* (1768).

Sadie, Stanley (ed.) [1992]. *The New Grove Dictionary of Opera* (London: Macmillan, 1997), 4: 491-518.

Safranszki R. *Goethe: Life as a Work of Art.* Trans. Dollenmayer (London: Liveright Publishing Corporation, 2017), 477.

Schiller, Friedrich von. *On the Sublime* in *Two Essays.* Trans. J.A. Elias (New York, 1966), 198.

Schopenhauer, Arthur. *The World as Will and Idea* (London: J M Dent, 1995).

Schubert, Franz Peter. *Der Doppelgänger* (1828). One of the six songs from *Schwanengesang,* words by Heinrich Heine.

Scott, Sir Walter. See Pittock, Murray (ed.).

Scruton, Roger. *Death-Devoted Heart: Sex and the Sacred in Wagner's Tristan and Isolde* (Oxford: Oxford University Press, 2004), 192.

—. "Music and the Transcendental". *The Imaginative Conservative* (27 January 2020), 84.

—. *Music as an Art* (London: Bloomsbury Continuum, 2018), 71-84, 109-124.

—. *Understanding Music* (London: Continuum International Publishing, 2009), 13.

—. *Wagner and German Idealism*. Based on *Wagner's Philosophers*, BBC Radio 3 (20 May 2013), 151.

Shakespeare, William. *The Complete Works of William Shakespeare* (New York: Barnes and Noble, 1994).

Shelley, Mary. *Frankenstein; or The Modern Prometheus* (Oxford: Oxford World Classics, 1994).

Shelley, Percy Bysshe. 'Music'. *The Complete Poetic Works of Percy Bysshe Shelley* (Oxford: Oxford University Press, 1914), 651.

Sigmund, R. *My Time in Heaven* (New Kensington, PA: Whitaker House, 2009).

Smith, Patrick J. *The Tenth Muse: A historical study of the opera libretto* (London: Victor Gollancz Ltd., 1971).

Sno, H.N.; Linszen, D.H. and De Jonghe, F. "Art Imitates Life: Déjà vu experiences in Prose and Poetry", *British Journal of Psychiatry,* 160 (1992): 511-518.

Sorgner, S. L. and Fürbeth, O. *Music in German Philosophy: An introduction* (Chicago: University of Chicago Press, 2010), 38.

Spaethling, R. "Folklore and Enlightenment in the Libretto of Mozart's *Magic Flute*". *Eighteenth-Century Studies*, 9:1 (1976): 45-68.

Spencer, S. and B. Millington (trans.). *Wagner's Ring of the Niblung: A Companion*. Trans. S. Spencer and B. Millington (London: Thames and Hudson, 2000).

Steane, J. B. *The Grand Tradition: Seventy Years of Singing on Record, 1900 to 1970* (London: Duckworth, 1974, 1978).

Steege, B. *Helmholtz and the Modern Listener* (Cambridge: Cambridge University Press, 2012), 228.

Steiner, George. *Tragedy re-considered.* In *Rethinking Tragedy.* Ed. R. Felski (Baltimore: Johns Hopkins University Press, 2008), 8.

Stendhal. *Naples and Florence: A Journey from Milan to Reggio*. Quoted in Magherini, G, *La Sindrome di Stendhal* (Florence: Ponte Alle Grazie, 1989).

Stendhal. *On Love*. Forward A. C. Grayling. Trans. S. Lewis (London: Hesperus Press Ltd., 2009), 11, 15, 30, 34, 52, 52.

Stevens, Nicholas. Review of Silvio J. dos Santos, *Narratives of Identity in Alban Berg's 'Lulu'* (Rochester, NY: University of Rochester Press, 2014).

Storr, A. *Music and the Mind* (London: Harper Collins, 1997).

Stott, R. *Darwin's Ghosts. The Secret History of Evolution* (New York: Spiegel & Grau Trade Paperbacks, 2012).

Swann, J. *A four-note cell as a basic element in Wagner's 'Ring'.* The Juilliard School (April 1980), 188 and personal communication.

—. "Wagner and Proust". *The Wagner Journal*, 12 (2018): 34-55.

Taylor, M. *Last Works: Lessons in Leaving* (New Haven: Yale University Press, 2018), 150-151.

Temkin O. *The Falling Sickness: A History of Epilepsy from the Greeks to the Beginnings of Modern Neurology.* 2nd Edition (Baltimore: Johns Hopkins Press 1971) 15, 92-96.

Tennyson, Alfred Lord. *The Works of Alfred Tennyson*. (London: MacMillan & Co, 1907) (Quotes and poems, *In Memorium),*

Thurman, D. J.; Hesdorffer, D.C.; French, J. A. et al. "Sudden unexpected death in epilepsy: assessing the public health burden". *Epilepsia*. 55(10):1479-85; doi: 10.1111/epi.12666. Epub 2014.

Tillyard, E. M. W. "*Shakespeare's Last Plays"* (London: The Athlone Press,1938), 68.

Trimble Michael Robert. *The Intentional Brain Motion, Emotion and the Development of Modern Neuropsychiatry* (Baltimore: John Hopkins Press, 2016).

—. *The Soul in the Brain: The Cerebral Basis of Language, Art and Belief* (Baltimore: Johns Hopkins Press, 2007).

—. *Why Humans Like to Cry: Tragedy, Evolution and the Brain* (Oxford: Oxford University Press, 2012).

Trimble, M. R.; Hesdorffer, D.; Letellier, R. I. "In Wagner's Eyes: Casting Light on a disputed portrait." *The Wagner Journal*, 13 (2019):20-31.

—. "The Mystery of Wotan's Missing Eye." *The Wagner Journal*, 7 (2013): 26-37.

Vaughan, H. *The World. Selected Poems* (London: Society for Promoting Christian Knowledge, 2004),113.

Vickhoff, B.; Malmgren, H.; Aström, R. et al. "Music Structure Determines Heart Rate Variability of Singers". *Front Psychol.* 4 (2013): 334. doi: 10.3389/fpsyg.2013.00334.

Vikram, Seth. *An Equal Music* (London: Phoenix, 1999), 484.

Vivien Green, *Mystical Symbolism: The Salon de la RoseXcroix in Paris 1892-1897* (New York: The Solomon Guggenheim Foundation, 2017).

Wagner, Cosima. *Cosima Wagner's Diaries, Volume One, 1869-1877*. Trans. G. Skelton (London: St James's Place, 1978), 1007 [21 Dec. 1877]; *Volume Two, 1878-1883*, 813-1010.

Wagner, Wilhelm Richard. *Wagner's Ring of the Nibelung: A Companion*. Trans. S. Spencer and B. Millington (London: Thames and Hudson, 2000), 217-218.

—. *A communication to my friends*. Trans. William Ashton Ellis (Wokingham: Dodo Press, 1851).

—. *Religion and Art*. Trans. William Ashton Ellis [1897] (Lincoln: University of Nebraska Press 1994), 213).

—. *Opera and Drama,* Trans. William Ashton Ellis [1900] (Lincoln: University of Nebraska Press, 1995),156

—. *Richard Wagner's Letters to August Röckel* (Forgotten Books, 2012), 84, 85, 90, 99.

—. Letters:

—. Letter to Franz Liszt (16 December 1854).

—. Letters to Mathilde Wesendonck (7 April and 3 September 1858).

—. Letter to Mathilde von Wesendonck 1858. In Garten, H. F. *Wagner the Dramatist*. (London: Overture Publishing, 1977), 137.

—. Diary or letter to Mathilde von Wesendonck, 1 Oct.1858. See Garten, H.F. Ibid., 149

—. *Opera and Drama,* Trans. W. Ashton Ellis (London: University of Nebraska Press,1995).

Watson, P. *The German Genius: Europe's Third Renaissance, The Second Scientific Revolution and the Twentieth Century* (London: Simon Schuster, 2010).

Watts, Isaac. "When I survey the wondrous cross". *Hymns and Spiritual Songs: The Three Books of Great Christian Bible Hymns— Complete* (CreateSpace Independent Publishing Platform 2018).

Wedekind, Frank.*The Lulu Plays.* Trans. Stephen Spender (London: Alma Classics, 2015).

Whitehead, Alfred North. *Process and Reality* (New York: Free Press, 1979), 39.

Whitman, Walt. See Brasher, T.

—. Essay *The Poetry of the Future* (1881). In *Prose Works* 1892. Vol. 2: *Collect and Other Prose*. Ed. Floyd Stovall (New York: New York University Press 1964).

—. *Leaves of Grass* (New York: Brooklyn, 1855),

Williams, S. *Wagner and the Romantic Hero* (Cambridge: Cambridge University Press, 2004).

Woolf, Virginia. "Impressions at Bayreuth", *The Times* (August 1909).

Wroe, A. '*Orpheus: The Song of* Life' (London: Jonathan Cape 2011).

Wu, W. *Romanticism: An Anthology*. Third Ed. (Oxford: Blackwell Publishing, 2006).

Yeats, William Butler. 'Wild Swans at Coole'. *Collected Poems of W. B. Yeats* (London: Macmillan and Co Ltd, 1955), 147.

Yovel, Y. *Kant's Philosophical Revolution: A Short Guide to the Critique of Pure Reason* (Princeton: Princeton University Press, 2018).

Zucker, Stefan, *Franco Corelli and a Revolution in Singing: Fifty-Four Tenors Spanning 200 Years* (New York: Bel Canto Society, 2015).

Zweig, S. *The Burning Secret and other stories* (London: Pushkin Collection, 1989).

INDEX

1. General (Authors and Scholars)

2. Artists

3. Ballets

4. Librettists

Translators consulted

5. Operas

6. Singers

7. Staging

Theatres

Designers and Producers

Lighting